SEDIMENT TRANSPORT

Edited by **Silvia Susana Ginsberg**

INTECHOPEN.COM

Sediment Transport
Edited by Silvia Susana Ginsberg

Published by InTech
Janeza Trdine 9, 51000 Rijeka, Croatia

Publishing Process Manager Ana Nikolic

Technical Editor Teodora Smiljanic

Cover Designer InTech Design Team

Image Copyright Richard Schramm, 2010. Used under license from Shutterstock.com

Additional hard copies can be obtained from orders@intechopen.com

Sediment Transport, Edited by Silvia Susana Ginsberg
p. cm.
ISBN 978-953-307-189-3

For EU product safety concerns:
IN TECH d.o.o., Prolaz Marije Krucifikse Kozulić 3, 51000 Rijeka, Croatia,
info@intechopen.com or visit our website at intechopen.com.

Meet the editor

Dr. Silvia Susana Ginsberg is a marine geologist at the Instituto Argentino de Oceanografia in Bahía Blanca, Argentina. Her general research area is in the field of geomorphology and sedimentology of coastal and estuarine environments. She completed the degree in oceanography with orientation in marine geology and her doctorate in Geology, both from Nacional del Sur University of Bahía Blanca city, Argentina. She is a Senior Research at the Consejo Nacional de Investigaciones Científicas (CONICET) y Técnicas and professor at the Department of Ingeniería Civil, University of Tecnológica Nacional, Facultad Regional Bahía Blanca. She has supervised the graduate students and post-graduate fellows focusing especially on estuarine processes. Her research interest includes aspects on sediment transport processes, hydrodynamic and morphodynamics on marine areas. She has authored numerous international and national journal articles and leads several projects on sediment transport.

Contents

Preface

Sediment Transport is a book that covers a wide variety of subject matters. Physicists, engineers, mathematicians, statisticians, and geologists have been investigating sediment transport for several decades and keep on doing so to learn further about it in an attempt to give solutions to problems resulting from human intervention in nature. In this respect, research on sediment transport has gained importance due to the increasing use of water resources. On the other hand, fully understanding erosion and sedimentation processes and their impact on sediment transport plays a key role in the growing knowledge on the importance of fluvial, lake and/or marine sediments over a highly heterogenous variety of environmental problems. In this respect, this book is a doubly significant contribution, firstly because of the growing interest of society in environmental matters and secondly because of the updated issues it deals with. It reports results from studies carried out in aqueous systems and in laboratory flumes by researchers whose expertise focuses on transportable solid particles and related problems.

It is with great satisfaction that I write the preface to this book by InTech, which includes 16 chapters written by researchers from different parts of the world on hydrodynamics and its relationship with sediment transport and morphological development. The main subject matter is presented in the light of different approaches such as sediment transport modeling in non-alluvial beds and hyperconcentrated flows; coastal area modeling for harbor planning; littoral zone modeling, boat traffic-induced sediment resuspension modeling; and settling flux modeling. Other contributions included in the book focus on cohesive sediment modeling; sediment dynamics in stream confluence or river diversion; analytical modeling applied to studies on fluid-bed interfaces in meandering channels; changes in sediment transport under fine materials, cohesive materials and ice cover; environmental remediation of contaminated fine sediments; bedform field evolution over sediment transport; and sediment circulation at interconnected tidal channels system. Authors are responsible for their views and conclusions.

In short, by fully accomplishing its objectives, this book is an excellent source of information which opens new paths worth of exploring. It is an invaluable interdisciplinary textbook and an important contribution to the field. Let me also gratefully thank each of the authors for their contributions and strongly recommend this textbook to those in charge of conducting research on engineering issues or wishing to deal with equally important scientific problems.

Silvia Susana Ginsberg
Instituto Argentino de Oceanografía, Bahía Blanca,
Universidad Tecnológica Nacional, FRBB,
Departamento de Ingeniería Civil, Bahía Blanca,
Argentina

1

Advanced Topics in Sediment Transport Modelling: Non-alluvial Beds and Hyperconcentrated Flows

B. Dewals, F. Rulot, S. Erpicum, P. Archambeau and M. Pirotton
Hydrology, Applied Hydrodynamics and Hydraulic Constructions, University of Liège, Chemin des Chevreuils 1, B52/3+1 - B-4000 Liège, Belgium

1. Introduction

Civil and environmental engineers frequently face sediment transport issues such as local scouring, sedimentation in reservoirs, erosion after floods or dam breaching flows as well as long term aggradation or degradation of riverbed (Dewals et al. 2010b; van Rijn 2007; White 2001). Such sediment related problems are of huge importance in most projects of river engineering, calling for structures to be designed considering sediment transport issues from the very early stages of project development. Sustainable operation rules also need to be developed, both in short and long term perspectives. As a result of the complexity of the governing physical processes and the significant uncertainties affecting input data, modeling tools with a genuine predictive capacity, such as comprehensively validated numerical models, constitute key elements to provide quantitative decision-support in project design and developments.

Sediment transport has been studied from a physical point of view for almost two hundred years but is not yet fully understood (Frey and Church 2009). In particular, while the Navier-Stokes and continuity equations represent a generally accepted mathematical description of fluid flow, there is no comparable model for the complete interaction of flow, sediment transport and bed evolution (Spasojevic and Holly 2008). Therefore, sediment transport remains a challenging topic of research today, since a unified description of processes is still to be achieved. In this quest, both new experimental approaches and more advanced numerical models have a part to play; the former providing new insight into fundamental processes while the later enabling to upscale the results for real-life applications.

In this chapter, we first present an original two-phase flow model for the water-sediment mixture, acting as a unified basis for all our subsequent developments. Next, we focus on two topics in which we have made original contributions, namely sediment routing on alluvial and non-alluvial beds and modelling of transient hyperconcentrated flows. In both cases, we use our original two-phase flow modelling framework to derive specific governing equations, for which we detail an appropriate finite volume numerical scheme and demonstrate their validity through a number of test cases.

2. Two-phase flow model for water-sediment mixtures

In this section, we subsequently review existing mathematical models for sediment-laden flows, including two-phase flow theories, present our own two-phase flow morphodynamic model and eventually detail its mathematical properties together with its numerical discretization.

2.1 Existing mathematical models

Cao and Carling (2002a) have provided a comprehensive discussion of existing approaches for mathematical modelling of sediment-laden flows. They have emphasized a number of shortcomings, particularly with respect to the turbulence closure models and the bottom boundary conditions. The former fail to reproduce the intricate interactions between the sediments and the flow turbulence; while the formulation of the later is affected by a high level of uncertainty as detailed in their companion paper (Cao and Carling 2002b). Existing numerical models also generally rely on simplifications in the water-sediment mixture and global sediment continuity equations: respectively, ignored time derivative of the bottom elevation and neglected sediment storage in the water column. These assumptions become questionable when significantly transient processes take place.

In contrast with other fields of hydraulic engineering such as aerated flows (Kerger et al. 2010b), the governing equations underlying most existing models for sediment-laden flows stem from single-phase flow theory. They involve continuity and momentum equations for clear water, combined with a continuity equation for sediments (Spasojevic and Holly 2008). They are therefore only valid for low sediment concentrations (<0.1 in volume) (Greimann and Holly 2001). Although water-sediment mixtures constitute obviously two-phase media, only very few attempts to account explicitly for this multiphase nature have been reported in sediment transport models and hardly none in morphodynamic models.

Two-phase flow models for sediment transport

In a two-phase formulation, Cao et al. (1995) derived suspended sediment concentration profiles valid for both low and high concentrations. Importance was stressed on the influence of the closure relations for turbulent viscosity and diffusivity. Similarly, Greimann et al. (1999) explained the increased diffusive flux of large particles and measurable velocity lag of particles (drift velocity), two observed phenomena but theoretically unexplained so far. A generalized mathematical model for the liquid-solid mixture was derived by Wu and Wang (2000) based on the two-fluid model, but validation and application were not reported. Greimann and Holly (2001) accounted for both particle-particle interactions and particle inertia in the expression of equilibrium concentration profiles for suspended sediments. Still, empiricism was necessary to formulate the turbulence quantities and they highlighted the need for further experimental and analytical work to develop improved models for the fluid eddy viscosity, relative magnitude of the particle turbulence intensity and boundary conditions applicable to loose beds. Criteria were given to determine if particle-particle interactions and particle inertia are significant. Recently, Bombardelli and Jha (2009) showed that both a standard sediment-transport model and a two-fluid-model predict accurately the velocity field, whereas only the later provides satisfactory predictions for the concentration profiles and the turbulence statistics. Extending to two-phase flows the well-known results of single-phase flows in open channels, they also found that the Reynolds stress model does not improve the predictions beyond the accuracy of the standard $k-\varepsilon$ model, at least for dilute mixtures. Values of the Schmidt number that fit the datasets indicate that the eddy viscosity is smaller than the diffusivity of sediment (Bombardelli and Jha 2009), which is in agreement with part of the literature but not all (Cao and Carling 2002a; Cao and Carling 2002b).

Two-phase morphodynamic models

Two-phase formulations for complete morphodynamic models are hardly available in literature. Recently, Greco et al. (2008) presented a 1D single layer two-phase morphodynamic model, which they successfully applied for dam break flow over an erodible topography.
A more rigorous theoretical derivation of a general two-phase flow model for flows in hydraulic environmental engineering has also been presented by Kerger, Dewals et al. (2011), but validation and applications have not yet been reported.

Double-averaged models

The double-averaging concept was recently introduced in hydraulic engineering by Nikora et al. (2007) and current research suggests that it may become a standard tool for fluvial applications. By means of explicit consideration of roughness mobility and form-induced stresses stemming from a rigorous derivation, the approach provides an enhanced treatment of the bed shear stress formulation, which will prove valuable for morphodynamic modelling. Nonetheless, using the double-averaged hydrodynamic equations for developing numerical models is still in its infancy and closures for subgrid scale effects remain to be developed (Walters and Plew 2008).

2.2 Derivation of an original two-phase flow model

In the following paragraphs, we detail the derivation of a two-phase mathematical model for flow, sediment transport and morphodynamics. The finite volume numerical technique developed to solve the set of governing equations is also detailed, together with a comparative discussion of synchronous vs. sequential resolution of the flow, sediment transport and morphodynamic models.

Conservation laws

Following an Eulerian description of the flow, we formulate conservation laws for the flow in a Cartesian system of coordinates (x, y, z), as sketched in Fig. 1.

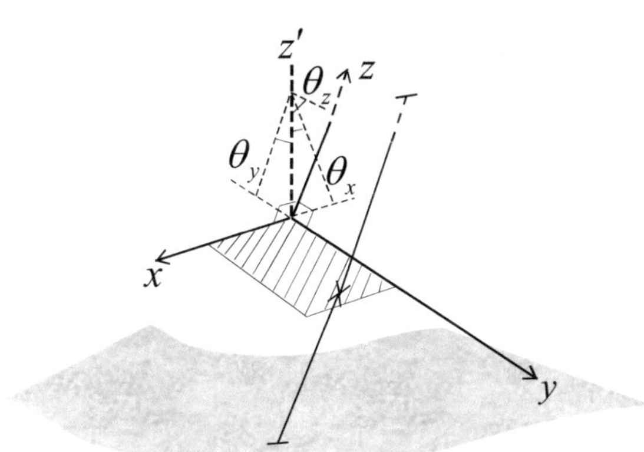

Fig. 1. Axis of reference.

The x and y axes are chosen in such a way that the plane they define corresponds to the main flow direction. Axis z is simply set normal to this plane. The axes x, y and z are inclined of angles θ_x, θ_y and θ_z with respect to the vertical direction (axis z'). In the particular case where the x-y plane is horizontal the angles become: $\theta_x = \theta_y = 0$ and $\theta_z = \pi / 2$.
The Reynolds-averaged mass and momentum conservation equations for the water-sediment mixture read:

$$\frac{\partial \rho}{\partial t}+\frac{\partial(\rho u)}{\partial x}+\frac{\partial(\rho v)}{\partial y}+\frac{\partial(\rho w)}{\partial z}=0 \tag{1}$$

$$\begin{aligned}
\frac{\partial(\rho u)}{\partial t}+\frac{\partial(\rho u^2)}{\partial x}+\frac{\partial(\rho uv)}{\partial y}+\frac{\partial(\rho uw)}{\partial z} &= -\frac{\partial p}{\partial x}+\frac{\partial \sigma_x}{\partial x}+\frac{\partial \tau_{xy}}{\partial y}+\frac{\partial \tau_{xz}}{\partial z}+\rho g \sin\theta_x \\
\frac{\partial(\rho v)}{\partial t}+\frac{\partial(\rho uv)}{\partial x}+\frac{\partial(\rho v^2)}{\partial y}+\frac{\partial(\rho vw)}{\partial z} &= -\frac{\partial p}{\partial y}+\frac{\partial \tau_{yx}}{\partial x}+\frac{\partial \sigma_y}{\partial y}+\frac{\partial \tau_{yz}}{\partial z}+\rho g \sin\theta_y \\
\frac{\partial(\rho w)}{\partial t}+\frac{\partial(\rho uw)}{\partial x}+\frac{\partial(\rho vw)}{\partial y}+\frac{\partial(\rho w^2)}{\partial z} &= -\frac{\partial p}{\partial z}+\frac{\partial \tau_{zx}}{\partial x}+\frac{\partial \tau_{zy}}{\partial y}+\frac{\partial \sigma_z}{\partial z}-\rho g \sin\theta_z
\end{aligned} \tag{2}$$

with ρ = mixture density; u, v and w = velocity components along x, y and z; g = gravity acceleration; p = pressure; σ_i = Reynolds normal stresses ($i = x, y, z$); τ_{ij} = Reynolds shear stresses ($i, j = x, y, z$). Viscous stresses in the momentum equations have been neglected compared to Reynolds stresses since the later greatly excess the former in practical applications.
Conservation of a dispersed phase in the fluid, namely suspended sediments, is expressed by the following advection-diffusion equation:

$$\frac{\partial(\rho\phi)}{\partial t}+\frac{\partial(\rho u\phi)}{\partial x}+\frac{\partial(\rho v\phi)}{\partial y}+\frac{\partial}{\partial z}\left[\rho(w-w_s)\phi\right]=-\left(\frac{\partial \tilde{q}^{d}_{\phi,x}}{\partial x}+\frac{\partial \tilde{q}^{d}_{\phi,y}}{\partial y}+\frac{\partial \tilde{q}^{d}_{\phi,z}}{\partial z}\right)+\tilde{S}_{\phi}\cdot, \tag{3}$$

in which ϕ = mass concentration and w_s = settling velocity of sediments. Notations $\tilde{q}^{d}_{\phi,x}$, $\tilde{q}^{d}_{\phi,y}$ and $\tilde{q}^{d}_{\phi,z}$ refer to the mass fluxes induced by turbulence, which are usually evaluated as follows:

$$\tilde{q}^{d}_{\phi,x}=\frac{\nu_T}{\sigma_T}\frac{\partial \phi}{\partial x},\quad \tilde{q}^{d}_{\phi,y}=\frac{\nu_T}{\sigma_T}\frac{\partial \phi}{\partial y}\quad \text{and}\quad \tilde{q}^{d}_{\phi,z}=\frac{\nu_T}{\sigma_T}\frac{\partial \phi}{\partial z}, \tag{4}$$

with σ_T the Schmidt number taking typical values in-between 0.8 and 1.2 (e.g. Hervouet 2003). The notation $\tilde{S}_{\phi}$ designates the production rate within the flow layer, which is usually zero for non-reactive flows.

Depth-averaging concept

Most flows of interest in civil and environmental engineering are characterized by significantly larger length scales in a reference plane (often almost horizontal) compared to the characteristic depth of the flow. This motivates the use of depth-averaged models, which require far less intricate numerical resolution procedures than needed for general three-dimensional free surface flows. In addition, besides reducing the complexity of the model, such a depth-averaged description of the flow better fits with available data and outputs of interest for most applications in civil and environmental engineering.

Bottom and free surface boundary conditions

Deriving a depth-averaged model from equations (1)-(3) requires boundary conditions to be prescribed at the bottom ($z = z_b$) and at the free surface ($z = z_s$). These include kinematic boundary conditions expressed as follows:

$$\frac{\partial z_b}{\partial t} + u\frac{\partial z_b}{\partial x} + v\frac{\partial z_b}{\partial y} - w = r_b \, , \tag{5}$$

$$\frac{\partial z_s}{\partial t} + u\frac{\partial z_s}{\partial x} + v\frac{\partial z_s}{\partial y} - w = 0 \, , \tag{6}$$

with r_b (m/s) the exchange flux with the bed; z_b and z_s the bed and surface elevations respectively. Note that $\partial z_b/\partial t$ has not been set to zero in (5) since we deal here with flows over erodible beds.

Since wind effects are not considered here, dynamic boundary conditions at the free surface simply state that pressure remains equal to the atmospheric pressure and that both normal and shear stresses are zero. Dynamic boundary conditions at the bottom link the components of the stress tensor with the bottom shear stress τ_b per unit horizontal surface:

$$\begin{aligned}
\tau_{bx}\,\Delta\Sigma &= \sigma_x \frac{\partial z_b}{\partial x} + \tau_{xy}\frac{\partial z_b}{\partial y} - \tau_{xz} \\
\tau_{by}\,\Delta\Sigma &= \tau_{xy} \frac{\partial z_b}{\partial x} + \sigma_y\frac{\partial z_b}{\partial y} - \tau_{yz} \\
\tau_{bz}\,\Delta\Sigma &= \tau_{xz} \frac{\partial z_b}{\partial x} + \tau_{yz}\frac{\partial z_b}{\partial y} - \sigma_z \, ,
\end{aligned} \tag{7}$$

where notation $\Delta\Sigma$ stands for $\sqrt{1 + \left(\partial z_b/\partial x\right)^2 + \left(\partial z_b/\partial y\right)^2}$ (Hervouet 2003).

Boundary conditions for the sediment advection-diffusion equation express the exchange rate $\tilde{S}^d_{\phi,b}$ of sediments between the bed and the flow layer:

$$\tilde{S}^d_{\phi,b}\,\Delta\Sigma = \left[\tilde{q}^d_{\phi,x}\right]_b \frac{\partial z_b}{\partial x} + \left[\tilde{q}^d_{\phi,y}\right]_b \frac{\partial z_b}{\partial y} - \left[\tilde{q}^d_{\phi,z}\right]_b . \tag{8}$$

while this exchange rate is simply zero at the free surface.

Scaling and shallow flow assumption

Standard scaling analysis of equations (1)-(3) proves useful to further simplify the derivation of the depth-averaged model. If the characteristic flow depth H is assumed much smaller than the characteristic length scale L in the x-y plane, then the depth-averaged z momentum balance is shown to reduce to the hydrostatic equilibrium:

$$\frac{\partial p}{\partial z} = -\rho g \sin\theta_z \quad \Rightarrow \quad p(z) = g \sin\theta_z \int_z^{z_s} \rho \, dz' . \tag{9}$$

This in turn provides an explicit relationship between the pressure and other flow variables such as density and water depth.

General depth-averaged model

Integration of the three-dimensional equations (1)-(3) over the local flow depth, accounting for the boundary conditions (5)-(8), results in the following set of two-dimensional equations:

$$\frac{\partial}{\partial t}\left(h\overline{\rho}\right) + \frac{\partial}{\partial x}\left(h\overline{\rho u}\right) + \frac{\partial}{\partial y}\left(h\overline{\rho v}\right) = -\rho_b r_b \tag{10}$$

$$\begin{aligned} &\frac{\partial}{\partial t}\left(h\overline{\rho u}\right) + \frac{\partial}{\partial x}\left(h\overline{\rho u^2}\right) + \frac{\partial}{\partial y}\left(h\overline{\rho u v}\right) \\ &= -\frac{\partial h\overline{p}}{\partial x} - \rho_b u_b r_b - [p]_b \frac{\partial z_b}{\partial x} + \frac{\partial h\overline{\sigma_x}}{\partial x} + \frac{\partial h\overline{\tau_{xy}}}{\partial y} + \tau_{bx}\Delta\Sigma + h\overline{\rho} g \sin\theta_x \end{aligned} \tag{11}$$

$$\begin{aligned} &\frac{\partial}{\partial t}\left(h\overline{\rho v}\right) + \frac{\partial}{\partial x}\left(h\overline{\rho u v}\right) + \frac{\partial}{\partial y}\left(h\overline{\rho v^2}\right) \\ &= -\frac{\partial h\overline{p}}{\partial y} - \rho_b v_b r_b - [p]_b \frac{\partial z_b}{\partial y} + \frac{\partial h\overline{\tau_{yx}}}{\partial x} + \frac{\partial h\overline{\sigma_y}}{\partial y} + \tau_{by}\Delta\Sigma + h\overline{\rho} g \sin\theta_y \end{aligned} \tag{12}$$

$$\begin{aligned} &\frac{\partial}{\partial t}\left(h\overline{\rho\phi}\right) + \frac{\partial}{\partial x}\left(h\overline{\rho u\phi}\right) + \frac{\partial}{\partial y}\left(h\overline{\rho v\phi}\right) \\ &= -\rho_b\phi_b r_b - \left[\frac{\partial}{\partial x}\left(h\overline{\tilde{q}^{\mathrm{d}}_{\phi,x}}\right) + \frac{\partial}{\partial y}\left(h\overline{\tilde{q}^{\mathrm{d}}_{\phi,y}}\right)\right] + \tilde{S}^{\mathrm{d}}_{\phi,\mathrm{s}} - \tilde{S}^{\mathrm{d}}_{\phi,\mathrm{b}} + h\overline{\tilde{S}_\phi} \end{aligned} \tag{13}$$

Overbars denote depth-averaged quantities. So far, no particular velocity or concentration profile has been assumed and the set of equations holds whatever the velocity and concentration distributions across the flow layer.

Bed-load mass balance equation

Besides the flow and transport equations (10)-(13), an additional mass balance equation for the bed-load is necessary to constitute the complete morphodynamic model. As sketched in Fig. 2, two layers may be distinguished in the bed material. The lower layer consists of bed material at rest, while bed-load takes place in the upper layer. Sediment continuity in the bed-load layer is expressed by Exner equation:

$$(1-p)\frac{\partial z_b}{\partial t} + \frac{\partial q_{bx}}{\partial x} + \frac{\partial q_{by}}{\partial y} = -e_b \tag{14}$$

Notations q_{bx} and q_{by} denote the bed-load unit discharges along x and y respectively. The sediment flux normal to the bed $e_b = E - D$ represents the net sediment exchange rate between the bed and the flow layer.

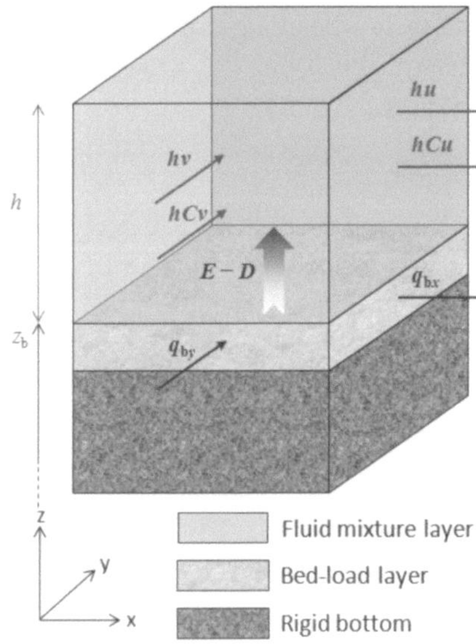

Fig. 2. Schematic configuration underlying the depth-averaged morphodynamic model.

Characteristics of flow-sediment mixtures

To derive our two-phase morphodynamic model from the general depth-averaged continuity and momentum equations (10)-(13) for the sediment-laden flow and for the dispersed phase (suspended load), adequate density, velocity and concentration profiles must be we incorporated in equations (10)-(13). For a water-sediment mixture, density and mass concentration are given by:

$$\rho = \rho_w (1 - C) + \rho_s C = \rho_w \left[1 + (s-1)C\right] = \rho_w (1 + \Delta s C) \tag{15}$$

$$\phi = \frac{\rho_s C}{\rho} = \frac{\rho_s C}{\rho_w (1 + \Delta s C)} = \frac{sC}{1 + \Delta s C} \tag{16}$$

with $s = \rho_s / \rho_w$ the relative density of sediments and the notation Δ is defined as $\Delta s = s - 1$. The real concentration profile is highly case-dependent and, for transient and varied flows, it may considerably deviate from computed equilibrium distributions, such as Rouse-type profiles. Therefore, we simply assume here a uniform concentration profile, which is found close to observations for highly transient flows accompanied by high transport rates. For similar reasons, the velocity profile is also simply assumed to be uniform.

We neglect the possible effects of differential advection of momentum due to a non-uniform velocity profile. Indeed, inclusion of a correction factor, so-called Boussinesq coefficient (Hervouet et al. 2003), appears unnecessary for a wide range of applications, in which the shear layer remains localized near the bed.

Based on the shallow flow assumption, the pressure distribution may be deduced from the simplified form (9) of the momentum equation along z, the direction normal to the main flow plane:

$$\frac{\partial p}{\partial z} = -\rho g \sin\theta_z = -\rho_w (1 + \Delta s C) g \sin\theta_z \, . \tag{17}$$

Assuming a uniform concentration profile along z and using the atmospheric pressure at the free surface as reference pressure, the following expressions involved in the governing equations (11)-(12) may be evaluated:

$$h\overline{p} = \rho_w\left(1+\Delta s\overline{C}\right)g\sin\theta_z h^2/2 \quad \text{and} \quad p_b = \rho_w\left(1+\Delta s\overline{C}\right)g\sin\theta_z h\,. \tag{18}$$

Diffusive fluxes in the sediment advection-diffusion equation are evaluated as follows:

$$\overline{q^{d}_{C,x}} = -\frac{\nu_T}{\sigma_T}\frac{\partial\overline{C}}{\partial x} \quad \text{and} \quad \overline{q^{d}_{C,y}} = -\frac{\nu_T}{\sigma_T}\frac{\partial\overline{C}}{\partial y}, \tag{19}$$

where the Schmidt number σ_T accounts for the difference between eddy viscosity and turbulent diffusivity of sediments. Although a generally accepted evaluation of this parameter is still lacking (Cao and Carling 2002a), it is reported to vary between 0.8 and 1.2 and is often simply set to unity.

Governing equations for flow-sediment mixtures

Introducing the results (15)-(19) into equations (10)-(13) leads to the following formulation of the governing equations for two-phase mixtures of water and suspended sediments. The mixture continuity equation becomes:

$$\frac{\partial}{\partial t}\left[h\left(1+\Delta s\overline{C}\right)\right]+\frac{\partial}{\partial x}\left[h\left(1+\Delta s\overline{C}\right)\overline{u}\right]+\frac{\partial}{\partial y}\left[h\left(1+\Delta s\overline{C}\right)\overline{v}\right]=\left[1+\Delta s(1-p)\right]\frac{e_b}{1-p} \tag{20}$$

whereas, using expressions (18), the mixture momentum equations may be written as:

$$\begin{aligned}&\frac{\partial}{\partial t}\left[h\left(1+\Delta s\overline{C}\right)\overline{u}\right]+\frac{\partial}{\partial x}\left[h\left(1+\Delta s\overline{C}\right)\overline{u}^2\right]+\frac{\partial}{\partial y}\left[h\left(1+\Delta s\overline{C}\right)\overline{u}\,\overline{v}\right]\\&\quad+\frac{\partial}{\partial x}\left[\left(1+\Delta s\overline{C}\right)g\sin\theta_z\frac{h^2}{2}\right]+h\left(1+\Delta s\overline{C}\right)g\sin\theta_z\frac{\partial z_b}{\partial x}\\&\quad=\left[1+\Delta s(1-p)\right]\frac{e_b-|e_b|}{2(1-p)}\overline{u}\beta+\frac{\tau_{bx}}{\rho_w}\Delta\Sigma+h\left(1+\Delta s\overline{C}\right)g\sin\theta_x+\frac{1}{\rho_w}\left(\frac{\partial h\overline{\sigma_x}}{\partial x}+\frac{\partial h\overline{\tau_{xy}}}{\partial y}\right),\end{aligned} \tag{21}$$

$$\begin{aligned}&\frac{\partial}{\partial t}\left[h\left(1+\Delta s\overline{C}\right)\overline{v}\right]+\frac{\partial}{\partial x}\left[h\left(1+\Delta s\overline{C}\right)\overline{u}\,\overline{v}\right]+\frac{\partial}{\partial y}\left[h\left(1+\Delta s\overline{C}\right)\overline{v}^2\right]\\&\quad+\frac{\partial}{\partial y}\left[\left(1+\Delta s\overline{C}\right)g\sin\theta_z\frac{h^2}{2}\right]+h\left(1+\Delta s\overline{C}\right)g\sin\theta_z\frac{\partial z_b}{\partial y}\\&\quad=\left[1+\Delta s(1-p)\right]\frac{e_b-|e_b|}{2(1-p)}\overline{v}\beta+\frac{\tau_{by}}{\rho_w}\Delta\Sigma+h\left(1+\Delta s\overline{C}\right)g\sin\theta_y+\frac{1}{\rho_w}\left(\frac{\partial h\overline{\tau_{xy}}}{\partial x}+\frac{\partial h\overline{\sigma_y}}{\partial y}\right).\end{aligned} \tag{22}$$

Expressing the density ρ as in (15) and the mass concentration ϕ as a function of the volume concentration C according to (16), the continuity equation for the dispersed phase writes:

$$\frac{\partial}{\partial t}\left(h\overline{C}\right)+\frac{\partial}{\partial x}\left(h\overline{u}\overline{C}\right)+\frac{\partial}{\partial y}\left(h\overline{v}\overline{C}\right)=-\left[\frac{\partial}{\partial x}\left(h\frac{\nu_T}{\sigma_T}\frac{\partial\overline{C}}{\partial x}\right)+\frac{\partial}{\partial y}\left(h\frac{\nu_T}{\sigma_T}\frac{\partial\overline{C}}{\partial y}\right)\right]+e_b. \tag{23}$$

Finally, the morphodynamic evolution is governed by Exner equation expressing the bed-load continuity as in (14):

$$\frac{\partial}{\partial t}\left[(1-p)z_{\mathrm{b}}\right]+\frac{\partial q_{\mathrm{b}x}}{\partial x}+\frac{\partial q_{\mathrm{b}y}}{\partial y}=-e_{\mathrm{b}}, \tag{24}$$

where $e_{\mathrm{b}} = E - D$ is the *net* erosion rate, evaluated as the difference between the erosion flux E (m/s) and the deposition flux D (m/s). A detailed discussion of the formulation of all terms involving the net erosion rate in (20)-(24) is given by Dewals (2006). The solid unit discharges $q_{\mathrm{b}x}$ and $q_{\mathrm{b}y}$ include both the flow-induced and the gravity-induced (slope failures) components of bed sediment transport.

Closure of the set of equations (20)-(24) requires a resistance formula to compute $\tau_{\mathrm{b}x}/\rho_{\mathrm{w}}$ and $\tau_{\mathrm{b}y}/\rho_{\mathrm{w}}$, a turbulence model, as well as an exchange model for evaluating $E - D$.

Instead of a uniform concentration profile, a piecewise uniform profile may also be assumed, leading to only slight modifications in the governing equations. Indeed, if the concentration is assumed to take a uniform value C_{b} for $z_{\mathrm{b}} \le z \le z_{\mathrm{b}} + h_1$ and zero above, only two changes are necessary in the equations: $h\bar{C}$ is replaced by $h_1 C_{\mathrm{b}}$ and the pressure term $\left(1+\Delta s\bar{C}\right)g\sin\theta_z\left(h^2/2\right)$ becomes $\left(h^2/2+\Delta s C_{\mathrm{b}} h_1^2/2\right)g\sin\theta_z$. Previous authors have used such approximations keeping C_{b} constant and computing the evolution of h_{s} (Fraccarollo and Capart 2002) or using appropriate empirical relations (Leal et al. 2003).

The set of governing equations (20)-(24) may be recast in the following vector form, which simplifies the formulation of the numerical scheme detailed in section 2.3:

$$\frac{\partial \mathbf{s}}{\partial t}+\frac{\partial \mathbf{f}_{\mathrm{a}}}{\partial x}+\frac{\partial \mathbf{f}_{\mathrm{d}}}{\partial x}+\mathbf{A}_{\mathrm{NC}}^{-}\frac{\partial \mathbf{s}}{\partial x}+\frac{\partial \mathbf{g}_{\mathrm{a}}}{\partial x}+\frac{\partial \mathbf{g}_{\mathrm{d}}}{\partial x}+\mathbf{B}_{\mathrm{NC}}^{-}\frac{\partial \mathbf{s}}{\partial x}=\mathbf{r} \tag{25}$$

with

$$\mathbf{s}=\left[h\left(1+\Delta s\bar{C}\right)\quad h\left(1+\Delta s\bar{C}\right)\bar{u}\quad h\left(1+\Delta s\bar{C}\right)\bar{v}\quad h\bar{C}\quad (1-p)z_{\mathrm{b}}\right]^{\mathrm{T}} \tag{26}$$

$$\mathbf{f}_{\mathrm{a}}=\left[h\left(1+\Delta s\bar{C}\right)\bar{u}\quad h\left(1+\Delta s\bar{C}\right)\bar{u}^2+gh^2\left(1+\Delta s\bar{C}\right)\sin\theta_z\quad h\left(1+\Delta s\bar{C}\right)\overline{uv}\quad h\bar{u}\bar{C}\quad q_{\mathrm{b}x}\right]^{\mathrm{T}} \tag{27}$$

$$\mathbf{g}_{\mathrm{a}}=\left[h\left(1+\Delta s\bar{C}\right)\bar{v}\quad h\left(1+\Delta s\bar{C}\right)\overline{uv}\quad h\left(1+\Delta s\bar{C}\right)\bar{v}^2+gh^2\left(1+\Delta s\bar{C}\right)\sin\theta_z\quad h\bar{v}\bar{C}\quad q_{\mathrm{b}y}\right]^{\mathrm{T}} \tag{28}$$

$$\mathbf{f}_{\mathrm{d}}=\left[0\quad \frac{h\bar{\sigma}_x}{\rho_{\mathrm{w}}}\quad \frac{h\bar{\tau}_{xy}}{\rho_{\mathrm{w}}}\quad h\frac{\nu_{\mathrm{T}}}{\sigma_{\mathrm{T}}}\frac{\partial\bar{C}}{\partial x}\quad 0\right]^{\mathrm{T}};\quad \mathbf{g}_{\mathrm{d}}=\left[0\quad \frac{h\bar{\tau}_{xy}}{\rho_{\mathrm{w}}}\quad \frac{h\bar{\sigma}_x}{\rho_{\mathrm{w}}}\quad h\frac{\nu_{\mathrm{T}}}{\sigma_{\mathrm{T}}}\frac{\partial\bar{C}}{\partial y}\quad 0\right]^{\mathrm{T}} \tag{29}$$

$$\mathbf{A}_{\mathrm{NC}}^{-}=\begin{pmatrix}0&0&0&0&0\\0&0&0&0&h\left(1+\Delta s\bar{C}\right)\dfrac{g\sin\theta_z}{1-p}\\0&0&0&0&0\\0&0&0&0&0\\0&0&0&0&0\end{pmatrix};\quad \mathbf{B}_{\mathrm{NC}}^{-}=\begin{pmatrix}0&0&0&0&0\\0&0&0&0&0\\0&0&0&0&h\left(1+\Delta s\bar{C}\right)\dfrac{g\sin\theta_z}{1-p}\\0&0&0&0&0\\0&0&0&0&0\end{pmatrix} \tag{30}$$

and $\mathbf{r}$ gathers all source terms.

2.3 Numerical discretization

The set of governing equations (20)-(24) is solved using a finite volume technique. Details of the time and space discretizations are provided here, together with a discussion on sequential vs. synchronous resolution procedures of the morphodynamic model.

Space and time discretization

The computation domain is discretized by means of a multiblock grid, in which each block consists in a locally Cartesian mesh. Since this multiblock structure enables refined meshes in critical areas (high gradients, complex geometry), it compensates for the lower flexibility of Cartesian grids, while keeping the benefits of regular grids in terms of order of accuracy, computation time and memory requirement.

The space discretization of the divergence form of equations (20)-(24)is performed by means of a finite volume scheme. Within each block, variable reconstruction at cells interfaces can be performed by constant or linear extrapolation, combined with a slope limiter, leading respectively to first or second order accuracy. Variables at the borders between adjacent blocks are extrapolated linearly, using additional ghost points. The value of the variables at the ghost points is evaluated from the value of the subjacent cells. Moreover, to ensure conservation properties at the border between adjacent blocks and thus to compute accurate continuity and momentum balance, fluxes related to the larger cells are computed at the level of the finer ones.

Advective fluxes are computed by a Flux Vector Splitting (FVS) method developed by the authors. According to this FVS, the upwinding direction of each term of the fluxes $\mathbf{f}_a$ and $\mathbf{g}_a$ is simply dictated by the sign of the flow velocity reconstructed at the cells interfaces. It has thus the advantage of being completely Froude independent and of facilitating a satisfactory adequacy with the discretization of the bed elevation gradient (Erpicum et al. 2010a). It can be formally expressed as follows:

$$\mathbf{f}_a^+ = \left[h\left(1+\Delta s\overline{C}\right)\overline{u} \quad h\left(1+\Delta s\overline{C}\right)\overline{u}^2 \quad h\left(1+\Delta s\overline{C}\right)\overline{uv} \quad h\overline{u}\overline{C} \quad q_{bx} \right]^T \tag{31}$$

$$\mathbf{f}_a^- = \left[0 \quad gh^2\left(1+\Delta s\overline{C}\right)\sin\theta_z \quad 0 \quad 0 \quad 0 \right]^T \tag{32}$$

$$\mathbf{g}_a^+ = \left[h\left(1+\Delta s\overline{C}\right)\overline{v} \quad h\left(1+\Delta s\overline{C}\right)\overline{uv} \quad h\left(1+\Delta s\overline{C}\right)\overline{v}^2 \quad h\overline{v}\overline{C} \quad q_{by} \right]^T \tag{33}$$

$$\mathbf{g}_a^- = \left[h\left(1+\Delta s\overline{C}\right)\overline{v} \quad h\left(1+\Delta s\overline{C}\right)\overline{uv} \quad h\left(1+\Delta s\overline{C}\right)\overline{v}^2 + gh^2\left(1+\Delta s\overline{C}\right)\sin\theta_z \quad h\overline{v}\overline{C} \quad q_{by} \right]^T \tag{34}$$

where the exponents + and − refer to, respectively, an upstream and a downstream evaluation of the corresponding terms. A Von Neumann stability analysis has demonstrated that this FVS leads to a stable spatial discretization of the terms $\partial \mathbf{f}_a/\partial x$ and $\partial \mathbf{g}_a/\partial y$ in (25) (Dewals 2006). Due to their diffusive nature, the fluxes $\mathbf{f}_d$ and $\mathbf{g}_d$ are legitimately evaluated by means of a centred scheme.

Since the model is applied to compute steady-state solutions, the time integration is performed by means of a three-step first order accurate Runge-Kutta algorithm, providing adequate dissipation in time. For stability reasons, the time step is constrained by the Courant–Friedrichs–Levy (CFL) condition. A semi-implicit treatment of the bottom friction term (3) is used, without requiring additional computational cost.

where δ_a is the maximum thickness of alluvium for which the non-erodible layer still affects the sediment transport, δ is the actual thickness of alluvium (Fig. 3) and ψ is a function of δ/δ_a which limits the bed-load transport flux due to the proximity of non-erodible areas. The thickness value δ_a is reported to be approximately equal to half of the bed form height. For $\delta = 0$, sediment transport is not possible anymore ($\psi = 0$); while for $\delta = \delta_a$, the presence of the non erodible layer is considered to have no influence on the bed-load discharge ($\psi = 1$). Thus, the function ψ monotonously increases between these two extreme values. The modified bed-load transport formula is then used in the Exner equation. A disadvantage of this method may rise from some lack of generality because δ_a remains a calibration parameter which may vary depending on topography, water depth, flow regime...

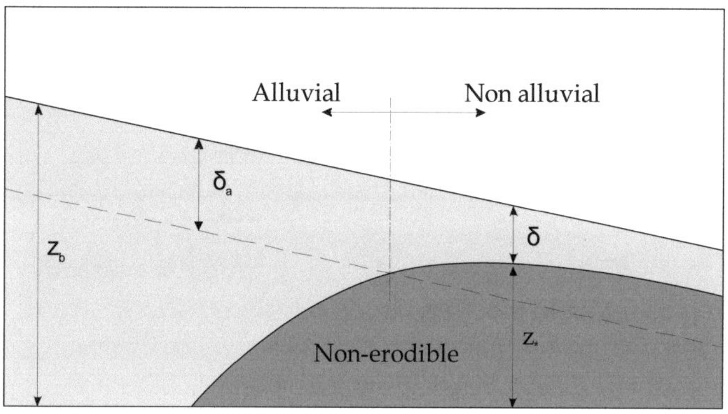

Fig. 3. Definition of the concept of alluvial and non-alluvial areas

Original procedure

Our original method introduced here uses the depth-averaged two-phase flow model presented in section 2. We focus here on bed-load transport on a partially non-alluvial bottom. Thus, the concentration is set to zero in equation (23) and the set of equations (20)-(24) becomes:

$$\frac{\partial h}{\partial t} + \frac{\partial h\overline{u}}{\partial x} + \frac{\partial h\overline{v}}{\partial y} = 0 \tag{36}$$

$$\frac{\partial h\overline{u}}{\partial t} + \frac{\partial h\overline{u}^2}{\partial x} + \frac{\partial h\overline{u}\,\overline{v}}{\partial y} + \frac{\partial}{\partial x}\left(g\sin\theta_z \frac{h^2}{2} \right) + hg\sin\theta_z \frac{\partial z_b}{\partial x} = \frac{\tau_{bx}}{\rho}\Delta\Sigma + hg\sin\theta_x + \frac{1}{\rho}\left(\frac{\partial h\overline{\sigma_x}}{\partial x} + \frac{\partial h\overline{\tau_{xy}}}{\partial y} \right), \tag{37}$$

$$\frac{\partial h\overline{v}}{\partial t} + \frac{\partial h\overline{u}\,\overline{v}}{\partial x} + \frac{\partial h\overline{v}^2}{\partial y} + \frac{\partial}{\partial y}\left(g\sin\theta_z \frac{h^2}{2} \right) + hg\sin\theta_z \frac{\partial z_b}{\partial y} = \frac{\tau_{by}}{\rho}\Delta\Sigma + hg\sin\theta_y + \frac{1}{\rho}\left(\frac{\partial h\overline{\tau_{xy}}}{\partial x} + \frac{\partial h\overline{\sigma_y}}{\partial y} \right), \tag{38}$$

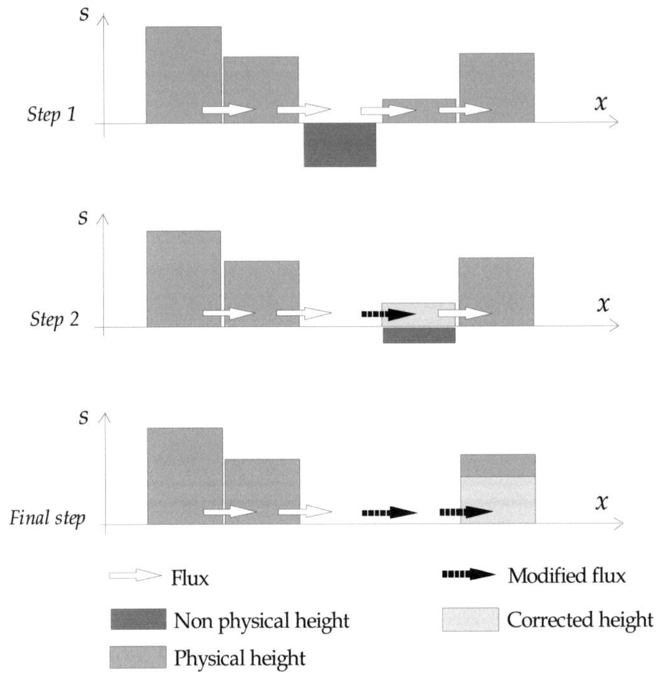

Fig. 4. Three-step procedure

$$\frac{\partial}{\partial t}\left[(1-p)z_b\right]+\frac{\partial q_{bx}}{\partial x}+\frac{\partial q_{by}}{\partial y}=0\,. \tag{39}$$

As emphasized at the beginning of this section, the difficulties in the numerical treatment of drying cells and non-alluvial beds are highly similar. This is notably due to the formal similarity in the mathematical formulation of Exner equation (39) and of the flow continuity equation (36). Our original approach consists thus in developing a single procedure to correct in a similar way the non-physical sediment level and flow depth. The general mathematical form for the continuity equations (i.e. continuity equation for the mixture and Exner equation for bed-load) can be written as:

$$\frac{\partial s}{\partial t}+\frac{\partial f_1}{\partial x}+\frac{\partial f_2}{\partial y}=0 \tag{40}$$

where s can be the sediment level or the water depth; f_1 and f_2 are fluxes in the two directions (sediment bed-load unit discharge or flow unit discharge). Thanks to an efficient iterative resolution of the continuity equations, correct mass and momentum conservation are ensured using a three-step procedure at each time step:

- Equation (40) is evaluated (step 1 in Fig. 4).
- Algorithm detects whether the current height as given by Equation (40) is under the level of the reference height (zero water depth or fixed bottom level). Thus, it highlights the occurrence of computed non physical configurations such as negative water depth and erosion of non-erodible bottom. Then, in cells with non physical configurations, the

outflow discharge is reduced (step 2 in Fig. 4 ; dashed arrow) such that the computed height becomes strictly equal to the reference height ($f_{1,new}^{out} = f_1^{out}.\alpha$ and $f_{2,new}^{out} = f_2^{out}.\alpha$).

- Since these flux corrections may in turn induce another non physical configuration in neighbouring cells, the two points above are repeated iteratively. At the end, this leads to a configuration in which the heights are all in their physical range, as shown in final step in Fig. 4.

3.2 Model verification

The depth-averaged two-phase flow model combined with the algorithm of flow depth and sediment level correction has been verified using several benchmarks leading to configurations with negative water depth or sediment transport over non-erodible bottoms. After a clear-water flow standard benchmark (dam break flow travelling on a sharp bump), scouring of a trench initially filled with sediments as well as the migration of a trench passing over a fixed bump have been tested.

Dam break flow travelling on a sharp bump

Water at rest is assumed to be initially stored in a prismatic reservoir 15.5 m long, 0.75 m wide and 0.75 m high (Fig. 5). In the downstream part of the channel, a sharp symmetric bump is located 13 m downstream of the dam, followed by water at rest with a maximum depth of 0.15 m. The bump is made of two 13.33 % slopes, with a maximum height of 0.4 m. The computational domain is a 38-m long straight channel limited by two fixed walls (Fig. 5). At time $t = 0$ s, the dam is assumed to break, releasing the previously stored water which flows downstream leading eventually to wetting and subsequently drying of the bump crest.

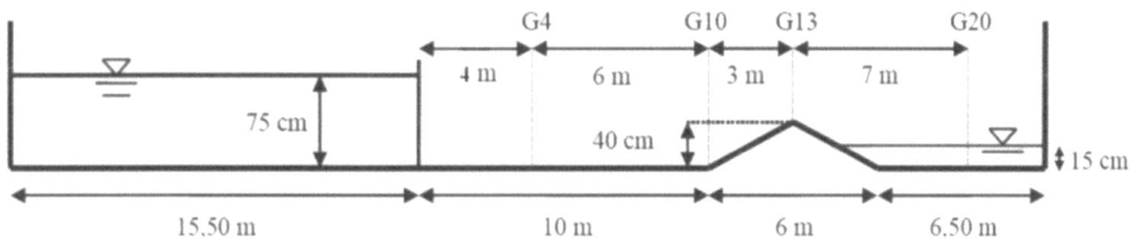

Fig. 5. Sketch of the experimental setup and location of the gauges.

The numerical simulation was carried out with a cell size of 0.1 m and a Manning coefficient equal to 0.005 s/m$^{1/3}$. The computed results are compared with experimental results (Morris and Galland 2000) obtained at four gauges (G4, G10, G13, G20 in Fig. 5). This experiment has been repeated twice (Experimental 1 and 2 in Fig. 6).

Fig. 6 shows that the computed depth remains always positive and that the numerical predictions match measurements throughout the considered time range. In particular, the reflexion wave reaching gauge G4 after approximately 13.5 seconds in the experiments is accurately reproduced in the numerical simulation; both in terms of wave velocity and wave height. The slightly noticeable temporal shift between experimental data and computed results may be attributed to the non-instantaneous collapse of the dam in the experimental setup. Finally, mass is conserved in the simulations, as verified by comparing water volumes between initial and final times: $\Delta V_{\text{water}} \approx 6 \times 10^{-10}$ m^3/m.

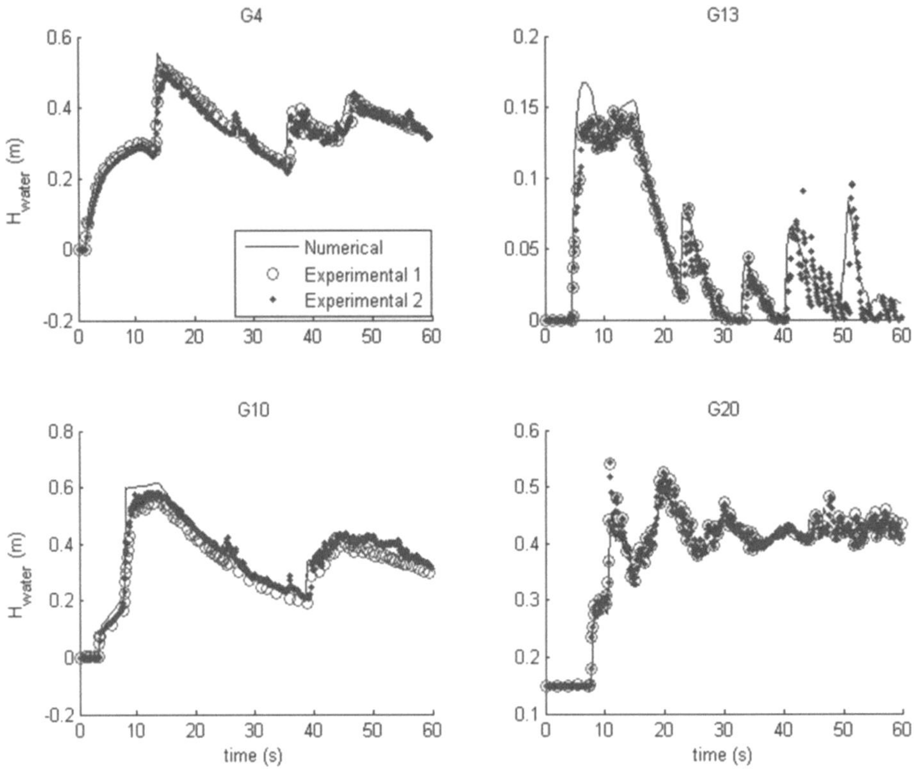

Fig. 6. Evolution of the water depth at gauges.

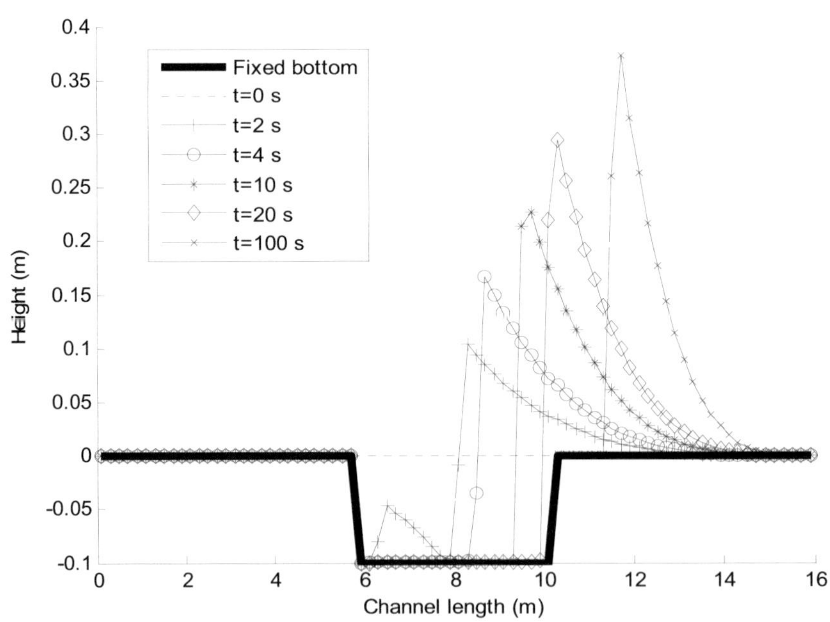

Fig. 7. Bed evolution in the hypothetical test case.

Scour of a trench initially filled with sediments

A prismatic channel is considered here, with a 1.1 m-wide rectangular cross section. The length of the channel is 16 m. The cell size used in the longitudinal direction is 0.2 m. The rigid bottom corresponds to the level zero throughout the channel, except in a 4.5 m-long trench located at mid-length of the channel, where the rigid bottom elevation is set to - 0.1 m (Hervouet et al. 2003). This trench is filled with sediments up to the level zero (see Fig. 7). In this hypothetical test case, the flow conditions are kept artificially constant in time. The flow discharge is assumed to increase linearly from 0 to 10 m^2/s between the abscissa x = 0m and the abscissa x = 8m; and decrease linearly from 10 to 0 m^2/s between the abscissa x = 8m and the abscissa x = 16m. The water depth is set everywhere equal to 2 m. The Manning coefficient is taken equal to $n = 0.04\ s/m^{1/3}$. The Engelund-Hansen bed-load transport formula is used in this case, in which the grain diameter is assumed equal to d_{50} = 0.3 mm, the specific gravity of sediments is s = 2.65 and the porosity of bed material is p = 0.375.
The computed time evolution of the bed profile is shown in Fig. 7. After 100 seconds, the bed profile does not move anymore. Although no direct comparison data are available, the model is found to perform satisfactorily since the computed bed level always remains above the level of the non erodible bottom and no mass conservation error is found in the computational results.

Evolution of a trench over a fixed bump

This test case considers the evolution of a trench passing over a non erodible bump. The length of the straight channel is 11.5 m and its width is 0.2 m. A bump is located in the middle of the domain while an approximately 0.04 m deep and 2 m long trench is excavated in the alluvial bed upstream. The grain diameter is taken equal to 0.45 mm. The cell size is 0.1 m. The computed bed-load transport law is assumed to be a power function of the water velocity: $q_b = m\, u^5$. The hydrodynamic and morphodynamic conditions are detailed in Table 1. Coefficient m in Table 1 was used as a tuning parameter to reproduce the propagation of the front of the trench over the first two meters in the upstream section (Struiksma 1999). Two experiments were carried out. In Test n°1, the thickness of alluvium on the bump is small while in test n°2, the thickness is zero. We can also observe that the two fixed bumps and the two trenches have different shapes.
Comparisons between numerical and experimental results are shown in Fig. 8. For both test-cases, experimental data are scattered but the overall agreement with numerical predictions is found satisfactory. Our new algorithm performs well since the sediment level is never computed under the level of non erodible bottom, with a mass conservation error of the order of the floating-point accuracy.

Quantity	Unit	T1	T2
Discharge	l/s	9.2	9.2
Mean water depth	m	0.106	0.106
Sediment transport (including pores)	l/h	4.0	4.4
Coefficient of sediment transport formula (m)	$10^{-4}\ s^4/m^3$	3.6	4.0
Water surface slope	mm/m	1.75	1.75
Chézy coefficient	$m^{1/2}/s$	31.8	31.8

Table 1. Hydrodynamic and morphodynamic conditions for two tests T1 and T2.

Computations however overpredict erosion depth downstream of the non erodible bump. This may result from the simplified transport capacity formula used, accounting neither for an explicit threshold for transport inception nor for gravity-induced sediment transport. Vertical accelerations might also play a part in this region. Results of T1 and T2 also reveal that the computed sediment level on the bump is underpredicted. The deeper sediment layer found experimentally may result from the medium gravels used to build the bump (non erodible under considered hydraulic conditions) leading to a high bed roughness.

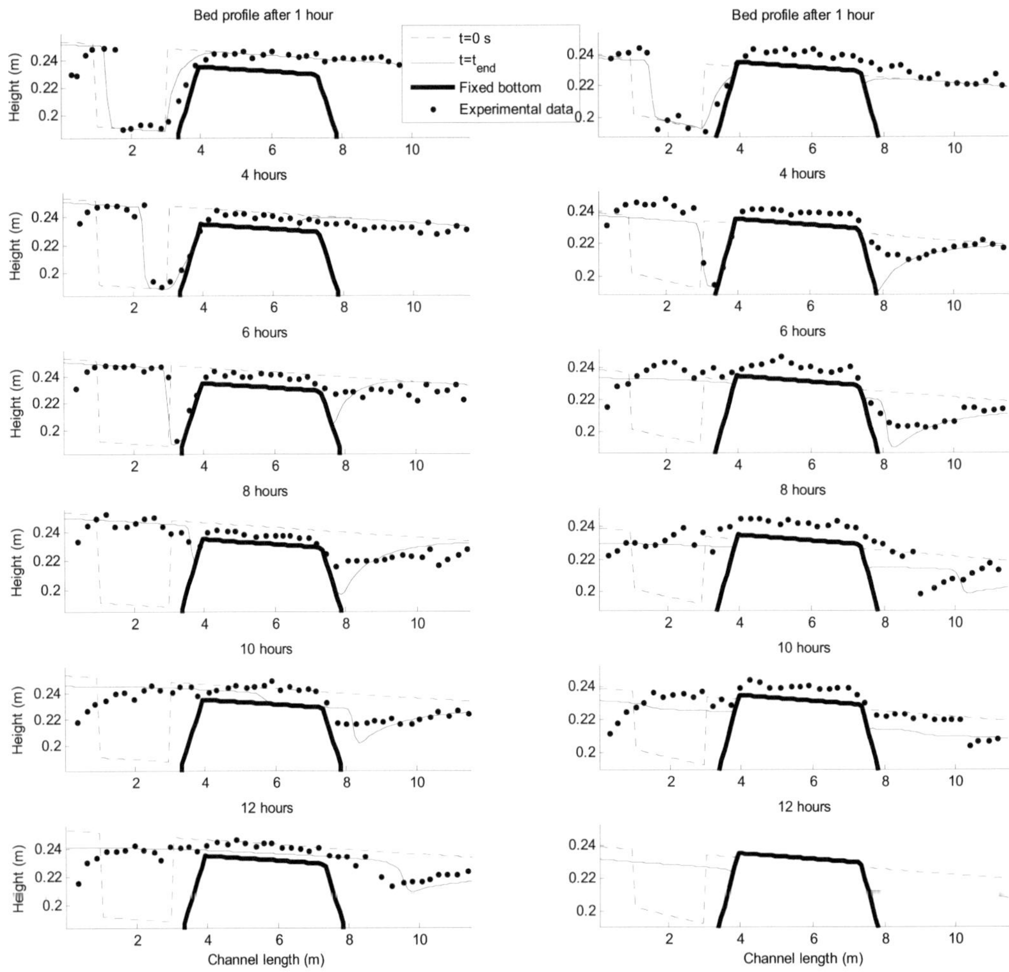

Fig. 8. Time evolution of the longitudinal bed profiles (T1, left and T2, right)

4. Application of the two-phase flow model for hyperconcentrated flows

The two-phase flow model for fluid-sediment mixtures presented in section 2 has also been used to study hyperconcentrated flows, including granular flows induced by mass failures or collapse of tailing dams on a rigid basal surface. Hyperconcentrated flows exhibit non-Newtonian behaviour and shear stress may highly depend on the concentration, properties and dimensions of the solid particles. Two different rheological models, involving both a yield stress, are investigated, namely Bingham and frictional fluid models. In the former case, also referred to as linear viscoplastic model, once the yield stress is exceeded, the shear stress is proportional to the shear rate like in viscous flows. In the later model, yield stress depends on the pore pressure following a Coulomb-type friction law.

This section focuses first on the necessary rheological models (subsection 4.1), including their appropriate formulation for inclusion into a depth-averaged flow model. Next, the set of governing equations is recast and the numerical treatment of the yield stress is discussed (subsection 4.2). The model has eventually been validated by comparisons with analytical solutions, previous numerical results and field observations, as detailed in subsection 4.3.

4.1 Rheological models for hyperconcentrated flows

In contrast with flows generally encountered in fluvial hydraulics, dominant stresses in hyperconcentrated flows usually stem not from turbulence but mainly from collisional and frictional interactions between particles. Generalized rheological fluid models are presented hereafter, both in their 3D formulation and adapted for depth-averaged modelling.

General formulation

We first briefly introduce examples of rheological models in the particular case of a simple shear flow, and then we address general three-dimensional configurations.

When a Newtonian fluid undergoes simple shear, the internal shear stress (τ) evolves linearly with the shear rate ($\dot{\gamma} = du / dz$): $\tau = \mu\, du/dz$, where μ denotes the dynamic viscosity of the fluid. Contrarily, shear stress in non-Newtonian fluids evolves non-linearly with the shear rate. Besides, the fluid may additionally be characterized by a yield stress τ_0 (Fig. 9), below which the fluid does not move, despite the application of shear stress. Bingham model is an example of yield stress fluid model, with shear stress evolving linearly as a function of the shear rate beyond the threshold (linear visco-plastic model). This corresponds to a particular case of the more general Herschel-Bulkley formulation (Fig. 9).

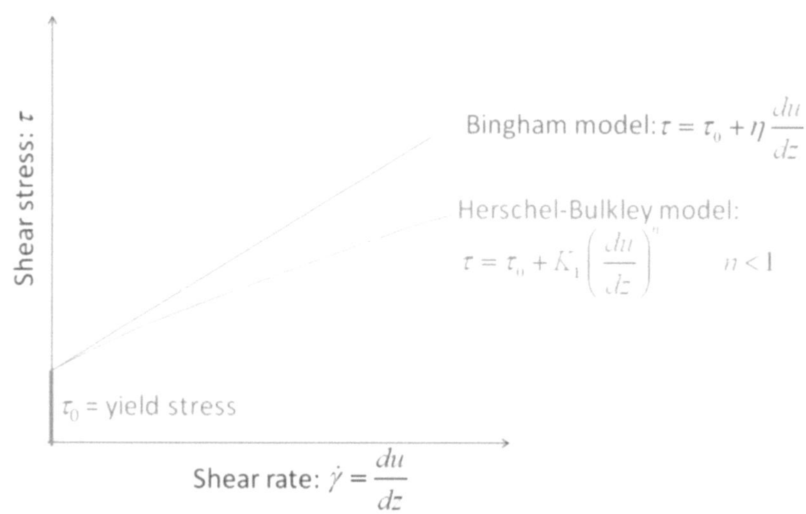

Fig. 9. Shear stress as a function of shear rate for yield stress fluids.

More generally, a rheological fluid model takes the following tensor form:

$$\boldsymbol{\sigma} = -p\mathbf{I} + \mathbf{F}(\mathbf{D}) \tag{41}$$

with $\boldsymbol{\sigma}$ the stress tensor within the fluid, p the pressure, $\mathbf{D}$ the shear rate tensor, and $\mathbf{F}$ a function to be specified depending upon the characteristics of the water-sediment mixture. The shear rate tensor is defined as:

$$D_{ij} = \frac{1}{2}\left(\frac{\partial u_i}{\partial x_j} + \frac{\partial u_j}{\partial x_i}\right) \quad \text{with} \quad i, j = x, y, z, \tag{42}$$

where x_i and u_i respectively designate spatial coordinates and velocity components.
A standard simplification for isotropic and incompressible fluids consists in specifying the general functional relationship $\mathbf{F}$ in the form of a scalar function ϕ_1 (Quecedo et al. 2004), depending only on the second invariant of the shear rate ($I_{2,\mathbf{D}}$), the first being zero as a result of the fluid incompressibility ($I_{1,\mathbf{D}} = \text{tr}\,\mathbf{D} = 0$):

$$\boldsymbol{\sigma} = -p\mathbf{I} + \phi_1\left(I_{2,\mathbf{D}}\right)\mathbf{D} \quad \text{with} \quad I_{2,\mathbf{D}} = \frac{1}{2}\text{tr}\left(\mathbf{D}^2\right). \tag{43}$$

Generally speaking, the function depends upon multiple factors, such as solid concentration, possible cohesive effects, pore pressures... The Herschel-Bulkley model (non-linear viscoplastic model) is a specific case where the function involves three parameters: yield stress τ_0, dynamic viscosity μ, and the Herschel Bulkley exponent n ($n \leq 1$):

$$\boldsymbol{\sigma} = -p\mathbf{I} + \left(\frac{\tau_0}{\sqrt{I_{2,\mathbf{D}}}} + 2\mu\left|4I_{2,\mathbf{D}}\right|^{\frac{n-1}{2}}\right)\mathbf{D} \tag{44}$$

and can in turn be particularized by choosing $n = 1$ to obtain the linear viscoplastic model:

$$\boldsymbol{\sigma} = -p\mathbf{I} + \left(\frac{\tau_0}{\sqrt{I_{2,\mathbf{D}}}} + 2\mu\right)\mathbf{D}. \tag{45}$$

Also referred to as Bingham fluid model, it involves only two parameters, assumed constant: viscosity μ and yield stress τ_0. In particular, models (44) and (45) have been applied to simulate, respectively, debris flows (e.g. Kaitna and Rickenmann 2007) and mudflows (e.g. Laigle and Coussot 1997), but also waste dump failures (e.g. Jeyapalan et al. 1903...).

Depth-averaged formulations for Bingham fluids

Similarly to all standard hydraulic resistance formulae developed for uniform flows, the depth-averaged formulation of the Bingham model is derived here assuming simple shear flow, consistently with Pastor et al. (2004).
Following notations from Fig. 1, the velocity field in a simple shear flow may be written:

$$u = U(z)\cos\alpha \qquad v = U(z)\sin\alpha \qquad w = 0 \tag{46}$$

where α represents the angle of the flow direction with respect to x- axis. Accounting for this particular velocity field, definition (42) enables to write out in full the shear rate tensor:

$$\mathbf{D} = \begin{pmatrix} 0 & 0 & \frac{1}{2}\frac{dU}{dz}\cos\alpha \\ 0 & 0 & \frac{1}{2}\frac{dU}{dz}\sin\alpha \\ \frac{1}{2}\frac{dU}{dz}\cos\alpha & \frac{1}{2}\frac{dU}{dz}\sin\alpha & 0 \end{pmatrix} \tag{47}$$

as well as its second invariant:

$$I_{2,\mathbf{D}} = \frac{1}{2}\mathrm{tr}\left(\mathbf{D}^2\right) = \frac{1}{4}\left(\frac{dU}{dz}\right)^2 . \tag{48}$$

Direct application of the rheological model (45) provides the stress tensor:

$$\boldsymbol{\sigma} = -p\mathbf{I} + \left(\frac{\tau_0}{\frac{1}{2}\frac{dU}{dz}} + 2\mu\right) = \begin{pmatrix} -p & 0 & \left(\tau_0 + \mu\frac{dU}{dz}\right)\cos\alpha \\ 0 & -p & \left(\tau_0 + \mu\frac{dU}{dz}\right)\sin\alpha \\ \left(\tau_0 + \mu\frac{dU}{dz}\right)\cos\alpha & \left(\tau_0 + \mu\frac{dU}{dz}\right)\sin\alpha & -p \end{pmatrix} \tag{49}$$

The depth-averaged flow model used hereafter (subsection 4.2) involves the following expressions, derived from the deviatoric part $\boldsymbol{\sigma}'$ of the stress tensor:

- depth-averaged normal stresses $\overline{\sigma'_{xx}}$ and $\overline{\sigma'_{yy}}$, as well as shear stress $\overline{\sigma'_{xy}}$;
- components τ_{bx} and τ_{by} of the bed shear stress $\boldsymbol{\tau}_b$, obtained from: $\boldsymbol{\tau}_b = \boldsymbol{\sigma}\cdot\mathbf{n}$.

The deviatoric part $\boldsymbol{\sigma}'$ of the stress tensor is defined by: $\boldsymbol{\sigma} = -p\,\mathbf{I} + \boldsymbol{\sigma}'$.

In the particular case of simple shear flow on a plane ($\mathbf{n} = [0\ 0\ 1]^T$), relation (49) leads to the following results:

- depth-averaged stresses are equal to zero: $\overline{\sigma'_{xx}} = \overline{\sigma'_{yy}} = \overline{\sigma'_{xy}} = 0$,
- bed shear stress is given by:

$$\tau_b = \tau_0 + \mu\left.\frac{dU}{dz}\right|_{z_b} . \tag{50}$$

This latter result, in particular dU/dz, must be expressed as a function of depth-averaged velocity, the primitive unknown of the depth-averaged model. To this end, integrating twice (50) over the flow depth, enables to obtain successively the velocity profile and the depth-averaged velocity, as detailed below.

In a simple shear flow, shear stress within the fluid varies linearly with depth:

$$\tau(z) = \tau_b\left(1 - \frac{z - z_b}{h}\right). \tag{51}$$

Consequently, two flow layers may be distinguished:

- the lower layer: $z \leq z_b + h\,(1-\tau_0/\tau_b)$, in which shear stress exceeds yield stress τ_0 and a velocity profile develops,

- the upper layer: $z > z_b + h\,(1-\tau_0/\tau_b)$, where stress remains below the yield stress, so that the fluid moves like a rigid body.

Integrating first the following combination of relations (50) and (51):

$$\tau_b\left(1-\frac{z-z_b}{h}\right)=\tau_0+\mu\frac{dU}{dz} \quad\Leftrightarrow\quad \frac{dU}{dz}=\frac{\tau_b-\tau_0}{\mu}-\frac{\tau_b}{\mu}\frac{z-z_b}{h}, \tag{52}$$

leads to:

$$\begin{aligned} U&=\frac{\tau_b-\tau_0}{\mu}(z-z_b)-\frac{\tau_b}{\mu}\frac{(z-z_b)^2}{2h} && \text{for } z\le z_b+h\left(1-\frac{\tau_0}{\tau_b}\right) \\ U&=\frac{\tau_b h}{2\mu}\left(1-\frac{\tau_0}{\tau_b}\right)^2 && \text{for } z> z_b+h\left(1-\frac{\tau_0}{\tau_b}\right) \end{aligned}. \tag{53}$$

The depth-averaged velocity may subsequently be deduced from (53):

$$\bar{u}=\frac{\tau_b h}{2\mu}\int_0^{\frac{z_0-z_b}{h}}\left[2\left(1-\frac{\tau_0}{\tau_b}\right)\eta-\eta^2\right]d\eta+\frac{\tau_b h}{2\mu}\int_{\frac{z_0-z_b}{h}}^{1}\left(1-\frac{\tau_0}{\tau_b}\right)^2 d\eta \quad \text{with } z_0=z_b+h\left(1-\frac{\tau_0}{\tau_b}\right). \tag{54}$$

which eventually leads to the following relationship between bed shear stress τ_b and depth-average velocity $\bar{u}$:

$$\xi^3-(3+a)\xi+2=0 \quad \text{with } \xi=\frac{\tau_0}{\tau_b} \quad \text{and } a=\frac{6\mu\bar{u}}{h\tau_0}. \tag{55}$$

In this depth-averaged formulation of Bingham rheological model, bed shear stress is evaluated numerically from the root of a third order polynomial, using a Newton-Raphson procedure. Indeed, no convenient analytical solution may be found since it corresponds to a *casus irreducibilis* according to Cardano's formulae (Quecedo et al. 2004).

The Newton-Raphson iterative process is made as effective as possible by appropriately choosing the first iterate as the root between 0 and 1 of the following second degree polynomial (Pastor et al. 2004):

$$\frac{3}{2}\xi^2-\left(\frac{114}{32}+a\right)\xi+\frac{65}{32}=0\,, \tag{56}$$

which constitutes the best possible approximation of polynomial (55).

The Bingham model applies if variations in pore pressure remain low, which is verified in two extreme cases: either high permeability of the mixture (pore pressures dissipate fast due to a long runout time compared to the consolidation time) or low permeability of the mixture (pore pressures hardly vary during runout).

Depth-averaged formulations for frictional fluids

Compared to the Bingham model, in the pure frictional fluid model the viscous term cancels ($\mu = 0$) and the yield stress depends on effective pressure in the material through a Mohr-Coulomb type relation:

$$\tau_0 = p' \tan\phi' = (p - p_w)\tan\phi' \tag{57}$$

where p' denotes effective pressure, p_w pore-pressure and ϕ' the effective friction angle. Hence relation (57) directly provides the bed shear stress, with no double integration being required, given that neither mixture velocity nor velocity gradients intervene directly in the expression:

$$\tau_b = p'_b \tan\phi' = (p_b - p_{w,b})\tan\phi' . \tag{58}$$

The frictional fluid model may be combined with a simple consolidation model (Hutchinson 1986), applying an exponential decrease in pore-pressure over time:

$$\tau_0 = p'_b \tan\phi' = p_b\left(1 - r_u^0 e^{-\frac{t}{T_c}}\right)\tan\phi' = \rho g h\left(1 - r_u^0 e^{-\frac{t}{T_c}}\right)\tan\phi' , \tag{59}$$

where r_u^0 represents the ratio between pore-pressure and initial pressure and T_c represents the characteristic consolidation time (Hungr 1995), given by:

$$T_c = \frac{4h_s^2}{\pi^2 c_v} \qquad \text{with } c_v = \frac{k}{m_v \gamma} , \tag{60}$$

with c_v, k and m_v denoting, respectively, the consolidation coefficient, the permeability of the material and the compressibility coefficient. The frictional fluid model has been used notably to simulate waste dump failures (Pastor et al. 2002).

4.2 Governing equations

We present here a particularized form of model (20)-(22) suitable for simulating hyperconcentrated flows on rigid basal surfaces. Numerical implementation of rheological models involving a yield stress is also detailed.

Depth-averaged two-phase model for hyperconcentrated flows

Water density ρ_w, solid density ρ_s and solid fraction C are assumed constant; so that the continuity equation (20) becomes simply:

$$\frac{\partial h}{\partial t} + \frac{\partial h\bar{u}}{\partial x} + \frac{\partial h\bar{v}}{\partial y} = 0 , \tag{61}$$

in which the net erosion rate e_b has been set to zero since the basal surface is assumed non-erodible.

Similarly, momentum equations become:

$$\frac{\partial h\bar{u}}{\partial t} + \frac{\partial h\bar{u}^2}{\partial x} + \frac{\partial h\bar{u}\bar{v}}{\partial y} + \frac{\partial}{\partial x}\left(g\sin\theta_z \frac{h^2}{2}\right) + hg\sin\theta_z \frac{\partial z_b}{\partial x} = \frac{\tau_{bx}}{\rho_m}\Delta\Sigma + hg\sin\theta_x , \tag{62}$$

$$\frac{\partial h\bar{v}}{\partial t} + \frac{\partial h\bar{u}\bar{v}}{\partial x} + \frac{\partial h\bar{v}^2}{\partial y} + \frac{\partial}{\partial y}\left(g\sin\theta_z \frac{h^2}{2}\right) + hg\sin\theta_z \frac{\partial z_b}{\partial y} = \frac{\tau_{by}}{\rho_m}\Delta\Sigma + hg\sin\theta_y . \tag{63}$$

where the mixture density $\rho_m = \rho_w\left(1+\Delta s\bar{C}\right)$ has been introduced. Terms involving depth-averaged stresses have all been lumped into the flow resistance terms involving τ_b.

Numerical treatment of the yield stress

If a kinematic or diffusive wave model was used, the yield stress could be treated in a straightforward way through the algebraic relation providing velocity: when the yield stress exceeds the bed or surface gradient, velocity remains zero.

In contrast, in the case of a dynamic wave model such as used here, velocity components are evaluated from the numerical integration of partial differential equations and the accounting for the yield behaviour of the material is less straightforward, in particular in multidirectional configurations.

The main point in the numerical treatment of the yield stress consists in preventing this yield stress to cause velocity reversal, whereas such a reversal should not be prevented if it results from the action of other contributions in the equations such as adverse topographic slope. More precisely, evaluation of velocity components at each time-step is split into three stages:

1. firstly, a first velocity predictor $\mathbf{u}_{w/o}$ is evaluated without taking into account the flow resistance terms (including yield stress)
2. subsequently, a second predictor $\mathbf{u}_w$ is evaluated by resolving the complete equations (62) and (63), including the flow resistance terms.
3. finally, the value of the velocity finally retained depends on the relative position of vectors $\mathbf{u}_{w/o}$ and $\mathbf{u}_w$:
   a. if their scalar product is positive, the effect of the term of flow resistance corresponds to a deceleration in flow and the predictor $\mathbf{u}_{w/o}$ may be retained as the new velocity value;
   b. if their scalar product is negative, the term of flow resistance results in reversal of the flow, which is not physically sound because in reality the fluid would stop such a case; therefore the new velocity value is simply set to zero.

4.3 Model verification

Examples of model verification are presented here, namely a slump test, treated with both Bingham and frictional fluid rheological models, as well as a case of failure of a real tailing dam. More model verifications will be detailed in subsequent contributions, enabling to systematically validate all components of the model, including the purely viscous stresses, yield stress as well as their one- and two-dimensional implementations.

Slump test

The slump test consists in the sudden release of a cone of material which was previously confined. For a Bingham fluid, the one-dimensional profile of the material at the end of the test corresponds to an analytical solution of the system of equations (61)-(62), combined with the flow resistance formula (55). Indeed, when the flow stops, all velocity components are zero and only the following terms remain in equation (62), expressing the balance between the surface slope and the yield stress:

$$-\frac{\partial}{\partial x}\left(\rho g \frac{h^2}{2}\right) - \tau_0 = 0 \qquad \Rightarrow \qquad h = h_0\sqrt{1-\frac{2\tau_0}{\rho g h_0}\frac{x}{h_0}}\,. \tag{64}$$

This leads to a parabolic free surface profile when the material freezes (Fig. 10), the shape of which depends on the yield stress τ_0 and the initial height of the material h_0.
The model has been similarly verified for the frictional fluid rheological model. Numerical predications could be successfully compared with those from the model by Manganey-Castelnau (2005).

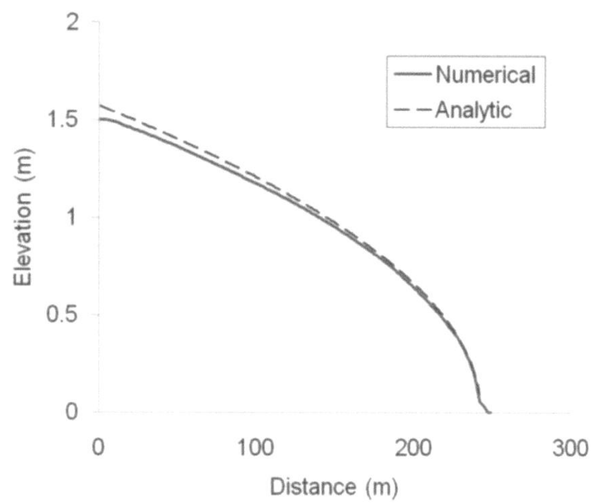

Fig. 10. Numerical predictions (Bingham model) vs. analytical solution for the slump test.

Gypsum tailing dam

Jeyapalan et al. (1983) described the flow of liquefied mine residues following the failure of the "Gypsum Dam" in Texas in 1966. The deposits were confined inside a rectangular reservoir with a depth of 11m at the time of the accident. Following seepage at the toe, a 140m breach opened up in the dike. The flow stretched to over 300m in length, with velocities in the order of 2.5 to 5 m/s. The material was characterised by a mean diameter of 70 microns and a density of 2,450 kg/m^3.
Consistently with Pastor et al. (2002), the simulation has been conducted using the Bingham fluid rheological model, with a yield stress $\tau_0 = 10^3$ Pa and a viscosity $\mu = 50$ Pa s. The density of the mixture was estimated at 1,400 kg/m^3.
Fig. 11 shows the agreement between the reference results by Pastor et al. (2002) and the predictions of the model developed here. In particular, a hydraulic jump appears around $t = 60$ s, following the stoppage of material situated along the breach axis, while the flow continues laterally until around $t = 120$ s in the more upstream part the wave.

5. Conclusion

Numerous issues remain challenging in current modelling capacities of flow, sediment transport and morphodynamics. In this chapter we have addressed two of them, namely handling mixed alluvial and non-alluvial beds and modelling hyperconcentrated flows.
Those two topics have been analyzed within an original modelling framework developed by the authors. It relies on a two-phase flow model set up to describe the flow of water-sediment mixtures. Increased inertia of the mixture as a result of the sediment concentration is accounted for in the momentum equations, which is hardly ever the case in currently

available morphodynamic models. To this end, the local continuity and momentum equations for the mixture have been depth-averaged without assuming straightaway particular concentration and velocity profiles, resulting in a generalized formulation of depth-averaged equations for water-sediment mixtures.

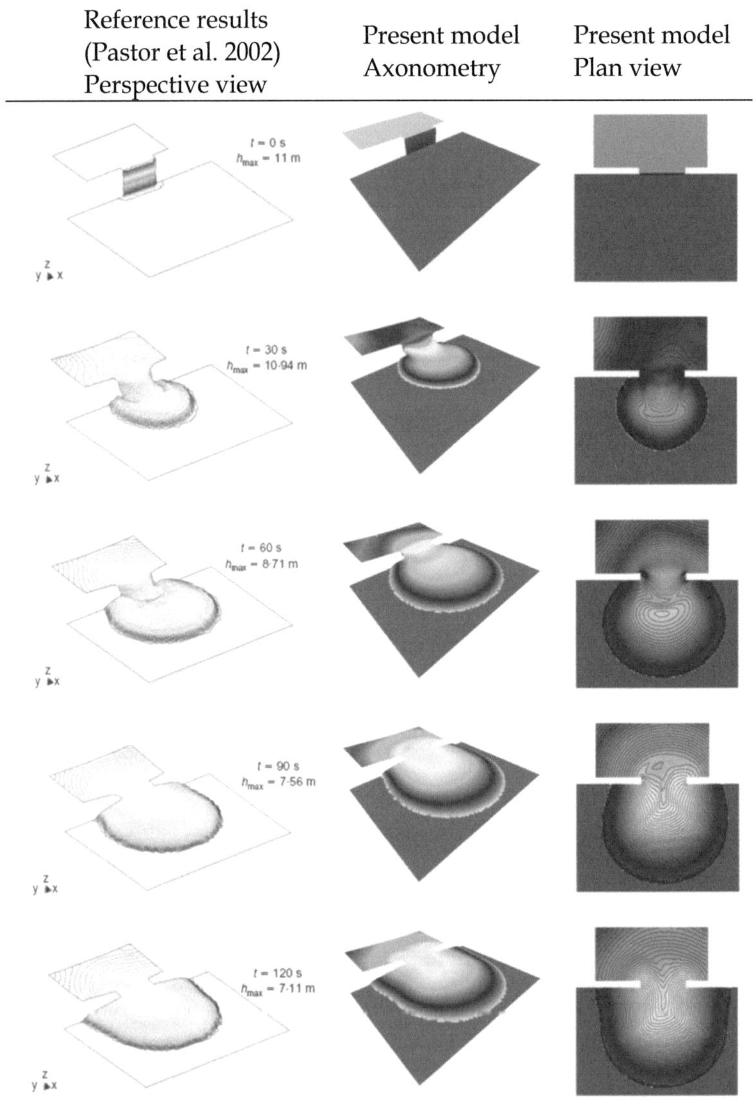

Fig. 11. Reference results in perspective (left), axonometric view and plan view of predictions of the model developed here for the "Gypsum dam" collapse.

In addition, an existing finite volume model for shallow flows has been accommodated to solve the generalized two-phase model for water-sediment mixtures. The stability of the extended scheme was demonstrated by Dewals (2006) and the resulting model succeeds in handling the wide range of time scales involved in practical sediment transport problems (Dewals et al. 2008a). Indeed, as a result of the flexibility offered in the levels of coupling between flow and sediment transport models, stable and accurate numerical solutions are obtained in a realistic CPU time for predictions of erosion and sedimentation patterns in the short, medium or long term, considering both bed-load and suspended load.

The set of governing equations has subsequently been particularized for two specific configurations, namely bed-load transport on partly non-alluvial beds and rapid runout of hyperconcentrated flows such as flowslides, mudflows or debris flows.

A unified algorithm with correction on the outward fluxes of the continuity equations in fluid mixture and sediment layer has been implemented in our two-phase depth-averaged flow model in order to deal with drying cells and sediment routing over partly non-alluvial beds. Our original contribution lies here in the unified mathematical treatment of these two issues. The new procedure has been successfully verified on three test cases, in which the flow and sediment mass conservation error has been shown to remain of the order of the floating-point accuracy.

Finally, our two-phase depth-averaged flow model has been adapted to account for the particular rheology of hyperconcentrated flows, including visco-plasticity and frictional behaviour influenced by pore pressure. A depth-averaged formulation of these rheological models has been derived. Based on mass and momentum conservation for the mixture of sediment and interstitial fluid, the resulting finite volume model has been shown to handle successfully flow initiation, propagation (including on dry areas) and stoppage consistently with the yield stress behaviour observed in nature and experiments. An original numerical treatment of the yield stress has been presented and applies for multidimensional problems, both for the Bingham fluid and the frictional fluid models. This newly elaborated model has been verified by comparison with a number of experimental, numerical and field data; and is readily available for practical applications such as flowslide hazard mapping and emergency planning. The feasibility and opportunity to develop a rheological model further integrating the "Bingham fluid" and "frictional fluid" approaches will be explored in future research.

6. References

Audusse, E., Bouchut, F., Bristeau, Klein, R., and Perthame, B. (2004). "A Fast and Stable Well-Balanced Scheme with Hydrostatic Reconstruction for Shallow Water Flows." *SIAM J. Sci. Comput.*, 25(6), 2050-2065.

Begnudelli, L., and Sanders, B. F. (2007). "Conservative Wetting and Drying Methodology for Quadrilateral Grid Finite-Volume Models." *Journal of Hydraulic Engineering*, 133(3), 312-322.

Bombardelli, F., and Jha, S. (2009). "Hierarchical modeling of the dilute transport of suspended sediment in open channels." *Environmental Fluid Mechanics*, 9(2), 207-235.

Cao, Z., and Carling, P. A. (2002a). "Mathematical modelling of alluvial rivers: reality and myth. Part 1: General review." *Water & Maritime Engineering*, 154(3), 207-219.

Cao, Z., and Carling, P. A. (2002b). "Mathematical modelling of alluvial rivers: reality and myth. Part 2: Special issues." *Water & Maritime Engineering*, 154(4), 297-307.

Cao, Z., Day, R., and Egashira, S. (2002). "Coupled and Decoupled Numerical Modeling of Flow and Morphological Evolution in Alluvial Rivers." *Journal of Hydraulic Engineering*, 128(3), 306-321.

Cao, Z., Wei, L., and Xie, J. (1995). "Sediment-Laden Flow in Open Channels from Two-Phase Flow Viewpoint." *Journal of Hydraulic Engineering*, 121(10), 725-735.

Dewals, B., Archambeau, P., Erpicum, S., Mouzelard, T., and Pirotton, M. (2002a). "Coupled computations of highly erosive flows with WOLF software." *Proc. 5th Int. Conf. on Hydro-Science & -Engineering*, Warsaw, 10 p.

Dewals, B., Archambeau, P., Erpicum, S., Mouzelard, T., and Pirotton, M. (2002b). "Dam-break hazard mitigation with geomorphic flow computation, using WOLF 2D hydrodynamic software." Risk Analysis III, C. A. Brebbia, ed., WIT Press, 59-68.

Dewals, B., Archambeau, P., Erpicum, S., Mouzelard, T., and Pirotton, M. (2002c). "An integrated approach for modelling gradual dam failures and downstream wave propagation." *Proc. 1st IMPACT Project Workshop*, Wallingford.

Dewals, B. J. (2006). "Une approche unifiée pour la modélisation d'écoulements à surface libre, de leur effet érosif sur une structure et de leur interaction avec divers constituants," PhD Thesis, University of Liege, Liège.

Dewals, B. J., Erpicum, S., Archambeau, P., Detrembleur, S., Fraikin, C., and Pirotton, M. (2004). "Large scale 2D numerical modelling of reservoirs sedimentation and flushing operations." *Proc. 9th Int. Symposium on River Sedimentation*, Yichang, Chine.

Dewals, B. J., Erpicum, S., Archambeau, P., Detrembleur, S., and Pirotton, M. (2006). "Depth-integrated flow modelling taking into account bottom curvature." *J. Hydraul. Res.*, 44(6), 787-795.

Dewals, B. J., Erpicum, S., Archambeau, P., Detrembleur, S., and Pirotton, M. (2008a). "Hétérogénéité des échelles spatio-temporelles d'écoulements hydrosédimentaires et modélisation numérique." *Houille Blanche-Rev. Int.*(5), 109-114.

Dewals, B. J., Erpicum, S., Detrembleur, S., Archambeau, P., and Pirotton, M. (2010a). "Failure of dams arranged in series or in complex." *Natural Hazards*, Published online: 27 Aug. 2010. DOI: 10.1007/s11069-010-9600-z.

Dewals, B. J., Kantoush, S. A., Erpicum, S., Pirotton, M., and Schleiss, A. J. (2008b). "Experimental and numerical analysis of flow instabilities in rectangular shallow basins." *Environ. Fluid Mech.*, 8, 31-54.

Dewals, B. J., Rulot, F., Erpicum, S., Archambeau, P., and Pirotton, M. (2010b). "Long-term sediment management for sustainable hydropower." Comprehensive Renewable Energy. Volume 6 - Hydro Power, A. Sayigh, ed., Elsevier, Oxford.

Ernst, J., Dewals, B. J., Detrembleur, S., Archambeau, P., Erpicum, S., and Pirotton, M. (2010). "Micro-scale flood risk analysis based on detailed 2D hydraulic modelling and high resolution land use data." Nat. Hazards, 55 (2), 181-209.

Erpicum, S., Dewals, B. J., Archambeau, P., Detrembleur, S., and Pirotton, M. (2010a). "Detailed inundation modelling using high resolution DEMs." *Engineering Applications of Computational Fluid Mechanics*, 4(2), 196-208.

Erpicum, S., Dewals, B. J., Archambeau, P., and Pirotton, M. (2010b). "Dam-break flow computation based on an efficient flux-vector splitting." *Journal of Computational and Applied Mathematics*, 234(7), 2143-2151.

Erpicum, S., Meile, T., Dewals, B. J., Pirotton, M., and Schleiss, A. J. (2009). "2D numerical flow modeling in a macro-rough channel." *Int. J. Numer. Methods Fluids*, 61(11), 1227-1246.

Fraccarollo, L., and Capart, H. (2002). "Riemann wave description of erosional dam-break flows." *J. Fluid Mech.*, 461, 183-228.

Frey, P., and Church, M. (2009). "How River Beds Move." *Science*, 325(5947), 1509-1510.

Gourgue, O., Comblen, R., Lambrechts, J., Kärnä, T., Legat, V., and Deleersnijder, E. (2009). "A flux-limiting wetting-drying method for finite-element shallow-water models, with application to the Scheldt Estuary." *Advances in Water Resources*, 32(12), 1726-1739.

Greco, M., Iervolino, M., Vacca, A., and Leopardi, A. (2008). "A two-phase model for sediment transport and bed evolution in unsteady river flow." River flow 2008, Altinakar, Kirkgoz, Kokpinar, Aydin, and Cokgor, eds., Izmir, Turkey.

Greimann, B. P., and Holly, F. M. (2001). "Two-Phase Flow Analysis of Concentration Profiles." *Journal of Hydraulic Engineering*, 127(9), 753-762.

Greimann, B. P., Muste, M., and Holly, F. M. J. (1999). "Two-phase Formulation of Suspended Sediment Transport." *Journal of Hydraulic Research*, 37(4), 479-491.

Hervouet, J.-M. (2003). *Hydrodynamique des écoulements à surface libre - Modélisation numérique avec la méthode des éléments finis*, Presses de l'école nationale des Ponts et Chaussées, Paris.

Hervouet, J., Machet, C., and Villaret, C. (2003). "Calcul des évolutions sédimentaires : le traitement des fonds rigides." *Revue européenne des éléments finis*, 12(2-3), 221-234.

Hungr, O. (1995), *A model for the runout analysis of rapid flow slides, debris flows, and avalanches*, Canadian Geotechnical Journal, 32 (4), 610-623.

Hutchinson, J. N. (1986), *A sliding-consolidation model for flow slides*, Canadian Geotechnical Journal, 23 (2), 115-126.

Jeyapalan, J. K., Duncan, and Seed. (1983). "Investigation of flow failures of tailings dams." *Journal of Geotechnical Engineering*, 109(2), 172-189.

Kaitna, R., and Rickenmann, D. (2007). "A new experimental facility for laboratory debris flow investigation." *Journal of Hydraulic Research*, 45(6), 797-810.

Kerger, F., Archambeau, P., Erpicum, S., Dewals, B. J., and Pirotton, M. (2010a). "A fast universal solver for 1D continuous and discontinuous steady flows in rivers and pipes." *Int. J. Numer. Methods Fluids*, Published online: 29 Dec 2009. DOI: 10.1002/fld.2243.

Kerger, F., Erpicum, S., Dewals, B., Archambeau, P., and Pirotton, M. (2010b). "1D Unified Mathematical Model for Environmental Flow Applicated to Aerated Mixed Flows." *Advances in Engineering Software*, In press.

Kerger, F., Dewals, B., Archambeau, P., Erpicum, S. and Pirotton, M. (2011), *A Multiphase Model for the Transport of Dispersed Phases in Environmental Flows: Theoretical Contribution*, European Journal of Mechanical and Environmental Engineering, In press.

Khuat Duy, B., Archambeau, P., Dewals, B. J., Erpicum, S., and Pirotton, M. (2010). "River modelling and flood mitigation in a Belgian catchment." *Proc. Inst. Civil. Eng.-Water Manag.*, 163(8), 417-423.

Kleinhans, M. G., Jagers, H. R. A., Mosselman, E., and Sloff, C. J. (2008). "Bifurcation dynamics and avulsion duration in meandering rivers by one-dimensional and three-dimensional models." *Water Resources Research*, 44, 31 PP.

Laigle, D., and Coussot, P. (1997). "Numerical modeling of mudflows." *Journal of Hydraulic Engineering*, 123(7), 617-623.

Leal, J. G. A. B., Ferreira, R. M. L., and Cardoso, A. H. (2003). "Dam-break wave propagation over a cohesionless erodible bed." Proc. 30rd IAHR Congress, J. Ganoulis and P. Prinos, eds., IAHR, Thessaloniki, Grèce, 261-268.

Mangeney-Castelnau, A., Bouchut, F., Vilotte, J. P., Lajeunesse, E., Aubertin, A., and Pirulli, M. (2005). "On the use of Saint Venant equations to simulate the spreading of a granular mass." *J. Geophys. Res*, 110, B09103.

Morris, M., and Galland, J. (2000). "CADAM, Dambreak Modelling Guidelines & Best Practice." European Commission.

Nikora, V., McEwan, I., McLean, S., Coleman, S., Pokrajac, D., and Walters, R. (2007). "Double-Averaging Concept for Rough-Bed Open-Channel and Overland Flows: Theoretical Background." *Journal of Hydraulic Engineering*, 133(8), 873-883.

Pastor, M., Quecedo, M., Gonzalez, E., Herreros, M. I., Merodo, J. A. F., Merodo, J. A. F., and Mira, P. (2004). "Simple approximation to bottom friction for Bingham fluid depth integrated models." *Journal of Hydraulic Engineering*, 130(2), 149-155.

Pastor, M., Quecedo, M., Merodo, J. A. F., Herrores, M. I., Gonzalez, E., and Mira, P. (2002). "Modelling tailings dams and mine waste dumps failures." *Geotechnique*, 52(8), 579-591.

Quecedo, M., Pastor, M., Herreros, M. I., and Merodo, J. (2004). "Numerical modelling of the propagation of fast landslides using the finite element method." *Int. J. Numer. Methods Engng*, 59, 755-794.

Roger, S., Dewals, B. J., Erpicum, S., Schwanenberg, D., Schüttrumpf, H., Köngeter, J., and Pirotton, M. (2009). "Experimental und numerical investigations of dike-break induced flows." *J. Hydraul. Res.*, 47(3), 349-359.

Spasojevic, M., and Holly, F. M. (2008). "Two- and thre-dimensional numerical simulation of mobile-bed hydrodynamics and sedimentation." Sedimentation engineering : processes, measurements, modeling, and practice, H. G. Marcelo, ed., American Society of Civil Engineers, 683-761.

Struiksma, N. (1999). "Mathematical modelling of bedload transport over non-erodible layers." *Proceedings of IAHR Symposium on River, Coastal and Estuarine Morphodynamics*, Genova, 6-10.

van Rijn, L. C. (2007). "Unified View of Sediment Transport by Currents and Waves. I: Initiation of Motion, Bed Roughness, and Bed-Load Transport." *Journal of Hydraulic Engineering*, 133(6), 649-667.

van Rijn, L. C., Walstra, R., and Ormondt, M. v. (2007). "Unified View of Sediment Transport by Currents and Waves. IV: Application of Morphodynamic Model." *Journal of Hydraulic Engineering*, 133(7), 776-793.

Walters, R., and Plew, D. (2008). "Numerical modeling of environmental flows using DAM: Some preliminary results." *Acta Geophysica*, 56, 918-934.

White, R. (2001). *Evacuation of sediments from reservoirs*, Thomas Telford.

Wu, W. M., and Wang, S. S. (2000). "Mathematical models for liquid-solid two-phase flow." *Int J Sediment Res*, 15(3), 288-298.

2

Modelling Coastal Sediment Transport for Harbour Planning: Selected Case Studies

Vincent Leys[1] and Ryan P. Mulligan[2]
[1]*CBCL Limited Consulting Engineers, Halifax NS*
[2]*East Carolina University, Greenville NC*
[1]*Canada*
[2]*USA*

1. Introduction

During the planning phase of coastal development projects, it is often necessary to determine potential sedimentation and erosion rates. This is particularly relevant at harbours where dredged channels are proposed, and accurate dredging projections are crucial for economic feasibility analyses. In addition, new structures that interfere with the natural processes may have major impacts on the adjacent shoreline.

In this chapter we consider a range of approaches for evaluating sediment transport for harbour planning studies (section 2), and present two detailed cases from Atlantic Canada. The sites described are representative of very different coastal environments. They include Saint John Harbour (section 3), a uniquely dynamic estuary on the Bay of Fundy with huge tides, a very large river outflow and significant sedimentation of silt and clay presenting various navigation and dredging challenges. The other site described is located on the sandy North coast of Prince Edward Island at Darnley Inlet, an exposed area where tides, storms and sea level rise are continuously reshaping the shoreline and navigation channels (section 4).

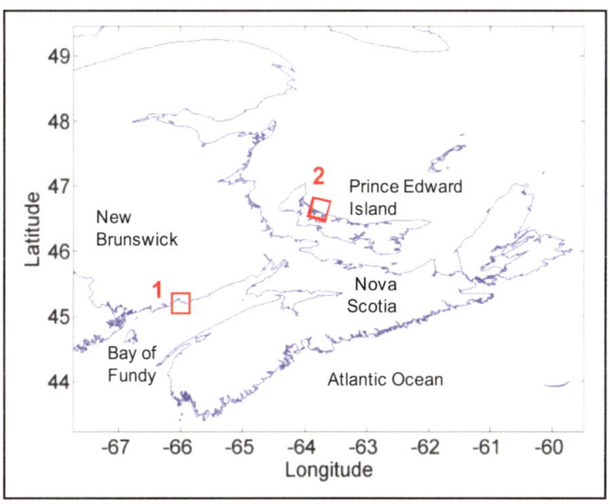

Fig. 1. Location of Saint John Harbour (1) , and Darnley Inlet (2) in Atlantic Canada

To put the case studies in perspective, a brief summary of approaches for evaluating coastal sediment transport processes is provided. The approaches include preliminary site investigations and data collection, basic sediment transport theory, and a range of numerical modelling techniques that can be applied to determine sediment erosion, transport and deposition.

2. Approaches for evaluating sediment transport

Engineering studies in natural environments have site specific conditions that require a unique approach to each problem. Therefore some or all of the following methods may need to be applied in order to determine the impacts of harbour structures on the sedimentological environment. Some adverse impacts may include interruption of the net wave-induced longshore transport causing downdrift erosion, scour at the base of breakwaters or jetties, silting-up of harbour basins requiring repeated dredging, increased agitation due to reflected waves, or increased currents through harbour openings. After the construction of new structures, sediment flows will adjust to a new equilibrium, typically over a timescale of years. Thus the effects of human-intervention on the coastal environment are not immediately obvious and coastal developments require careful planning.

Site investigations

Every harbour has a unique combination of structures, environmental forcing conditions, sediment sources and supply. Site investigations should include:

- Acquisition of bathymetry, water level, wind, wave and sediment properties information;
- Observation of shoreline features to identify erosional/depositional landforms;
- Examination of aerial photographs, which gives a larger scale view of the area and may allow other landforms to be identified. Analysing a sequence of historical aerial photography is the first (and oftentimes the most accurate) method to assess sediment processes and determine rates of change.

As a brief example, sediment flux at Arisaig, on Nova Scotia's North shore is dominated by wave-driven longshore transport supplied by sandy cliffs. The original harbour facing the direction of longshore transport became a natural sand trap. A new breakwater and extension of existing rock structures were recently considered. Some important aspects considered in the design process included impacts of episodic major storms, seasonal and annual climate variability, changes in water levels, and changes in up-drift shoreline use that affect the sediment supply from beaches, rivers or cliffs.

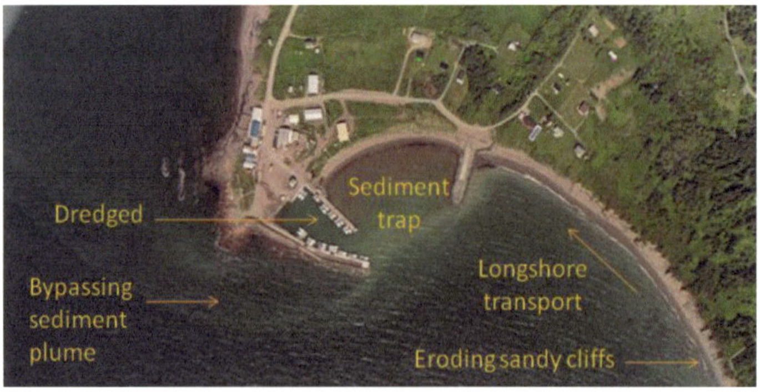

Fig. 2. Arisaig Harbour, Nova Scotia, 2003.

Shoreline contour models

Shoreline contour models simulate the evolution of one bathymetric contour (generally the shoreline at mean water level). They typically assume uniform grain size, beach profile shape and depth of closure (the seaward depth at which repeatedly surveyed profiles intersect). These models, developed for straight sandy coastlines, predate the full morphological models discussed next. However, within limitations, these models are very effective for long-term predictions of shoreline change when coastal structures are introduced or modified. As an example, consider the one-dimensional diffusion equation:

$$\frac{\partial y}{\partial t} - D\frac{\partial^2 y}{\partial x^2} = 0 \tag{1}$$

where y is the cross-shore coordinate, x is the alongshore coordinate, t is time and D (longshore diffusivity) is related to the sediment transport rate, beach profile shape and wave conditions. The equation can be solved analytically (Pelnard-Considere 1956, Dean 2002) and used to model the progressive shoreline evolution from an initially straight shoreline, assuming steady-state wave and sediment conditions and one structure perpendicular to the coast. As shown in Fig. 3, accretion against the up-drift side of the structure increases with time until the contour intersects the end of the structure, at which time bypassing begins.

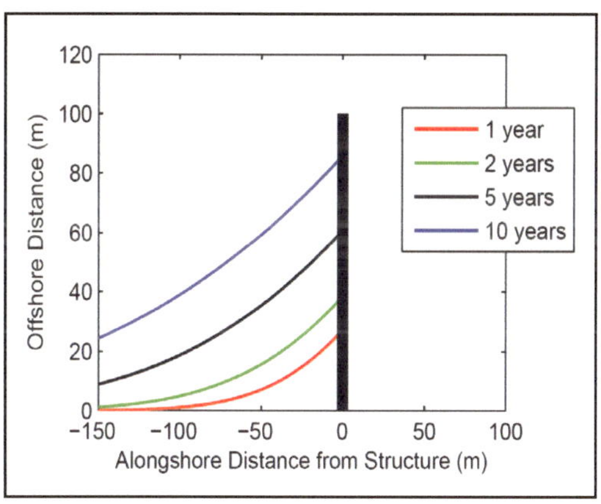

Fig. 3. Example sediment accretion along a groin estimated by a 1-D shoreline change model.

More sophisticated 1D models have been developed with the capability of simulating the beach response to the introduction of different coastal structures such as groins, detached breakwaters or seawalls. The shoreline models LITPACK (DHI 2008) and GENESIS (Generalized Model for Simulating Shoreline Change, (Gravens et al 1991)) simulate long-term averaged shoreline change produced by spatial and temporal differences in wave parameters and longshore sediment transport. The NLINE model (Dabees and Kamphuis, 2000) simulates beach evolution for multiple contour lines.

Hydrodynamic and morphological models

Morphodynamic models rely on numerical routines that explicitly predict the wave and hydrodynamic forcing, and sediment transport in two or three dimensions. The hydrodynamic models numerically solve the fluid momentum and continuity equations in order to predict water level changes, circulation and transport driven by winds, waves, tides, river discharge or density forcing. Some examples include Delft3D (Lesser et al, 2004), the Regional Ocean Modeling System (ROMS, Shchepetkin and McWilliams, 2005), Coupled MIKE21 (DHI 2009) and FVCOM (Chen et al, 2006). An example using Delft3D is shown on Fig. 4. Examples using MIKE by DHI are presented in the detailed case studies following this section. Each uses different numerical techniques, includes different features and operates on different types of computational grids (i.e. rectangular, curvilinear or unstructured).

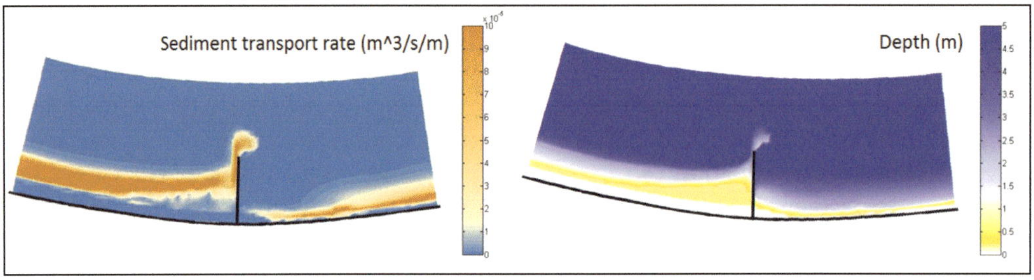

Fig. 4. Example of longshore sediment transport rate and resulting bathymetry along a sandy beach one year after introducing a groin, predicted using the Delft3D model.

Separately, a wave model is employed to predict wave transformation. The wave model is either a phase-averaging (spectral) or a phase-resolving (Boussinesq) model. The wave and hydrodynamic models typically operate on different timescales but are coupled such that they communicate at specified time steps. The wave model is used to propagate wave energy throughout the model domain, and predict changes to the wave energy distribution by refraction, diffraction, wind generation, non-linear energy transfers, dissipation (e.g. white-capping, bottom friction, and breaking) and interaction with currents. Examples of phase-averaged models include the SWAN model (Simulating Waves Nearshore, Booij et al, 1999), and MIKE21 SW (DHI 2009). Examples of phase-resolving models include CGWAVE (Demirbilek & Panchang 1998) and MIKE21 BW (DHI 2009). Phase-averaged models are typically more computationally efficient, since larger spatial resolution, larger time steps and simpler physics are used. Phase-resolving models are typically better at handling reflection and diffraction which become important processes near coastal structures and inside harbour basins.

The morphological models are coupled with the hydrodynamic models by including sediment equations to predict bottom shear stresses and track sediment concentrations through the model domain. Morphological models typically use a bed shear stress formulation in the form:

$$\tau_b = \rho C_D u_b |u_b| \tag{2}$$

where ρ is water density, u_b is the is the horizontal velocity above the bed, and C_D is a drag coefficient. The drag coefficient is proportional to von Karman's constant, the thickness of the bed layer, and the roughness length of the bed. The bed roughness length is used to parameterize sub-grid scale roughness features including bedforms and individual grains. Other sediment routines parameterize sediment processes, such as roughness in the bottom boundary layer, bedload and suspended-load transport, particle fall velocity and flocculation, with different formulations for cohesive and non-cohesive sediments. Sediment is eroded, transported, deposited and the bed morphology evolves with time. Examples are the Community Sediment Transport Model (CSTM, Warner et al, 2008), Delft3D and MIKE 3.

3. Saint John Harbour

Background

The marine physical environment of Saint John Harbour is very complex and dynamic. Key parameters such as water level, density and flow are highly variable in time and space due to the interaction of large semi-diurnal tides (of maximum range 8.9m) with strong freshwater discharge from the Saint John River, one of the largest rivers in Eastern Canada. The large tides are due to the fact that the Gulf of Maine-Bay of Fundy system is close to being in resonance with the semi-diurnal forcing from the North Atlantic (Greenberg 1990). The unusual characteristic of Saint John Harbour is that the large tides are being countered by particularly strong river outflow. The river discharges into the Harbour across a 200 m wide ridge and then through a narrow rock gorge (Fig. 6), creating spectacular rapids that

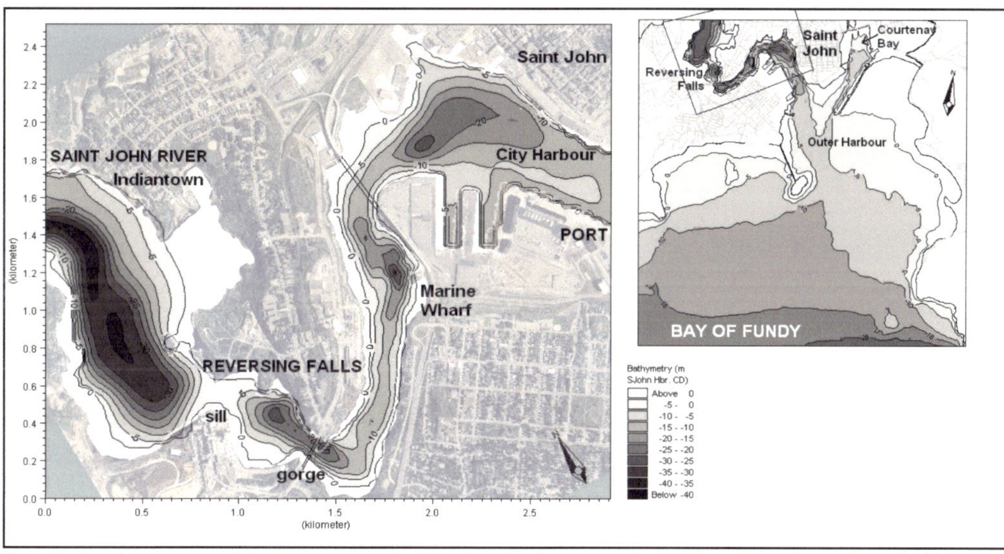

Fig. 5. Study area and bathymetry: Reversing Falls and adjacent channel reaches (left), Saint John Harbour on the Bay of Fundy (right).

Fig. 6. The Reversing Falls gorge at ebb tide, looking Northeast (16/11/2006).

reverse direction with the tides. The Reversing Falls only allow a relatively small volume discharge in and out over a tidal cycle. This hydraulic control causes a significant difference in the water levels on either side of the constriction and locally strong currents alternating in direction.

Dredging records

Downstream of the Falls, the Port of Saint John requires maintenance dredging of fine sediments settling along piers and in navigation channels. Target dredging areas for the Port of Saint John are shown in Fig. 7, along with summary grain size distributions. The dredging areas include channels in the Outer Harbour and Courtenay Bay, and deepwater berths in the City Harbour.

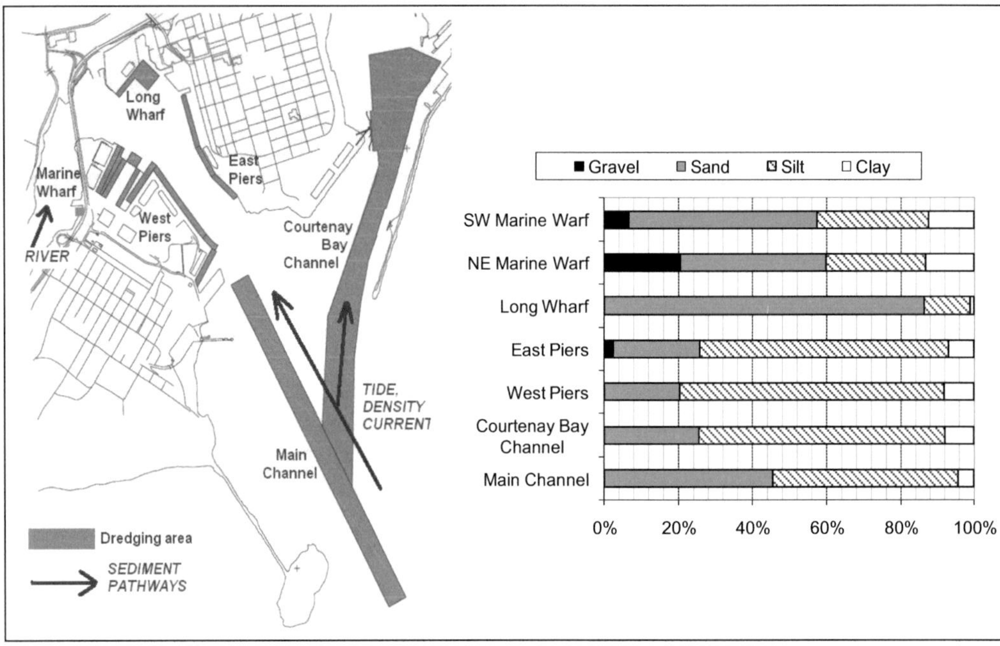

Fig. 7. Saint John Harbour dredging areas and grain size distributions (2006 data - source: Saint John Port Authority).

Measured dredging volumes represent the best available 'benchmark' data for estimating a mean annual sedimentation rate. However the extrapolation from scow-measured dredging volumes to sedimentation rate carries considerable uncertainty due to the bulking factor. The bulking factor is the ratio of dredged volume immediately after deposition in the scow, to the in-situ volume of the same mass of material. As a general rule of thumb, the smaller the grain size, the larger the bulking factor: sand can bulk up 1.0 to 1.2, silt 1.2 to 1.8 and clay 1.5 to 3.0 (USACE 2004). In addition, actual dredging areas vary from year to year, and may be less than target areas, possibly by a factor of 2 or 3. Ranges for sedimentation rates have been developed based on dredging records weighed with the above uncertainty factors. The calculated ranges represent averages in time and space, which could be exceeded in any given year or location. It is estimated that the sedimentation rates range from 0.2 to 1m/year, the higher end of the bracket applying to the deepwater berths in the City Harbour. The dredging records show considerable variability in the quantities from year to year, resulting in the wide range.

Description of sedimentation processes

The ample sediment supply combined with the hydrodynamic regime cause extensive sedimentation in dredging areas. These areas tend to decelerate sediment-bearing flows, and represent a departure from a natural equilibrium state where sedimentation is balanced by erosion. The primary local sediment sources include the River, the seabed of the Bay of Fundy and eroding coastlines. An extensive review of harbour sedimentation (or 'siltation') mechanisms is provided by Winterwerp (2005). Sediment transport modelling must resolve the following three major site-specific processes:

1. *Density currents* - The density difference between tidally-driven salt water inflow and freshwater river outflow causes an estuarine circulation pattern characterized by a mean seaward surface flow and a mean up-harbour bottom flow. The residual dense bottom flows carry silt that is deposited in the more stagnant areas. Sedimentation in dredging areas is due in most part to marine silt carried into the Harbour by the bottom density current (Neu 1960). Notably, yearly variations in river outflow and suspended concentrations in the river do not correspond to variations in the measured dredged quantities.
2. *Tidal exchange* - Water within a harbour basin is replaced by freshwater from the river water on the ebb tide, and then by saline water on the flood tide (with the exception of the near-surface where a layer of freshwater persists). This efficient and continuous exchange mechanism is caused both by the very large tidal range and by the large river discharge. Settling then occurs wherever weaker currents allow. The sedimentation rate depends on a complex array of variables including tidal prism (volume entering the harbour), trapping efficiency, suspended sediment concentrations, dry bed density and settling rates vs. local currents. Of these variables, settling rate probably has the highest variability and influence.
3. *Horizontal eddy exchange* - During peak flows, suspended sediments are transported in the lee of protruding wharf structures due to energetic residual eddies shed from the structures. Deposition occurs where weak currents and high settling rates allow.

Hydrodynamic modelling

The Danish Hydraulic Institute's MIKE3 finite-volume model was implemented to better understand flow patterns and sediment transport in the harbour, and to assist in the evaluation of maintenance dredging requirements for future harbour facilities. The hydrodynamic module solves the hydrostatic momentum and continuity equations, including the effects of turbulence and variable density, and the conservation equations for salinity in three dimensions together with the equation of state of sea water relating the local density to salinity, temperature and pressure. The model also features a coupled advection-diffusion algorithm to model the evolution of suspended sediment concentrations, which serve as input to the sediment transport module.

The model domain consists of an unstructured mesh of 2,590 triangular elements in each horizontal layer. In the vertical dimension, the model was set-up using up to 23 layers in the deepest areas. The upper three layers were defined as compressible 'sigma' layers following the oscillations in water level. Below a fixed depth of 1m below low tide, the model used 20 strictly horizontal 2m thick layers to better resolve the density stratification. The model domain (Fig. 8) was set-up to include all dredging areas, with its upstream boundary 500m downstream of the Reversing Falls. The upstream boundary conditions for this model (water level, salinity and suspended sediment concentrations) were obtained by extracting data from a calibrated, non-hydrostatic and higher-resolution MIKE3 hydrodynamic model of the Reversing Falls channel developed for a previous study (Leys 2007). At the Bay of Fundy boundary, tidal predictions were used as well as time-series of vertical salinity and suspended sediment concentrations constructed from field observations by Neu (1960).

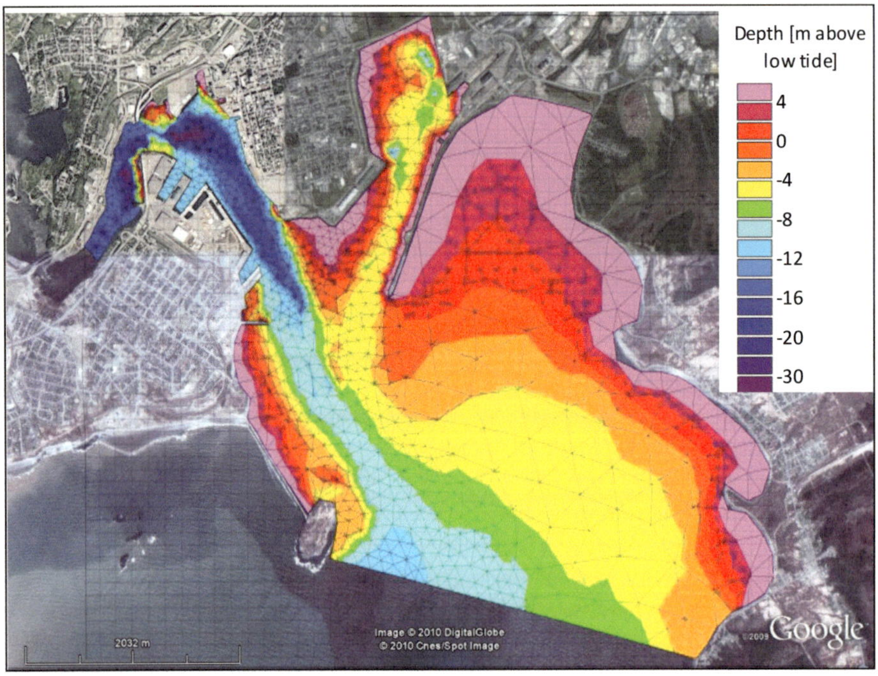

Fig. 8. Saint John Harbour Model mesh.

Vertical profiles

The evolution of vertical profiles for key model variables over a tidal cycle is shown on Fig. 9 at the intersection of the Main Channel and Courtenay Bay channel (see Fig. 7). The three dimensional circulation patterns are evidenced by salinity and Total Suspended Sediment (TSS) fields over a mean tidal cycle. During flood tide and at high tide, the bottom layer of denser, saline and sediment-laden tidal water extends into the channel and flows opposite the seaward surface current. On the ebb and low tide, saltwater is gradually replaced from surface to bottom by freshwater from the river which carries coarser and lower sediment content. The model results are consistent with field data, but show a stratified phase of shorter duration.

Residual current patterns

Residual bottom current patterns (i.e. averaged over a tidal cycle) are important as they govern the movement of the sediment-laden bottom layer. Residual currents over a mean tidal cycle during summer conditions are presented in Fig. 10. The results indicate that the mean currents along the bottom move in the up-harbour direction. The modelled bottom density current in the City Harbour and Courtenay Bay Channel is approximately 0.1m/s, slightly less than residual currents calculated from summer field measurement by Neu (1960). Modelled near-surface residual currents for summer conditions correspond well to past measurements, with values in the order of 0.3 to 0.4 m/s in the City Harbour.

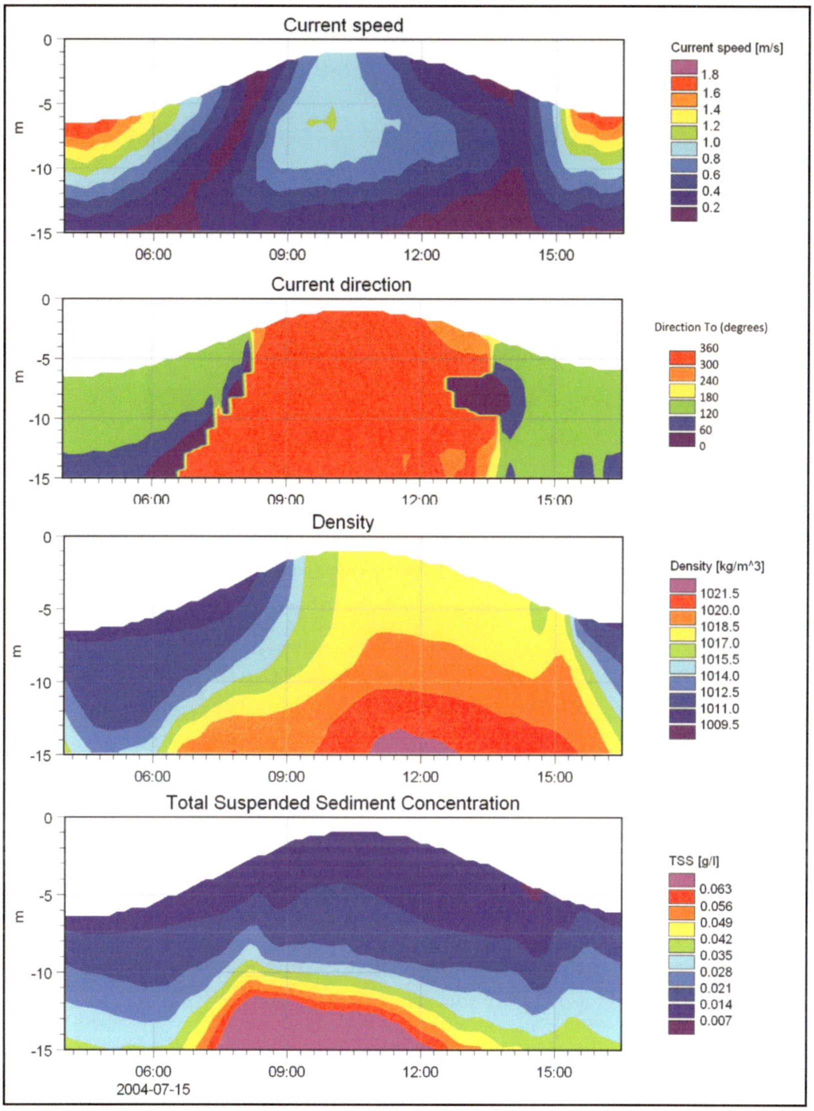

Fig. 9. Vertical profiles of model fields over one tidal cycle at the intersection of the Main Channel and Courtenay Bay channel.

Sediment transport model

The DHI MIKE3 Mud Transport Model calculates sediment transport of fine material in estuaries and coastal areas, for dredging and sedimentation studies. This model was used to simulate the erosion, transport and deposition of fine grained and cohesive material under the action of river, tidal and density currents calculated by the hydrodynamic model. Values for sediment parameters were adjusted within realistic ranges based on field data (suspended sediment concentrations and dredging records). Two fine sediment fractions were included in the model, which form the bulk of sediment deposits in dredged areas: sandy silt and clayey silt. The critical shear stress for deposition and settling velocity were treated as calibration parameters and the values adopted are listed in Table 1. At the open

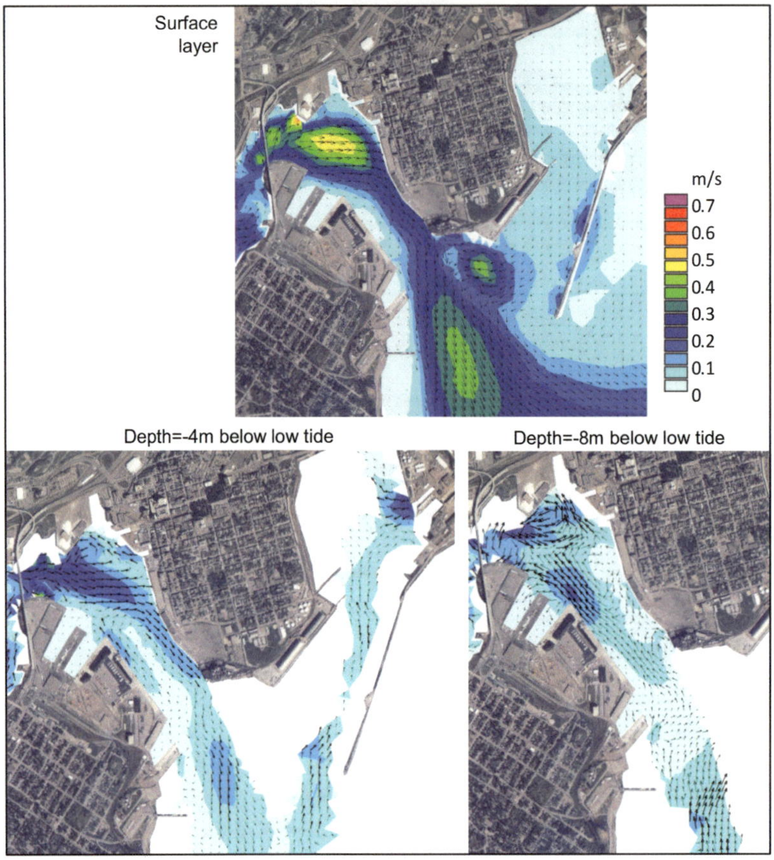

Fig. 10. Modelled residual flow patterns for summer conditions showing seaward surface flow from river (top panel) and up-harbour density current at the bottom (bottom panels).

boundaries, time-series of depth-varying suspended sediment concentrations were developed based on summer data from Neu (1960) and recent additional sampling (Leys 2007).

Parameter	Fraction 1 Clayey Silt	Fraction 2 Sandy Silt	Comments
Representative grain size [mm]	0.005	0.05	For reference - not actually used in computations
Critical shear stress for deposition [N/m2]	0.01	0.09	At each model time step, erosion or deposition algorithms are triggered if the bed stress is respectively above or below critical value
Settling velocity [mm/s]	0.1	1	For fine sediment the settling velocity can vary by several orders of magnitude. Flocculation, which generally starts at concentrations higher than observed in typical conditions, was not included.

Table 1. Input sediment characteristics for MIKE3 MT model.

The model accuracy can only be as good as the accuracy of dredging records used in the calibration process. Based on the Port dredging records, the estimated error range from the modelled sedimentation rates is -50% to +100% (i.e., in any given year the actual value could be twice as low or twice as high as predicted). Under existing conditions the finer fraction (clayey silt) deposits primarily in the calmer areas, i.e., Courtenay Bay and within rectangular mooring basins (Fig. 11). The coarser fraction (sandy silt) causes more deposition in the main channel, and there is still a significant amount that settles alongside piers. A comparison between deposition quantities for the two fractions over the entire model area indicates approximately 6 times more mass deposition for the coarser sandy silt fraction. This indicates that a large proportion of the fine fraction modelled remains in suspension in the more energetic areas.

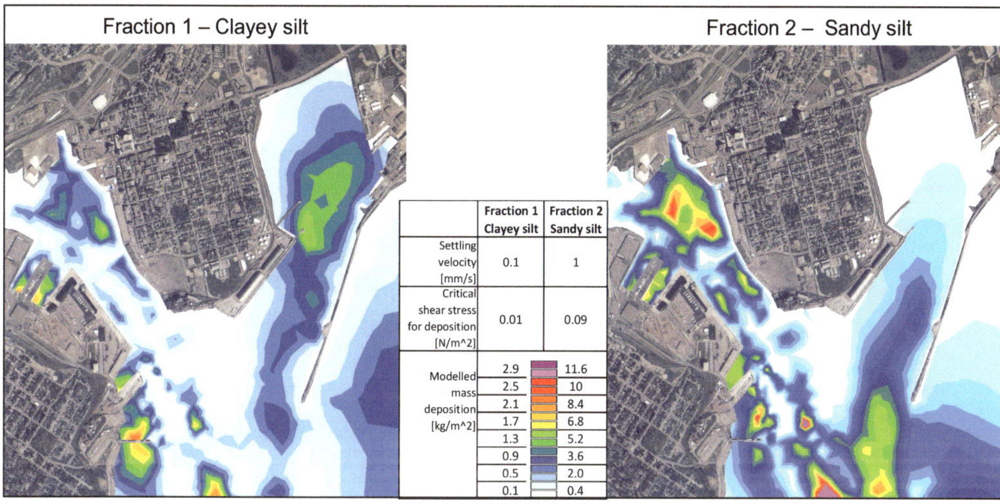

Fig. 11. Modelled mass deposition for two sediment fractions after 15 days.

The large difference in mass deposition between the two fractions is reduced when applying dry bed density to convert mass deposition into volume and determine the sedimentation rate. For a given mass accumulation calculated from settling, the dry bed density dictates the corresponding volume accumulation and therefore sedimentation rate.

Values were selected based on moisture contents of surficial sediment samples varying from 40% to 80% for freshly deposited silt. For example, a given mass deposition of clayey silt in Courtenay Bay (where dry bed density was assumed low at 200 kg/m^3) will result in a larger thickness than the same mass deposition of sandy silt in the centre of main channel where dry bed density is likely higher (assumed at 1000 kg/m^3).

Modelled sedimentation rates were extrapolated to a yearly basis based on a 2-week model run, which provides patterns that are generally consistent with annual dredging records for Port areas (Fig. 12). Sedimentation rates peak within man-made indentations in the shoreline, and in sections of the dredged channels where currents are lower such as the Courtenay Bay Channel along the eastern breakwater.

The model also helped understand why observed sedimentation rates in the Port were uncorrelated with the duration and the intensity of the spring freshet. It was previously thought that the source of the deposited sediment (River or Bay of Fundy) could influence the seasonal variations in sedimentation rates. Measurements from Neu (1960) indicate the suspended loads from the River during the spring freshet are much greater, on the order of 50 mg/l (comparable to typical bottom tidal loads) as opposed to 10 mg/l or less in the summer. Observations from the Port Authority indicate that yearly variations in river outflow do not correspond to observed variations in the dredged quantities. In an attempt to quantify the source region, modelled suspended sediment was divided into two fractions based on origin: the River or Harbour boundary. The results indicate that tidal currents account for more than 50% of the transport to and deposition within the harbour. Typical spring freshet conditions were also investigated. Conceptual model runs were conducted for non-reversing flow conditions (when river levels remain above high tide) and with increased suspended loads from the river boundary. Overall sedimentation rates were found to be in the same order of magnitude as during tidal, reversing flow conditions. Under these conditions the absence of prevailing tidal loads is compensated by the increase in river loads and settling. Continuous non-reversing flow conditions only occur during typically 2 weeks in the spring. Therefore, the influence of the spring freshet on harbour sedimentation is likely limited.

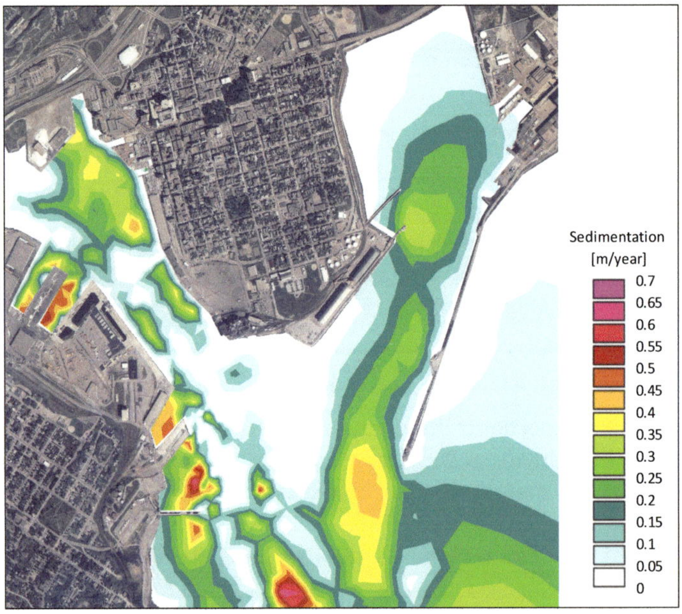

Fig. 12. Modelled annual sedimentation rate

The model is presently being used to optimize modifications to selected piers or mooring basins, taking into account challenges presented by current velocities and sedimentation. As demonstrated, accumulation rates in new dredging areas can be estimated, within a certain error range depending on the reliability of dredging records. Predicting yearly variations in sedimentation over dredging areas of the Port would also be useful for Port users, but would require a vast amount of additional field data and modelling efforts.
Saint John Harbour provides a good example for fine suspended sediment transport driven by tidal and density flows within a man-made port. The next case study describes a very different site where coarser sand transport prevails, driven by waves and tidal currents at a natural inlet without coastal structures.

4. Darnley inlet

Darnley inlet is located on the sandy North Shore of Prince Edward Island. The coast is formed of dunes, barrier islands and tidal inlets, and it is particularly sensitive to storms and sea level rise (Forbes et al 2004, McCulloch et al 2002). At various tidal inlets, dynamic coastal processes occurring on very different time-scales (tides, storms, long-term sea level rise) play an active role in sediment transport, presenting short-term navigation challenges and causing a long-term evolution in the geomorphology.

Malpeque Harbour is located within Darnley Basin (Fig 13). The long winding channel from the harbour exits through the inlet between two sand spits, then becomes considerably shallower as it fans out over a shallow sand bar 500 to 600m from the inlet throat. The bar is referred to as an 'ebb shoal' for the greater influence of the ebb tide on sediment transport. The ebb shoal represents the primary bypassing route for westward sediment transport, and bi-annual dredging is required at the ebb shoal to ensure navigational safety.

Considerations of the tidal prism (the volume exchanged over a tidal cycle) and longshore sediment transport have helped to understand historical changes in the inlet location and dune build-up. In-depth modelling analyses were conducted with a morphological model to better understand sedimentation at the ebb shoal, and to predict morphologic responses to channel-training structures such as jetties.

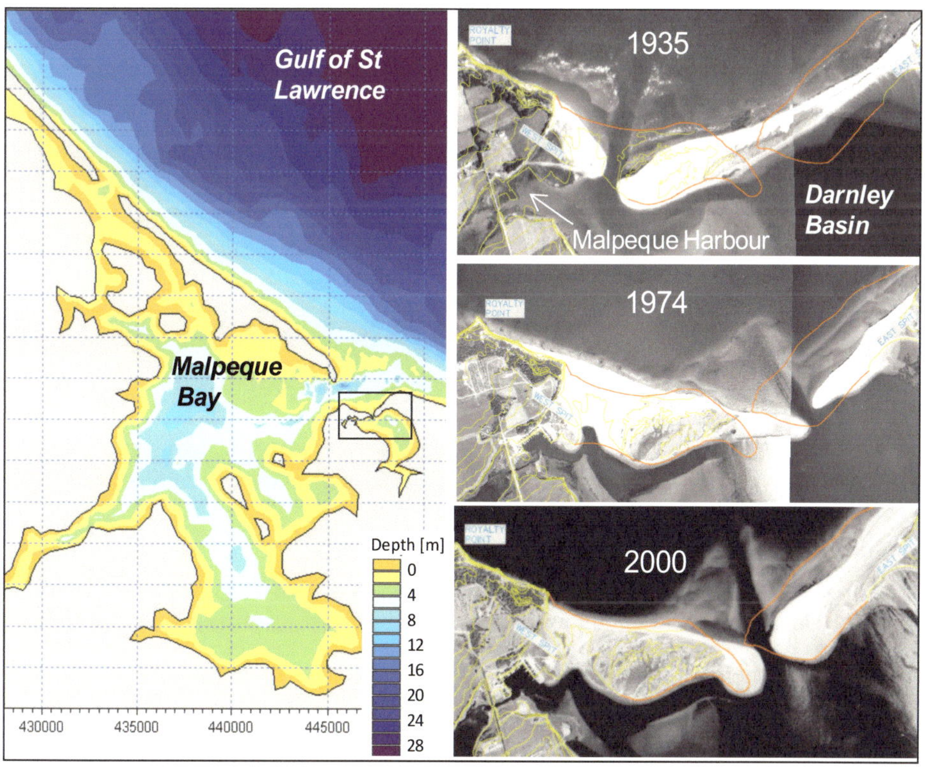

Fig. 13. Site location (left) and historical air photos at Darnley Inlet (right) showing beach evolution, inlet migration and locations of the East and West Spits. The 2008 shoreline is represented with the orange line.

Historical evolution

Darnley Inlet is an interesting example of an unstable natural inlet, where the inlet throat has moved considerably over the past century. In the first half of the 20th century, the inlet was located at the base of the present West spit (Fig. 13). The old inlet closed in the 1960's while the new inlet opened through the East spit. Due to westward net longshore transport decreasing toward the inlet, the East spit is accreting and the inlet location is presently migrating westward at a rate of about 14m/year.

Tidal hydrodynamics

The local tides are of mixed type, i.e. with diurnal and semi-diurnal influence. This creates asymmetries in the hydrodynamic regime (e.g. ebb or flood dominance) that have implications on sediment transport.

A hydrodynamic model of the inlet was developed using DHI MIKE21 to estimate key inlet variables that influence sediment dynamics, including tidal prism and ebb or flood dominance. MIKE21 is a two-dimensional, finite-volume model solving the momentum and continuity equations over a finite volume mesh (Fig. 14) to simulate hydrodynamic circulation.

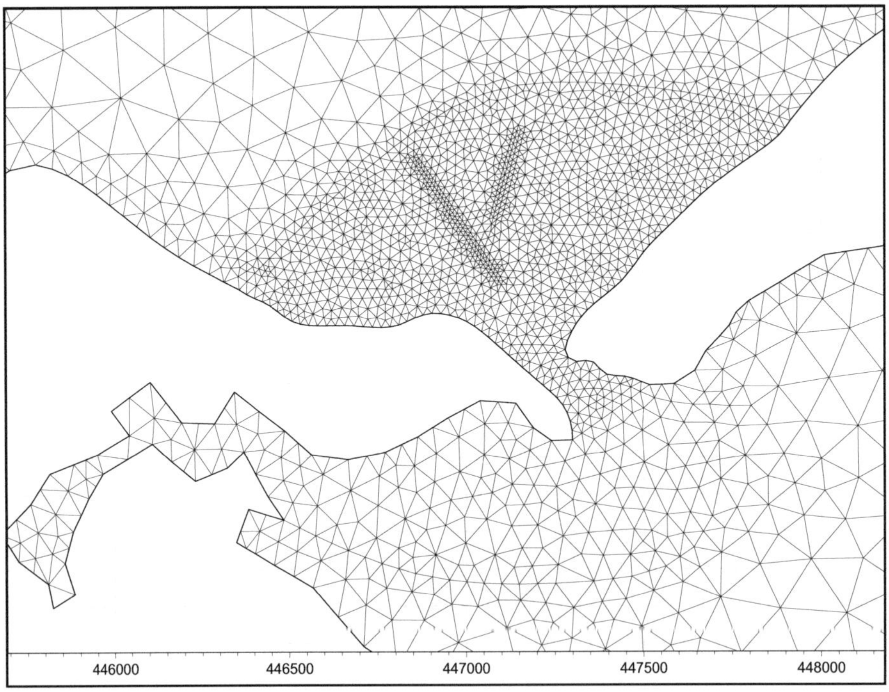

Fig. 14. High-resolution Darnley Inlet model mesh with higher resolution over dredged channels.

The process of ebb dominance on sediment transport at Darnley Inlet is shown on Fig. 15. In effect, during at least 60% of the month, there is one dominant ebb tide per day, meaning that on average, the sand transport direction through the channel is seaward. This is due to the ebb current being above the threshold for incipient sand motion for a longer period of time than the flood current. Sand transported away by the ebb tide deposits off the inlet, forming the ebb shoal.

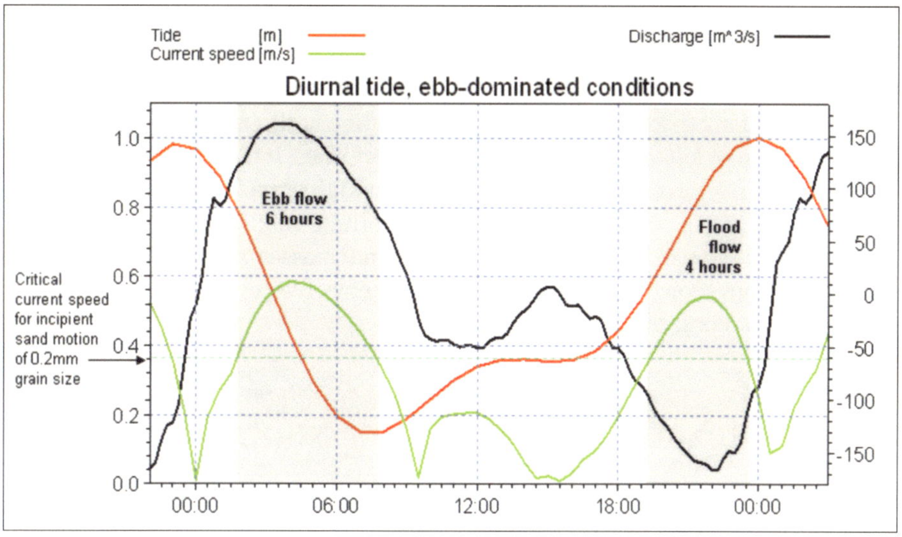

Fig. 15. Time-series of modelled current speed during ebb-dominated conditions

Wave-driven sediment transport

Prevailing Northeasterly offshore waves break on a shallow offshore bar (Malpeque Bay's ebb shoal) before reaching the inlet where storm wave heights are controlled by water level. Using spectral wave model MIKE21 SW, it was estimated that significant wave heights incident to the inlet and adjacent spits are less than 1m when the water level is at or below high tide. Northeasterly waves arrive at the East spit at a very oblique angle to the shoreline, causing significant Eastward longshore transport. In contrast, prevailing wave crests arrive near-perpendicular to the West spit shoreline, causing lower longshore sediment transport rates.

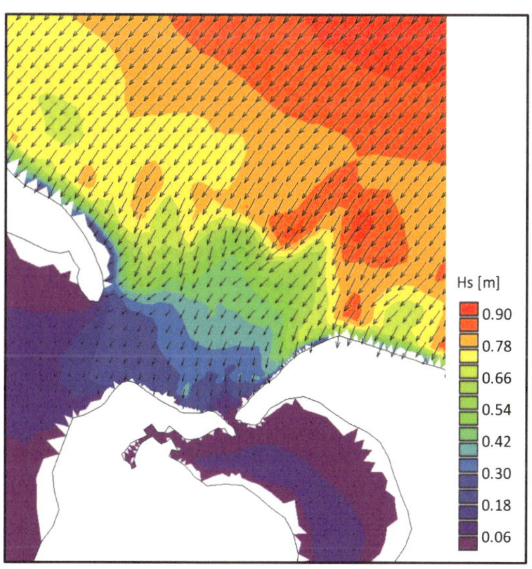

Fig. 16. Modelled near-shore wave height transformation for offshore waves of significant height 1m and peak period 8s.

Potential bulk longshore sediment transport rates were calculated based the standard CERC (1984) formula, using modelled breaking significant wave height parameters along representative shoreline reaches. The results are presented on Fig. 17. As a convention, left- and right-directed longshore transport rates are negative and positive, respectively. The net transport is the sum of left and right-directed transport rates. It is cautioned that:

- The effect of strong, reversing tidal currents from Malpeque Bay is not included; and
- The sediment transport values shown assume infinite sand supply, which is rarely the case. In the long-term, the beach sizes and sand masses are constantly adjusting to gradients in longshore transport rates, which themselves evolve as a function of sand supply and changing shoreline orientation.

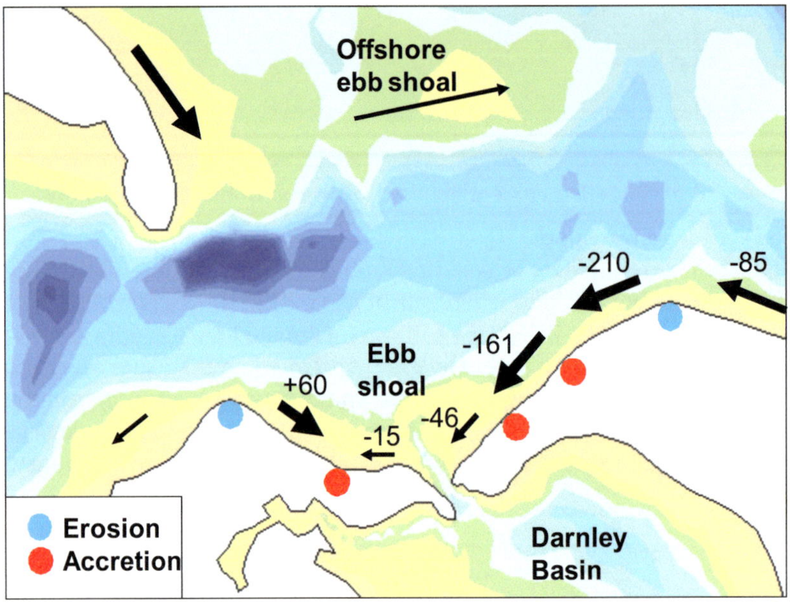

Fig. 17. Net potential longshore sediment transport rate (1000's m^3/year) based on wave forcing. The arrows indicate the direction of net transport. Erosion occurs in areas of increasing transport, while decreasing gradients in longshore transport results in accretion.

Gradients in potential longshore transport rates, caused by changing wave parameters or shoreline orientation, are used to identify areas of shoreline erosion or accretion and to explain observed trends:

- The net transport at the inlet is from East to West; the inlet throat location is therefore likely to continue migrating westward;
- The large differences in potential transport rates between the East and West spits would cause large deposition volumes into and near the inlet, notably at the ebb shoal;
- The growth of the East spit seen on aerial photographs (Fig. 13) can be explained by the along-shore gradient in transport potential due to diminishing wave heights;
- The location of the pre-1960's inlet at the base of the West spit is likely accreting due to converging shorelines on either side. This would have contributed to the original inlet closure; and
- Headlands represent eroding sediment sources.

Empirical tools for assessing inlet stability

Despite the complexities of the physical processes, morphologic characteristics of inlets such as cross-sectional area at the throat or ebb shoal dimensions have been successfully correlated to parameters such as tidal prism, wave climate and longshore sediment transport rate. Empirical relationships have been developed by researchers for sandy inlets worldwide, the largest available datasets being from the continental U.S. A review of these predictive tools is provided by Kraus (2008), and those found useful to understanding the sediment dynamics at Darnley Inlet are applied herein.

Inlet stability can refer to either the cross-sectional area of the channel across the inlet throat, or to the inlet location. With respect to the former, since the tidal prism must enter and exit through the relatively constricted inlet, flow increases and sediment is scoured until the inlet erodes to a stable channel cross sectional area. A natural channel is only stable when it is both large enough to allow full tidal passage yet small enough so constricted tidal currents flush the excess sand out. Jarrett (1976) examined the relationship between tidal prism and inlet throat cross sectional area, based on data at 108 inlets. The relationship derived from his data is plotted on Fig. 18.

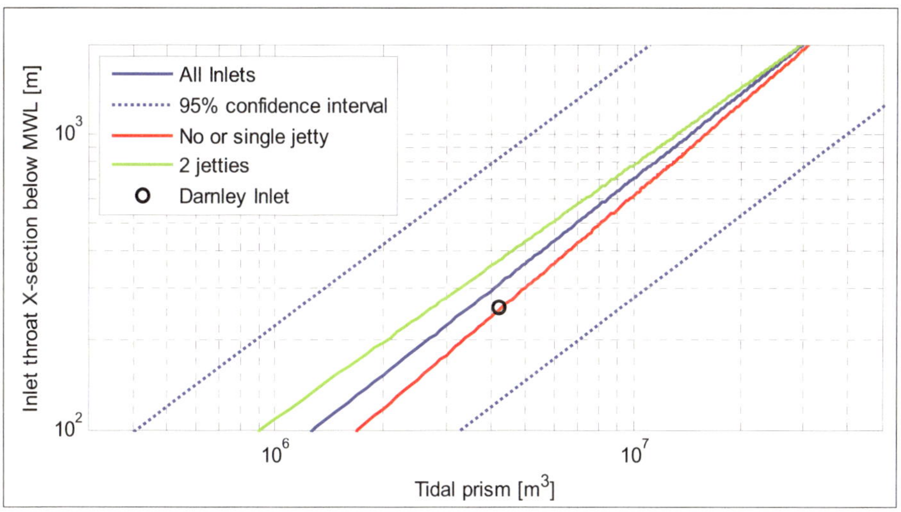

Fig. 18. Empirical equilibrium cross-sectional area at the inlet throat vs. tidal prism (after Jarrett 1976).

Darnley Inlet fits the trend line well, and the 270m^2 measured cross sectional area is at near-equilibrium dimensions. Jarrett's data suggests that parallel jetties may increase the cross-sectional area by about 40% at the throat (this does not consider channel size over the ebb shoal).

It is useful to predict the consequences of a departure from the equilibrium cross section. The question is whether the inlet would tend to close or scour itself back to its equilibrium shape after a sudden storm-induced deposition reduces the cross section. The analysis procedure was summarized by Dean and Dalrymple (2002), and is referred to as the Escoffier method. The results are graphically presented in Fig. 19. The maximum inlet velocity is calculated based on the Keulegan (1967) equation, accounting for inlet cross-section, upstream area, friction, entrance and exit losses using a semidiurnal tide.

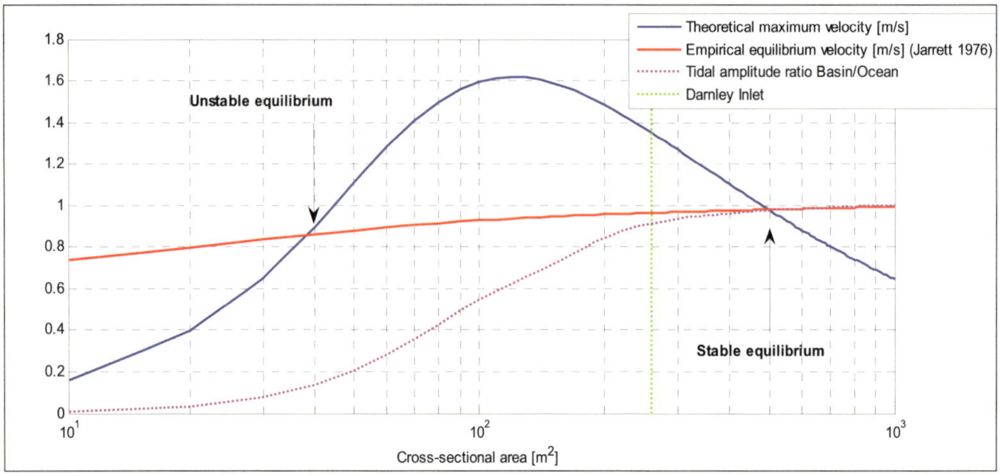

Fig. 19. Maximum and equilibrium inlet velocity vs. throat cross-sectional area.

Small entrance channels that limit tidal passage are dominated by friction, and their equilibrium is unstable. If a storm reduces the cross-section, friction increases, tidal passage decreases and the maximum velocity decreases, leading to more infilling, and ultimate closure of the inlet and relocation of the channel to a more hydraulically efficient location. This illustrates what likely happened at Darnley Inlet in the 1960's. The new inlet through the East spit grew towards stable equilibrium while the original inlet to the West became unstable and eventually closed. When the cross section was wide enough for most of the tide to go through (tidal passage over 70-80%), the influence of friction was much reduced. At stable inlets, storm-induced sedimentation increases maximum velocities by reducing the cross-sectional area, causing he channel to scour back to stable equilibrium. For a cross section of 270m^2, the maximum equilibrium velocity through the inlet throat is about 1m/s, which is consistent with numerical model results and other published estimates of maximum tidal current necessary for inlet channel stability (Kraus 2008). In summary, the tidal prism of Darnley Basin can only support one stable inlet. The inlet cross-section itself appears stable and not at risk of closure.

In terms of the inlet location, longshore sediment transport influences the tendency of un-jettied inlets to migrate along the shore, regardless of dredging schedule. Bruun and Gerritsen (1966) examined the influence of the tidal prism over annual longshore transport ratio P/Q, arguing that the longshore transport should play a role in inlet stability. In the present case, using P=4.2 10^6 m^3 (from the MIKE21 model) and Q=46,000 m^3/year, P/Q is below 100, which means the inlet position is more likely to be unstable and result in shifting channels with significantly large shoals. This conclusion is consistent with observations showing that the inlet throat position is migrating westward. This translation is supported by the strong westward longshore transport, and has been facilitated so far by the relatively lower elevation at the tip of the West spit between its well vegetated dune and the channel. As seen at other shifting tidal inlets, there is a possibility that the East spit overlaps the eroding tip of the West spit. This would create a more angled and less hydraulically efficient channel, compounding the navigational difficulties.

Morphological modelling

The DHI MIKE21 Coupled Model was used to investigate engineering alternatives to ebb shoal dredging. The model includes a dynamic coupling between DHI's sand transport model, the aforementioned hydrodynamic module and the spectral wave model MIKE21 SW. Feedback of the bed level changes on the waves and flow calculations is included, as well as dynamic coupling of flow and wave calculations. This model is typically used for investigating the morphologic evolution of the nearshore bathymetry due to the impact of engineering works (coastal structures, dredging works etc.). It is most suitable for short to medium-term investigations (a few weeks) over a limited coastal area. The computational effort becomes impractically large for long-term simulations, or for large areas. Fine sand of median grain size 0.2 mm was assumed throughout the domain based on field samples.

A morphological sediment transport model is ideally calibrated using a series of pre- and post storm bathymetries. However, with this site (as with many others), the temporal resolution of sounding surveys do not allow that level of detail. The model results can still be compared to dredging records for an order-of-magnitude validation. Two runs were carried out to validate the model for existing conditions:

- a 1-month, short-term simulation with varying wave height to investigate the infilling processes at the dredge channel, and
- a medium-term simulation with constant wave activity and using a speedup factor in the bed level changes to artificially accelerate the morphologic evolution. The medium-term simulation setup served as a template for modelling the morphologic effects of engineering alternatives to dredging.

Short-Term Simulation

The model was run for a period of 1 month including several representative wave events. Input time-series for the month of October 2004 were imported from an offshore wind and wave hindcast, which included several wave events of various magnitudes. The initial bathymetry included a typical 2m-deep, 20m-wide dredged channel through the ebb shoal.

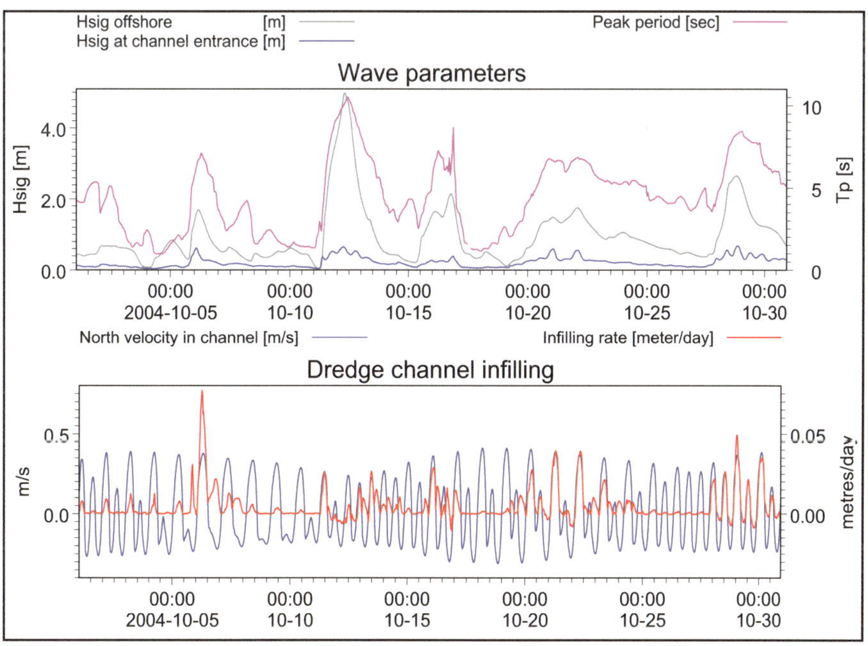

Fig. 20. Time-series of modelled dredge channel infilling rates vs. tidal current and wave parameters.

The mechanisms and occurrence of sedimentation events can be better understood with plots of infilling rates versus current and wave parameters, as shown in Fig. 20. The model indicates that sediment suspension occurs during wave events, and most of the infilling occurs during the ebb tide (on both spring and neap tides) when waves and currents oppose each other. Some channel scouring occasionally occurs, mostly on the flood tide, but only to be overcome by infilling on the next ebb tide. Sediment transport is typically triggered by local waves above 0.3 m, or offshore waves above 0.5 m, which occurs 38% of the year. During these events, the peak infilling rate is typically 0.02 to 0.04 m/day on the ebb tide, i.e. 0.0035 to 0.0070 m per tide. Based on a 38% yearly occurrence, this translates into an infilling rate of 0.9 to 1.8 m/year, which is consistent with dredging requirements.

The relatively low wave height threshold that triggers infilling explains why dredging is also required after the summer season. In fact, the waves incident to the inlet ebb shoal are depth limited due to the offshore ebb shoal, therefore the infilling rate is not proportional to the offshore wave height. In simple terms, channel infilling occurs gradually at each moderate wave event and not only during large storms.

Modelled sediment transport loads during a spring tide and a Northeasterly wave event are shown on Fig. 21. Longshore transport patterns along exposed shorelines near the open bounday are primarily wave driven. However, transport within the inlets where waves are depth limited, including Darnley Inlet, is very much influenced by tidal currents. For this particular event, the sediment transport direction varies with tidal currents. These effects add significant uncertainty to the longshore estimates shown on Fig 17. In a qualitative sense the wave-induced differences between left and right directed longshore transport are reduced by tidal currents.

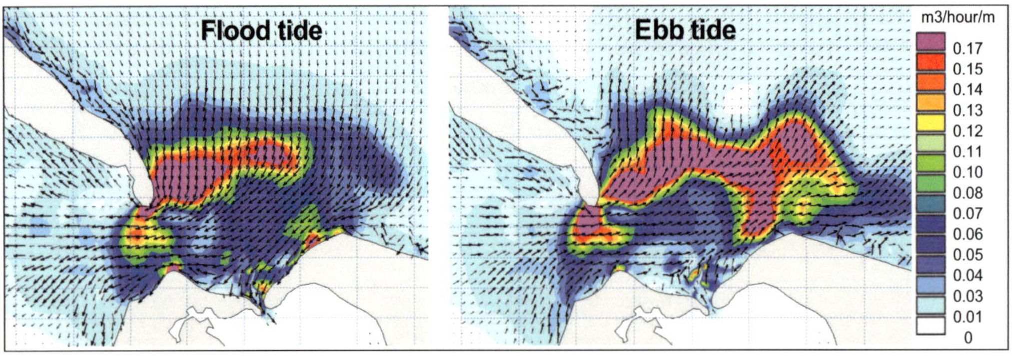

Fig. 21. Modelled sediment transport loads during a northeasterly wave event with significant wave height 1m and peak period 8s.

Long-term simulations (annual to decadal) would yield more accurate longshore transport estimates but would require a much larger computing effort than typically done for preliminary studies. However, medium-term morphologic evolutions can be investigated using artificial parameters to speed up the model, as presented below.

Medium-Term Morphological Simulation

Results from the short-term simulation indicated that the offshore significant wave height of 0.5 m was a trigger for sediment transport causing infilling in the dredge channel. To speed up the morphologic evolution in the medium-term simulations, significant wave heights at the offshore boundary were assigned a constant 1m value (the mean height for wave records greater than 0.5m). Assuming a 38% yearly occurrence of such waves, the modelled morphologic changes would occur 1/0.38=2.6 times faster than normal. For increase computational effeciency, a speedup factor of 72 was applied to the rate of bed level change computed every hour (speed up factors must be used cautiously because sudden bed level changes trigger numerical instabilities in the hydrodynamics). The resulting total morphological speedup is difficult to estimate, because the tidal variations cannot be artificially accelerated, and the areas of accelerated erosion versus deposition vary with the tidal stage. In any case, the simulations were stopped when the depth over the ebb shoal had levelled off to a near stable level. The results for existing conditions show that ebb shoal morphologic equilibrium is reached after 16 model days, which in real-time would take 6 months. This suggests that the ideal model speedup factor should be on the order of 10.

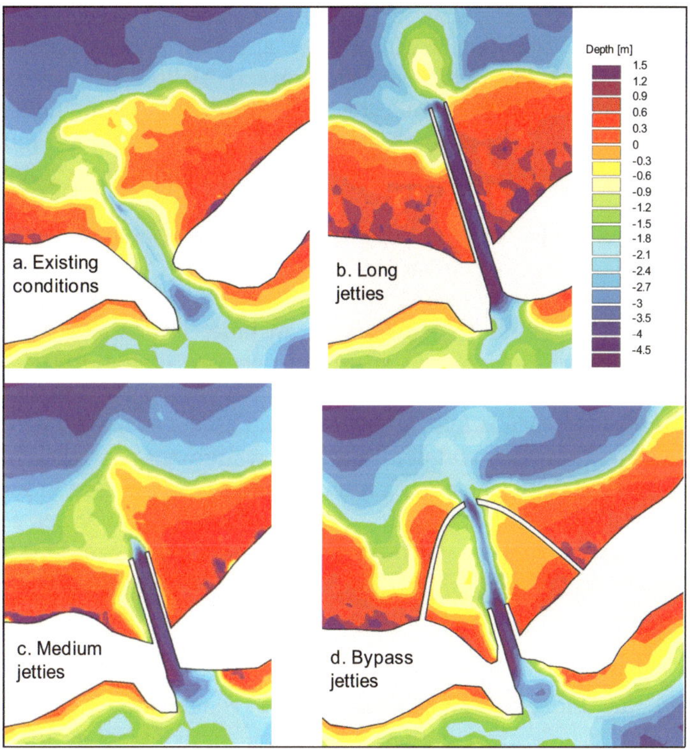

Fig. 22. Modelled morphologic evolution of Darnley Inlet under natural conditions (a) and with jetties of various configurations (b, c, d).

The model was used to investigate the effectiveness of inlet stabilization jetties to improve navigation over the ebb shoal. The bathymetry at the ebb shoal morphologic equilibrium stage for various scenarios is shown on Fig. 22. Inlet engineering presents a conflict between navigational requirements, which can only be met by interrupting longshore transport into the channel, and the necessity of sediment bypassing to ensure shoreline stability. Jetties would have to extend 600m from the inlet throat to provide a significant reduction in dredging costs. Shoreline impacts would include updrift accretion at the East spit and potential downdrift erosion on the West spit, which would be attenuated by transport from the west headland. Shoreline impacts can be mitigated by using curved 'bypass' jetties that improve sediment bypassing (details and examples are given by Mangor (2004)). Short jetties may be used to stabilize the inlet location, which is the primary purpose of many jetties at tidal inlets where maintenance dredging is still required. At this site the migrating inlet does not represent an erosion risk to any property or infrastructure, but the structures would help towards a more predictable channel. Short jetties would have a lesser impact on the adjacent shorelines, but would provide no significant reduction in dredging requirements. Finally, it is cautioned that morphologic response to an inlet has a long time-scale and great spatial extent. Long-term processes, such as the migration of the inlet throat under existing conditions, are not represented in the present modelling exercise.

4. Conclusion

Coastal sediment transport is of paramount importance for many harbour developments, as it dictates maintenance dredging requirements and shoreline response (erosion and accretion) to coastal structures. A site-specific harbour planning study should provide a thorough understanding of the local environment through various methods, some of which are presented herein. Traditional analysis methods have generally relied upon local experience, aerial photographs, bathymetric and oceanographic surveys, and simple analytical and/or empirical models. To complement these necessary first steps, advances in numerical models over the last decade make it possible to efficiently assess coastal sediment transport and its implications for infrastructure projects.

Over the next decade coastal morphology will be studied using new and improved modelling tools and techniques. These may include improvements to existing 3D morphological models such as better parameterizations of sediment processes, especially for fine-grained sediments (i.e. flocculation, fluid mud, turbulent interactions), and new methods such as smooth-particle hydrodynamic simulations (a mesh-free particle-tracking method) which depend on high-end computing. The different examples presented in this chapter underline the importance of using a site-specific approach and local observations in the application of numerical models. Improved predicting capabilities for morphologic models are, and will continue to be an important area of future research.

5. Acknowledgements

The authors would like to thank Public Works and Government Services Canada and the Small Craft Harbours Branch of the Department of Fisheries and Oceans for providing funding for the studies described herein. Valuable insights given by the Saint John Port Authority and John Hughes Clarke (University of New Brunswick) are also gratefully acknowledged.

6. References

Booij, N., R. Ris, and L. Holthuijsen (1999), A third-generation wave model for coastal regions: 1. Model description and validation, *J. Geophys. Res.*, 104(4), 7649- 7666.

Bruun, P., and Gerritsen, F. (1959). Natural bypassing of sand at coastal inlets. *J. Waterways and Harbors* Div. 85, ASCE, 401-412.

Chen, C, R. C. Beardsley and G. Cowles (2006). An unstructured grid, finite-volume coastal ocean model (FVCOM) system. *Special Issue entitled "Advance in Computational Oceanography", Oceanography,* vol. 19, No. 1, 78-89.

CERC (1984). Shore Protection Manual. Coastal Engineering Research Center, U.S. Army Corps of Engineers, Vicksburg.

Dabees M.A., Kamphuis J.W. (2000). NLINE: Efficient Modeling of 3-D Beach Change. *Proceedings of the 25th International Conference on Coastal Engineering*, Sydney Australia, ASCE pp.2700-2713.

Dean R.G. (2002). Beach Nourishment Theory and Practice. *Advanced Series on Ocean Engineerting* - Vol 18. World Scientific Publishing.

Dean, R.G., and Dalrymple, R.A. (2002). Coastal Processes With Engineering Applications. Cambridge U. Press, 475p

Demirbilek Z., Vijay Panchang V. (1998). CGWAVE: A Coastal Surface Water Wave

Model of the Mild Slope Equation. *U.S. Army Corps of Engineers Technical Report CHL-98-xx*

DHI Software (2009). MIKE, Coastal Hydraulics and Oceanography - User Guide. Danish Hydraulic Institute.

DHI Software (2008). LITPACK -An integrated modeling system for littoral processes and coastal kinetics - User Guide. Danish Hydraulic Institute.

Forbes, Donald L, George S. Parks, Gavin K. Manson and Lorne A Ketch, 2004. Storms and Shoreline Retreat in the Southern Gulf of St. Lawrence. *Marine Geology 210* (pp 169-204).

Gravens, M., Kraus, N. and Hanson, H. (1991). GENESIS: Generalized Model for Simulating Shoreline Change - *Report 1: Technical Reference, Technical Report CERC-89-19*, U.S. Army Waterways Experiment Station, Coastal Engineering Research Center, Vicksburg, MS.

Greenberg, D.A. (1990). The Contribution of Modelling to Understanding the Dynamics of the Bay of Fundy and Gulf of Maine. *Modeling Marine Systems*, Vol. 2, A.M. Davies, ed., CRC Press, Boca Raton., Fla., 107-140.

Jarrett, J.T. (1976). Tidal Prism-Inlet Area Relationships. *GITI Report 3*, U.S. Army Waterways Experiment Station, Vicksburg, MS.

Keulegan G.H. (1967). Tidal Flow in Entrances. U.S. Army Corps of Engineers, Committee on Tidal Hydraulics, Tech. Bull. 14, Vicksburg.

Kraus, N. C. (2008). Engineering of Tidal Inlets and Morphologic Consequences. In: Kim Y. (editor). *Handbook of Coastal Engineering*. World Scientific Press. Singapore, in press. 39p.

Lesser, G., J. Roelvink, J. van Kester, and G. Stelling (2004), Development and validation of a three-dimensional morphological model, *Coastal Eng*, 51, 883-915.

Leys, V. (2007). 3D Flow and Sediment Transport Modelling at the Reversing Falls - Saint John Harbour, New Brunswick. *Proceedings of MTS/IEEE Oceans Conference*, Vancouver BC, Canada, Sep 29-Oct 4 2007.

Mangor, K. (2004). Shoreline Management Guidelines. Published by DHI Water and Environment, Horsholm, Denmark.

McCulloch M. M., Forbes D.R., Shaw R.W and A041 Scientific Team. 2002. Coastal Impacts of Climate Change and Sea Level Rise on Prince Edward Island. Geological Survey of Canada. Open File 4261.

Neu, H. A. (1960). Hydrographic survey of Saint John Harbour, NB. National Research council of Canada, Mechanical Engineering Report MH-97.

Pelnard-Considere, R. (1956). Essai de Theorie de l'Evolution des Formes de Rivages en Plage de Sable et de Galets. *Fourth Journees de l'Hydralique, les Energies de la Mer*, Question III, Rapport No. 1, pp. 289–298

Shchepetkin, A.F. and J.C. McWilliams (2005). The Regional Ocean Modeling System (ROMS): A split-explicit, free-surface, topography-following coordinates ocean model, *Ocean Modelling*, 9, 347-404.

U.S. Army Corps of Engineers (2004). Status of Current Hopper and Bin Measurement Technologies. *Technical Note ERDC TN-DOER-T6*. September 2004.

Warner, J.C., Sherwood, C.R., Signell, R.P., Harris, C.K. and Arango, H.G. (2008). Development of a three-dimensional, regional, coupled wave, current, and sediment transport model. *Computers and Geosciences*, 34, 1284-1306.

Winterwerp, J. C. (2005). Reducing harbor siltation. I: Methodology. *J. Waterway, Port, Coastal and Ocean Engineering*. Vol. 131, No. 6, Nov 2005, pp. 258-266

3

Numerical Modelling of the Sediment Re-Suspension Induced by Boat Traffic

Hassan Smaoui[1,2], Abdellatif Ouahsine[2], Damien Pham Van Bang[1,3] and Philippe Sergent[1] and François Hissel[1]

[1] *Centre d'Études Techniques Maritimes et Fluviales (CETMEF). 2, Bd Gambetta, B.P. 60039, 60321 Compiègne Cedex*

[2] *LHN, Laboratoire Roberval UMR 6253 CNRS, Université de Technologie de Compiègne, B.P. 20529, 60206 Compiègne Cedex*

[3] *EDF/R&D, Laboratoire d'Hydraulique Saint Venant, 6, quai Watier, Bat K, B.P. 49, 78401 Chatou Cedex*

France

1. Introduction

Fluvial transport in inland waterways is used in France and Europe since millennia. It was the most effective transport mode for goods and merchandizes before the development of the terrestrial alternatives. As an illustration, the French network of the inland waterways counted 12800 *km* of both, rivers and channels, at the beginning of the twentieth century. The huge development of highways and railways encountered in France has progressively decreased the use of fluvial transport. By considering the duration for the transport, the terrestrial mode looks more attractive than the fluvial one. But, the actual context of climate change and the international or European commitments to strongly diminish CO_2 emissions are pleading for the reverse modal transfer, i.e. from highways and/or railways to waterways. Rebalancing the different transport modes are indeed compatible with the sustainable development concepts facing against road saturation, CO_2 emissions, visual or noise pollutions while ensuring a better safety.

Even through the shared awareness and the common efforts to increase the river traffic, this transport mode is still considered "under-developed" in the white paper of the European commission of 2001. For example, the fluvial transport in France has increased by 30% in ten years (1997 – 2007), but it only represents 3% of the means of transport. And the recent recommendations in "Grenelle de l'environnement" initiated by the French ministry of ecology may accelerate the rebalance between rails, roads and rivers, carrying the ambition of a 25% report from the road down to the river by 2020.

However, the development of this transport mode will certainly have economical and ecological consequences on the river management. The intensification of the river traffic by increasing the size and/or the frequency of passing boats will impact the actual management plan of the river. The waves induced by navigation could become stronger and bank erosion more frequent. The induced currents may be more intense and the resulting sediment transport more problematic. Indeed, it is accepted in aquatic environment that erosion

phenomenon have strong consequences on the ecosystem : destruction of habitats, increase in the turbidity, release of nutritive elements (phosphorus and nitrogenizes) which support the proliferation of the algae. With these economical and environmental considerations, the benefit/cost balance for the maintenance program of the inland waterways is changed and some practices could be renewed in the light of recent investigations.
To evaluate how different passing boats impact on the waterway, it was measured at two fields the hydraulic conditions (h, u) and the solid suspended matter (SSM, c) at different points. The measurements are related to the boat traffic that is obviously recorded or geo-localized. The two sites (canal de la Sensée and the Seine river: see figures 1 and 2) are investigated in November 2006. The water signals obtained close to the bank and the SSM concentrations close to the bed are measured at different points. The data description and analysis are fully presented in (Pham Van Bang et al., 2007). Most of the passing boats are impacting the waterway in modifying the water elevation and the turbidity that was frequency- or space-time analyzed by Pham Van Bang et al., 2008.
The field experiment aims to optimize maintenance program (dredging operations to maintain the nautical depth) and evaluate the effect of an increase in the boat frequentation. Due to the large scale of the site, it remains obvious the field measurements alone are not sufficient to quantify all the variables of the flow. For example, it is still not possible to measure *in situ* and non-intrusively the turbulent movement of both the water and the sediment. But during a boat episode, such measurements are missing, more particularly close to the bed or the bank of the studied river. These lacks of information is herein supplemented by the results of the mathematical/numerical models that enables a prediction, for inaccessible data field at least.

In this book chapter, we present the numerical approach for a 1DV model to predict the sediment re-suspension behind the boat passage. To proceed, we first start with the description of the mathematical model 1DV and justify the applicability to the sites of consideration. Then, we analyze the model results focusing both aspects, the TKE distribution along the vertical direction and the time-decrease in the fixed point measurement. In particular, we demonstrate the model is reliable to reproduce the field observed sediment re-suspension. Finally, we prove the numerical approach is effective to predict the sediment fluxes induced by the daily river traffic.

2. Presentation of the mathematical model

The model 1DV used in this study is part of a more general 3D model that is specifically devoted to free surface flows. The 1DV model is obtained by neglecting all the horizontal gradients, except the pressure gradient. It consists in these set of equations for the mean and turbulent movements and the continuity equation :

$$\frac{\partial u}{\partial t} + \frac{1}{\rho}\frac{\partial p}{\partial x} = \frac{\partial}{\partial z}\left[(\nu + \nu_t)\frac{\partial u}{\partial z}\right] \tag{1}$$

$$\frac{\partial k}{\partial t} = \nu_t \left(\frac{\partial u}{\partial z}\right)^2 + \frac{\partial}{\partial z}\left[\nu_{tk}\frac{\partial k}{\partial z}\right] - \frac{k\sqrt{k}}{B_1 L} \tag{2}$$

$$\frac{\partial kL}{\partial t} = A_1 L \nu_t \left(\frac{\partial u}{\partial z}\right)^2 + \frac{\partial}{\partial z}\left[\nu_{tk}\frac{\partial kL}{\partial z}\right] - \frac{L^2\sqrt{k}}{B_2\widetilde{w}} \tag{3}$$

$$\frac{\partial c}{\partial t} = w_s \frac{\partial c}{\partial z} + \frac{\partial}{\partial z}\left[\gamma_t \frac{\partial c}{\partial z}\right] \tag{4}$$

where t is the time, x and z are the horizontal and the vertical coordinates respectively, u the horizontal velocity, p the pressure, ρ the water density, ν the molecular viscosity, ν_t the turbulent viscosity, k the TKE (turbulent kinetic energy), L the mixing length, ν_{tk} the turbulent diffusivity of the TKE , $\widetilde{w}$ is the wall function, A_1, B_1, B_2 are empirical constants (Table 1), c the concentration of SSM, γ_t the turbulent diffusivity of SSM and w_s the settling velocity of sediment given by:

$$w_s = \begin{cases} \dfrac{g d_{50}^2 (s-1)}{18\nu} & \text{if } d_{50} < 63\mu m \\ \dfrac{-3\nu + \sqrt{9\nu^2 + g d_{50}^2 (s-1)(\alpha_1 + \alpha_2 d_{50})}}{\beta_1 + \beta_2 d_{50}} & \text{if } d_{50} \geq 63\mu m \end{cases} \tag{5}$$

Where ν is the molecular viscosity, g is the acceleration of gravity, d_{50} is the median diameter of the sediments, $s = \rho_s/\rho$, ρ_s is the density of the sediments and $\alpha_1, \alpha_2, \beta_1, \beta_2$ are empirical constants (Table 2).

A_1	B_1	B_2	S_h	S_k	S_m
1.80	16.6	12.48	0.49	0.20	0.39

Table 1. Closure empirical constants of $k - kL$ turbulence model

$\alpha_1(m)$	α_2	$\beta_1(m)$	β_2
3.8691×10^{-5}	2.4801×10^{-2}	1.1607×10^{-4}	7.4405×10^{-2}

Table 2. Empirical constants used to compute the settling velocity w_c

To close the system of equations (1, 2, 3, 4), the turbulent viscosity ν_t and diffusivities (ν_{tk}, γ_t) must be determined from TKE (k), the mixing length (L) and same stability functions S_m, S_k and S_h as :

$$\nu_t = L\sqrt{k}S_m \quad , \quad \nu_{tk} = L\sqrt{k}S_k \quad , \quad \gamma_t = \frac{S_h}{S_m}\nu_t \tag{6}$$

Values of the stability functions S_m, S_k and $S_h$are given by Blumberg & Mellor, 1987 in Table (1).

2.1 Bottom conditions

At the bottom, a non-slip condition is assumed for the velocity u $(u(z_0) = 0)$. Moreover, if we assume the equilibrium between the production and the dissipation of the TKE, the logarithmic profil of the velocity u and a linear mixing length, the following conditions can be imposed for k and L

$$k(z_0) = \lambda_b u_*^2 \quad , \quad L(z_0) = \kappa z_0 \tag{7}$$

where λ_b is an empirical coefficient, u_* is the bottom friction velocity, $(\lambda_b = 6.74)$ and κ is the von Karmàn constant $(\kappa = 0.41)$.
The bottom condition for the equation (4) dealing with sediment conservation is expressed by equation (8). At the fluid-sediment bed interface, equation (8) assume the equilibrium

between between the descending (or the deposition rate D) and the ascending (or the erosion rate E) sediment fluxes.

$$\left(\gamma_t \frac{\partial c}{\partial z} + w_s c\right)_{z=z_0} = D - E \quad (8)$$

The deposition rate D is considered proportional to the SSM concentration close to the bed with the settling velocity w_s acting as a proportionality factor ($D = w_s c(z_0)$). In addition, the erosion rate, E, is computed by the equation (9) given in the equilibrium condition by Celik & Rodi,1984; Celik & Rodi, 1988; Celik & Rodi, 1991 and Van Rijn, 1986.

$$E = w_s c_b \frac{\gamma_0 S}{1 + \gamma_0 S} \quad (9)$$

where c_b is the volume concentration of the sediment ($c_b = 0.65$ for non-cohesive sediment), γ_0 is an empirical constant for the sediment re-suspension and S is the non-dimensional shear stress excess given by:

$$S = \frac{\tau_b - \tau_{th}}{\tau_{th}} \quad (10)$$

$\tau_b = \rho u_*^2$ is the shear stress at the bottom and τ_{th} the critical threshold for erosion.

2.2 Free surface conditions

At this interface (air-fluid),as a first approximation and without the effects of boat propellers, we consider that the velocity boats are transmitted directly by the shear effect to the surface layer. The effect of a boat passage is therefore taken into consideration by using the navigation velocity of boats as hydrodynamic forcing: $u(H) = V_{boat}$. However this first approximation is relevant since the navigation velocities of boats are measured accurately in the field conditions by using differential GPS systems.
The equilibrium between production-dissipation of the TKE, and a linear law for the mixing length are also assumed at $z = H$. Thus, the conditions to be imposed on the free surface are:

$$k(H) = \lambda_s u_{*,s}^2 \quad , \quad L(H) = \kappa z_s \quad (11)$$

where λ_s is an empirical coefficient, $u_{*,s}$ is the surface friction velocity and z_s is the roughness of the boat surface.
Concerning the boundary condition for the sediment equation (4) at the free surface, it is specified a zero flux condition (i.e. no sediment flux across the free surface):

$$\left(\gamma_t \frac{\partial c}{\partial z} + w_s c\right)_{z=H} = 0 \quad (12)$$

2.3 Numerical solution

The flow modeled in this study is sheared flow at the surface and boundary layer flow at the bottom. The variables of this type of flow are characterized by large variations in the vicinities of the boundaries of the domain. To reproduce numerically these variations, we used a variable mesh size by splitting the column water into three layers (bottom, intermediate layer and surface layer). The surface and bottom layers were meshed very fine (about one millimeter). To limit interpolation procedures during the computation , the mean variables (u, c) and the turbulent variables (k, L, ν_t, γ_t) were calculated on two staggered grids. This grid shift allows us to express naturally the various gradients of each equation of system (1-4). Equations (1-4) with their boundary conditions was discretized by the finite volume method in space and finite differences in time. The spatial partial derivatives were approximated by a centered scheme except the convective term of equation (4) for which the upwind scheme was adopted. The implicit Euler scheme was adopted for the time derivatives of all model equations. Note that the discretization of equations has led to solving a tri-diagonal linear system by the Thoma's algorithm (TDMA: Tri-Diagonal Matrix Algorithm).

3. Application of the 1DV model

3.1 Field measurements

The impacts of passing boats on the bank erosion and/or the sediment re-suspension are specifically addressed during previous field experiments (Pham Van Bang et al., 2007 and 2008). Two sites (figures 1, 2) are considered: "Canal de la Sensée" (site 1) in northern France; Seine river (site 2). The two weeks (November 2006) measurements consist in records on the water elevations close to the bank and SSM concentrations with flow velocities in the vicinity of the bottom. The water fluctuations induced by the passing boats are measured by the mean of classical capacitive gauges in three points. This choice enables studies on the following aspects: the amplitude, the frequency and the direction of waves that are produced by the boat passage. The concentration and velocity measurements close to the bed are obtained by using Optical Backscattering Sensor, OBS, and electromagnetic velocity-meter, S4. Both sensors, OBS and S4, are attached to a beam that is anchored to the bed in the center of the navigation channel. The OBS is located at 10 *cm* above the sediment bed while the S4 is at 15 *cm*. This experimental set-up is completed by Acoustic Doppler Current Profiler (ADCP) located behind the passing boat to measure the vertical profile of flow velocity. The backscattered acoustic signal from ADCP transducer is also analyzed since its intensity provides qualitative information on the SSM concentration. As far as this chapter deals with sediment re-suspension induced by passing boats, the analysis from boat wake and ADCP measurements (detailed in Pham Van Bang et al., 2007) are not considered in the present study and only measurements close to the bed (from S4 and OBS transducers) are of concern. The information provided by these sub-aquatic sensors are indeed more relevant to characterize the re-suspension of sediment. It is generally observed from OBS measurements the production of concentration peaks during the boats passage. Thereafter, the peak tends slowly to decrease after the boat passage. This could be explained by the non-negligible inertia of the water mass which delays the recovering of equilibrium as explained later

3.2 Parameters and model forcing

During the period of the field investigations on the two sites, the daily traffic is recorded in order to correlate measurements to the navigation parameters. Of prior importance, it is recorded the direction, speed, trajectory and type (size, draught) of boats by the use of DGPS, radar and/or digital camera (Pham Van Bang et al., 2007). For example, the navigation speed with the boat length is used to form a chronicle that is finally exploited to force the velocity u at the free surface $z = H$.

The river bed and the SSM are sampled for the calibration of transducer, but also for the estimation of the bed roughness. Indeed, from grain size analysis of bed samples, it is determined a median diameter for the sediment particles, d_{50}, close to 20 μm (site 1) and 40 μm (site 2). This result enables the classification of the sediment as muddy sand for both sites.

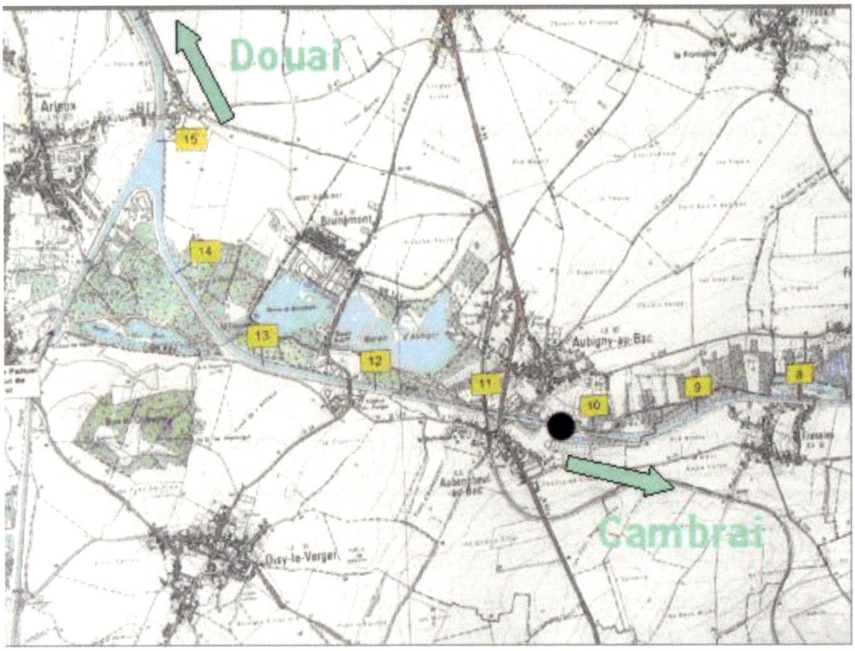

Fig. 1. Localization of the measurement site 1 "Canal de la Sensée" (North of France)

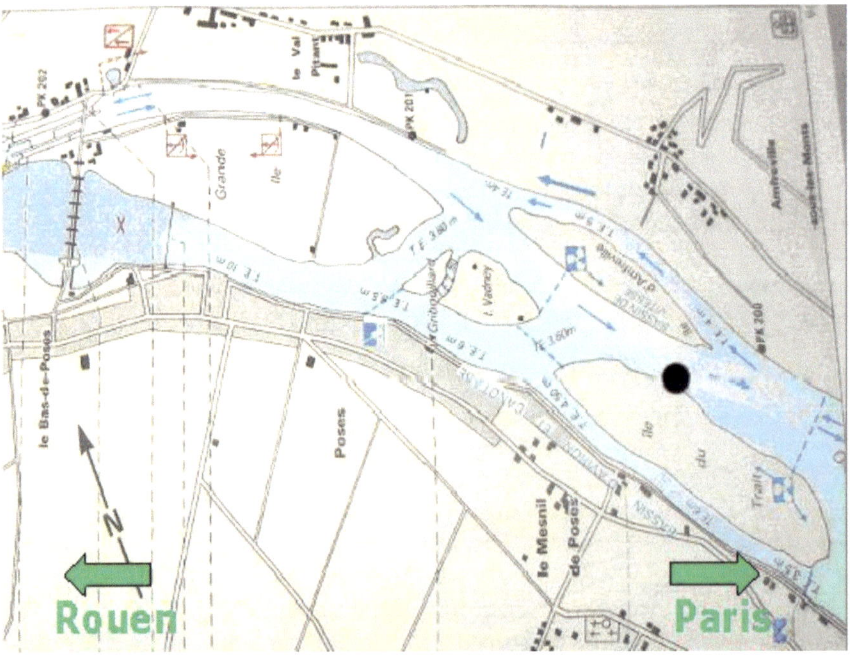

Fig. 2. Localization of the measurement site 2 "Seine river" (North of Paris, France)

The median diameter, d_{50}, is finally used to estimate the roughness of the bottom according to : $z_0 \cong 8d_{50}$.

4. Presentation of results simulations

As mentioned above, the measurement campaign was conducted on two different sites over a period of two weeks. However, we present here, for example, simulations for some days of both site 1 (11/08/2006 and 11/09/2006) and site 2 (11/15/2006).

4.1 The current induced by the boats

Figure 3 presents the velocity and the variables of turbulence obtained by the 1DV model. Two boat passages (namely D1 and D9, navigating down- and up-stream, see table (4)) are simulated for the day of the 11/08/2006. On this figure, it is first to notice that the obtained vertical profile of velocity is in good agreement with the flow characteristic expected from the theory of the surface-boundary layer. Moreover, the numerical model reproduces realistically the return currents induced by boats (inversion of the profile): the vertical profiles of velocity provided by our model are in good agreement with available data (mostly from physical model at laboratory) on the flow characteristics in the under keel clearance region.
Figures 3.b, 3.c and 3.d show that the model provides vertical profiles of turbulence variables (k, L , ν_t) in perfect agreement with the theory of the turbulent surface boundary layer structure.
The 1DV model also enables to simulate the velocity measurements performed by S4 sensors that are localized at 10 cm and 25 cm from the bottom. For the boat traffic recorded during the day of the 11/08/2006 (site 1), the chronicle presents strong and irregular peaks that is explained by the important inertia of water which delays considerably the return to equilibrium. The averaged peaks in the velocity for the day of the 11/08/2006 is 3.5 cm/s at $z = 10$ cm and 4.4 cm/s at $z = 25$ cm. These estimations are in close agreement with the field measurements (Pham Van Bang et al., 2007, 2008).
For the site 2 (day 11/15/2006), the model provides a velocity of 12 cm/s at $z = 10$ cm and 16 cm/s at $z = 25$ cm. This variation is undoubtedly due to the difference in the site conditions (geometry of the navigation channel, type of boat,..etc.). For example, the site 1 is localized at a narrow and straight channel (50 m wide, 4 m deep) with a negligible ambient current while the site 2 is related to a large river (200 m wide, 5 m deep) with an average ambient current of $10cm/s$.
Finally, it is useful to note the important advantage of our numerical model to provide direct estimation for the bottom shear stress, τ_b. This key parameter for sediment transport is not measured during the field experiment: the time resolution used for S4 records was limited due to the hard drive capacity of the data log; the measurement of velocity near the bed looks problematic if using such a technology. The bottom shear stress τ_b is however accessible from the numerical modelling. With the considered assumptions, the model estimates the bottom shear stress for the day of the 11/08/2006 (site 1) at 0.43 Nm^{-2} and at 1.14 Nm^{-2} for day of the 11/15/2006 (site 2). This last estimated value is very compatible with the values of τ_b founded between 1 and 2 Nm^{-2} from ADV measurements in the Seine estuary (Verney et al., 2007).

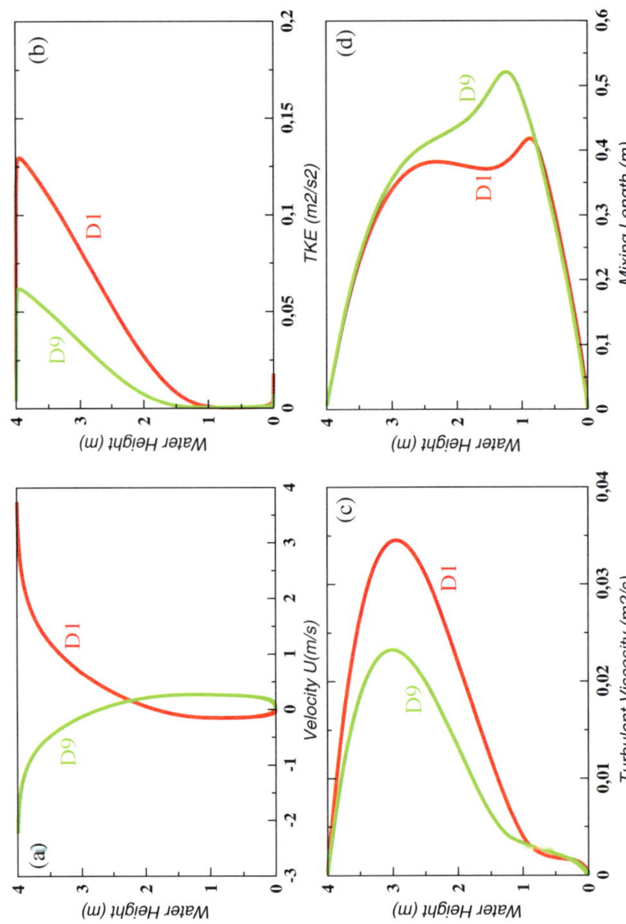

Fig. 3. Vertical profiles resulting from simulation: a) velocity (u), b) turbulent kinetic energy (k), c) turbulent viscosity (ν_t), d) mixing length (L)

4.2 Suspension of the sediments induced by boats

Figure 4 presents time series of SSM concentration measurements from OBS sensor located at 10 *cm* from the bottom on site 1. The confrontation between experiment and simulation presents a good agreement. In particular, the general shape and magnitude of concentration peaks are well reproduced by the computation. The numerical model is therefore able to capture dynamics, the fast mode of erosion and the following slow mode of sedimentation that are produced by a passing boat. As example, figure 4 focusing on the measurements of the 11/10/2006 proposes simulations of the sediment re-suspension peaks. For the peak values, results are in narrow agreement with the measurements performed close to the bed since the difference between the simulated and measured peaks does not exceed 3×10^{-3} g/l. However, it is also observed in figure 4 some discrepancies between simulation and experiment after the peaks of concentration, i.e. during the sedimentation of the re-suspended matter. This difference with measurements could be explained twofold. On one hand, the numerical model only considers an unique *'equivalent'* diameter (40 μm for the site of interest) for the sediment while OBS measurement concerns all the sediment particles whatever their size. As the particle settling velocity scales with the square of its diameter, sediment particles finer than 40 μm will remain longer in the suspension so that the simulation results underestimate the concentration during the sedimentation. On the other hand, the numerical model doesn't take account the added turbulence by the propellers effects. These effects could also maintain longer the sediment in suspension.

Despite of its simple assumptions, the actual version of the 1DV model provides practical information for the management plan of the waterway. For example, the amplitude and the time duration of a re-suspension peak are obtained from the model. The time relaxation of the concentration peak is estimated and could be used to avoid cumulative effect whilst increasing the boat frequentation. From figure 4.d, the D59 boat induced peak concentration is decreasing with a characteristic time (noted τ) between 10 and 12 mn. The following boat shouldn't imply cumulative effect on the D59 peak signature if it passes 15 mn later than D59.

Figure 5.a shows the SSM concentration chronic simulated by the model at the bottom (10 *cm* above the bottom) of site 1. This figure is used to establish a correspondence between the boats passage and the SSM peaks: the peaks succession of the SSM identify with precision the boats causing these peaks (figures. 5.b, 5.c, 5.d). In these figures, we also observe an asymmetry that reflects the rapid phase of erosion and the slow process of sedimentation. This asymmetry has been observed *in-situ* by others authors (Pham Van Bang et al., 2007; Parchure et al., 2001a; Parchure et al., 2001b). Note that the 1DV model can learn about the amplitudes and duration peaks of SSM without going through any simplifying assumption. Indeed, one can directly read from the SSM concentration figures 5.b, 5.c, 5.d), the characteristic time of decay (or sedimentation). As an example, we read $\tau_{D22} = 220\ s$, $\tau_{D24} = 222\ s$ et $\tau_{D36} = 185\ s$ respectively for boats D22, D24 and D36. These orders of magnitude are in agreement with the conclusions resulting from the measurements taken on site 1 by Pham Van Bang et al. (2007).

On figure 6.a, we present the vertical profiles of the SSM concentration induced by the passage of the D1 and D9 boats (site 1). It is first observed the vertical profiles match in good agreement with the Rouse's law (analytical solution at the equilibrium). It is also observed that the suspension takes predominantly place in the first fifty centimeters from the bottom. Let us remind that return currents due to the boat passage are localized in this zone. As a consequence, an accurate calculation of the return current near the bottom zone is strongly required in the numerical model for a good estimate of the SSM concentration.

The figure 6.b presents the same results as figure 6.a in term of vertical profile of sediment

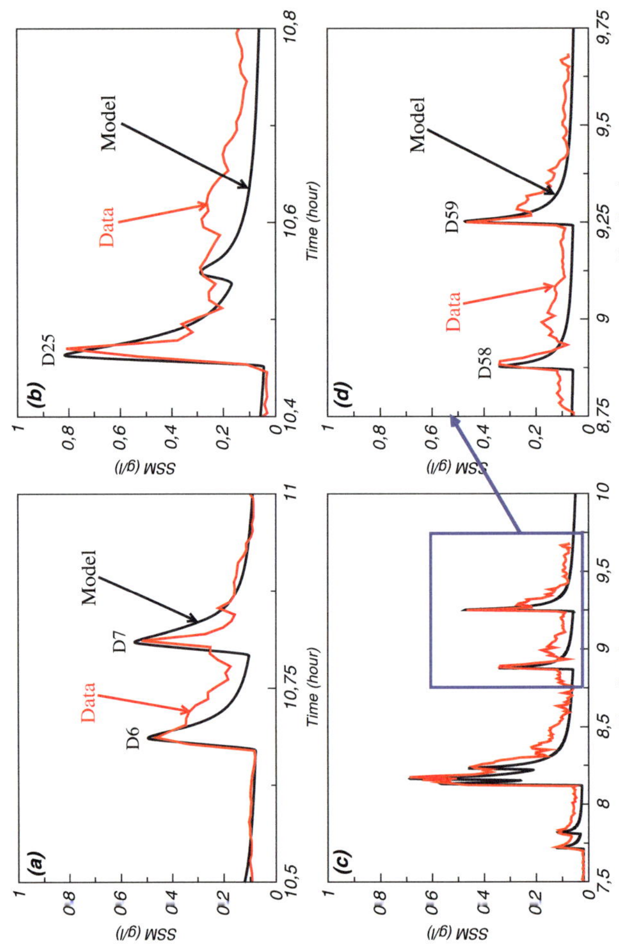

Fig. 4. The SSM chronic issued from simulation.
(a) the peaks of the SSM induced by boats D6 and D7 11/08/2006.
(b) the peaks of the SSM induced by boat D25 11/09/2006.
(c) the peaks of SSM induced by all river traffic 10/11/2006.
(d) zoom on the SSM induced by boats D58 and D59 11/10/2006

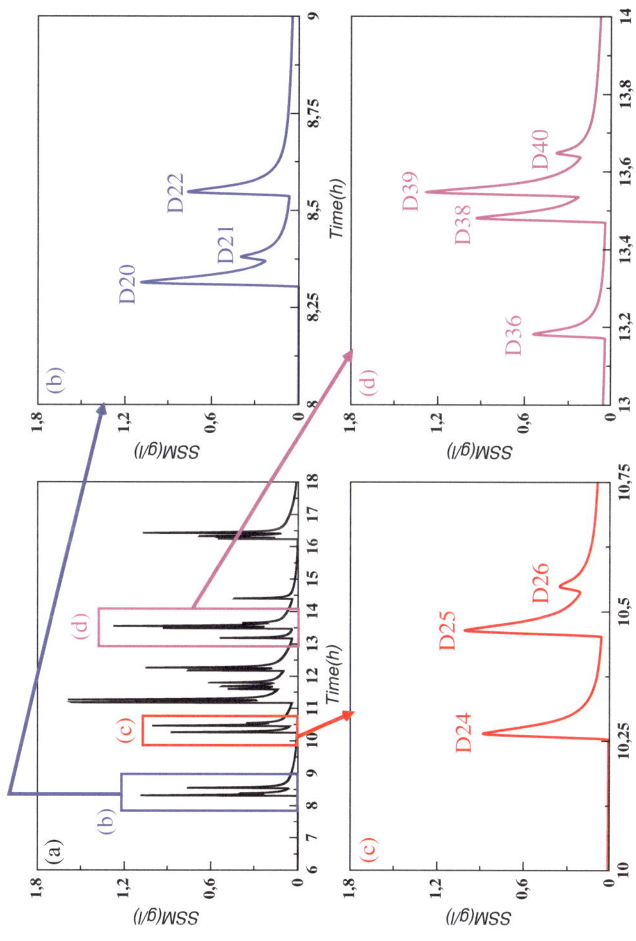

Fig. 5. The SSM chronic issued from simulation.
(a) the peaks of SSM induced by all river traffic of 11/09/2006.
(b), (c) and (d) identification of boats causing the SSM peaks

horizontal fluxes ($u.c$). As a consequence of the previously detailed velocity (u, in figure 3.a) and concentration profiles (c, in figure 6.a), the vertical distribution of sediment fluxes presents positive and negative values depending on the direction of the passing boat. More important values of the sediment fluxes occur close to the bottom and near the free surface. The results presented in figure 6.b provide again practical information for the management

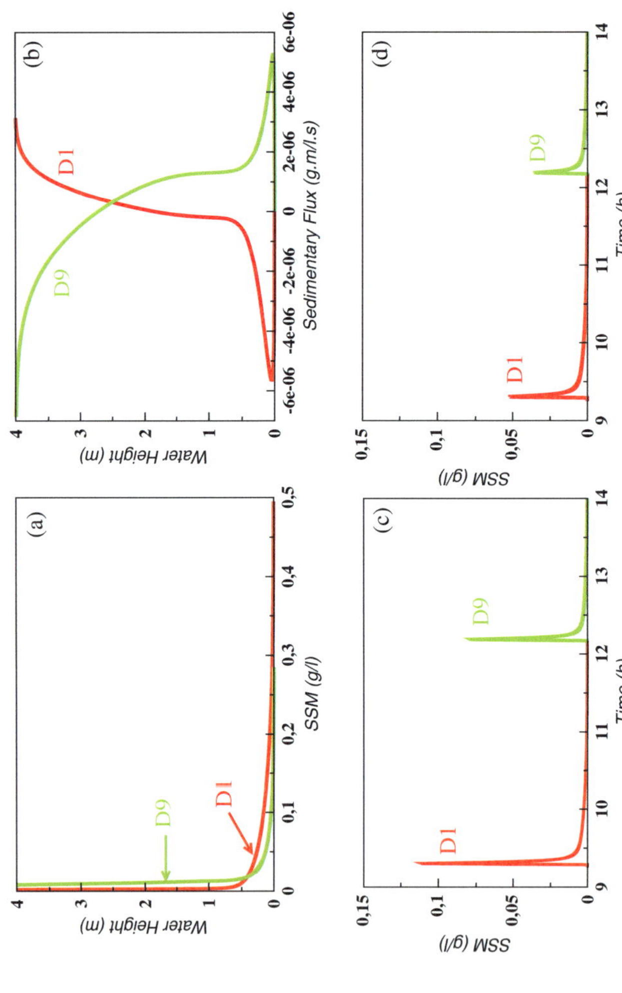

Fig. 6. The SSM chronic issued from simulation.
(a) the peaks of SSM induced by all river traffic of 11/09/2006.
(b), (c) and (d) identification of boats causing the SSM peaks

of the waterway. The net sediment transport induced by daily traffic could be estimated as follows. For the daily traffic recorded during the day of the 11/9/2006, the vertical profiles of the sediment concentration induced by each passing boat is integrated along the vertical. The net sediment mass transported per second and per width unit (figure 7) is therefore obtained. For the whole traffic (17 boats, site 1) of the 11/9/2008, the sediment transport is estimated by the model at 0.280 $kgm^{-1}s^{-1}$ and 2.940 $kgm^{-1}s^{-1}$ for the day of the 11/15/2006 (site 2). An extrapolation of these results at one year gives the estimate of 102.5 kg per year and per width unit for site 1 (and 1.1 *tons* per year and width unit for site 2). Finally, the estimate of the net mass transported due to navigation traffic is proposed in table 3 for other days.

Site 1		Site 2	
11/08/2006 :	1.720 Kgm^{-1}	11/14/2006 :	*missing data*
11/09/2006 :	0.280 Kgm^{-1}	11/15/2006 :	2.943 Kgm^{-1}
11/10/2006 :	0.124 Kgm^{-1}	11/16/2006 :	2.475 Kgm^{-1}

Table 3. Net mass of the sediments transported by suspension per day during de measurements period (November 2006)

5. Conclusions

Predicting the sediment transport in natural environment is a difficult task, but an essential exercise for ecological and economical issues. More difficult is the prediction of sediment transport in presence of navigation traffic. To study this peculiar case, a field experiment on two different sites is conducted with the measurement on the sediment re-suspension induced by passing boats.

In the continuity of this work, we developed a 1DV numerical model. Despite of its apparent simplicity, the model considers all the governing processes (hydrodynamic, turbulence, sediment transport) to compute the boat induced sediment transport. Although based on simplifying assumptions, this model recovers all the flow characteristics that are observed in the under keel clearance zone, meanwhile propeller effects are neglected.

The use of the developed model, this study demonstrates that the numerical model is able to predict reasonably all the measured variables. Moreover, the model provides precious information on the flow variables difficult to measure due to the complexity of their achievement in field conditions. More particularly, the 1DV model supplies information on turbulence, bottom stress shear, sedimentary flux. The model gives also estimation that could be used for an integrated management of the waterway.

Finally, let us note that the actual version of the model doesn't account for the total flow in the under keel clearance zone, nor for muddy sediment transport. A non-hydrostatic 2DV or 3D numerical model (hydrodynamic, turbulence, muddy sediment transport) is needed for a more realistic study of the flows including the propeller effects.

6. Acknowledgement

The authors acknowledged gratefully the Inland Waterways for France VNF (Voies Navigables de France), the University of Quebec and the Regional Laboratory of the Parisian East (LREP) for their financial and technical support during the field investigations

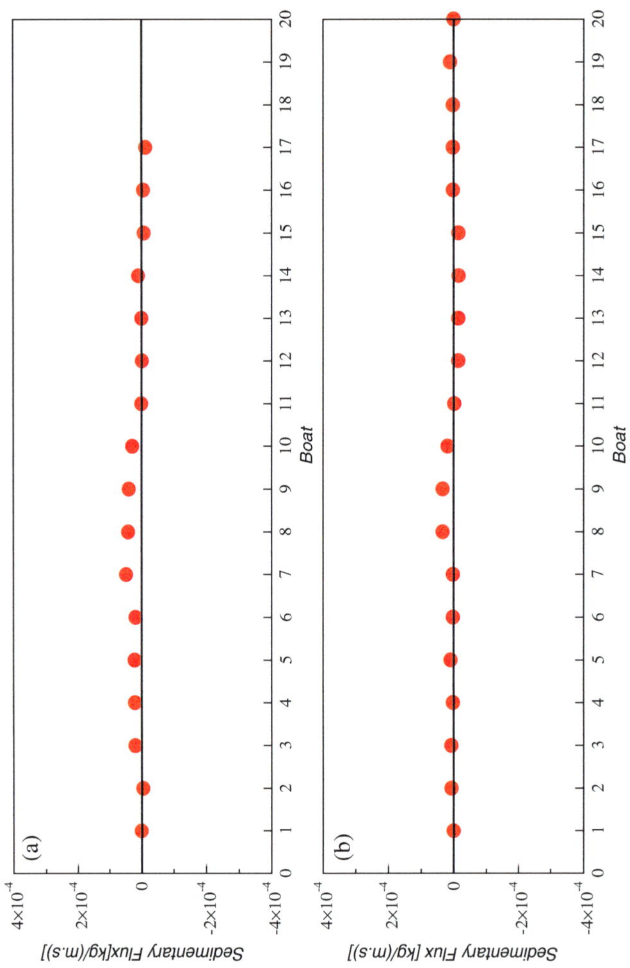

Fig. 7. Suspended sedimentary flux induced by boats at site 1.
a) daily river traffic of 11/08/2006
b) daily river traffic of 11/09/2006

Date	H (GMT)	Reference	C.E	Length (m)	Direction	Speed (km/h)
11/08/2006	09h45	D1	E	38	Douai	13.5
-	10h45	D6	C	45	Douai	08.5
-	10h55	D7	E	80	Cambrai	06.6
-	12h20	D9	E	2×38	Cambrai	08.0
11/09/2006	08h18	D20	C	-	Douai	11.0
-	08h22	D21	-	-	Cambrai	05.8
-	08h32	D22	C	38	Douai	10.0
-	10h15	D24	C	38	Douai	10.5
-	10h27	D25	E	80	Cambrai	10.0
-	10h32	D26	E	85	Cambrai	06.5
-	12h55	D36	C	67	Douai	09.7
-	13h28	D38	C	38	Cambrai	10.0
-	13h32	D39	C	90	Douai	11.0
-	13h38	D40	E	80	Cambrai	06.0
11/10/2006	08h05	D58	C	38	Douai	06.5
-	09h15	D59	C	79	Douai	09.7

Table 4. Information about some boats cited in the text:
C.E : Carrying or Empty boat.
Direction (see figure 1): Douai = downstrean; Cambrai = upstream

7. References

Blumberg A.F. & Mellor G.L. (1987). A description of three-dimensional coastal ocean circulation model has. In: *Three-Dimensional Coastal Models Ocean, Coastal and Estuarine Sciences, Volume 4 AGU* N.S. Heaps (Ed.), pp 1–16, Washington (D.C.).

Celik, I. & Rodi W. (1984). A deposition-entrainment model for suspended sediment transport. *Report SFB 210/T/6*, University of Karlsruhe (FRG), Karlsruhe, Germany, 55 p.

Celik, I. & Rodi W. (1988). Modeling suspended sediment-transport in nonequilibrium situations. *ASCE Journal of Hydraulic Engineering*, Vol. 114, No. 10, October 1988, 1157-1191.

Celik, I. & Rodi W. (1991). Suspended sediment-transport capacity for open Chanel flow. *ASCE Journal of Hydraulic Engineering*, Vol. 117, No 2, February 1991, 191-204.

Parchure, T.M.; McAnally W.H. & Teeter A.M. (2001a). Desktop method for estimate vessel-induced sediment suspension. *ASCE Journal of Hydraulic Engineering*, Vol. 127, No 7, July 2001, 577-587.

Parchure, T.M.; McAnally W.H. & Teeter A.M. (2001b). Wave-induced sediment resupension near the shorelines of the upper the Mississippi to rivet system. *(ENV Report 20)*, 99 p

Pham Van Bang, D.; Ouahsine, A.; Sergent, P. & Hissel, F. (2007). Impact of the passage of the boats on the transport of the sediments and the maintenance of the inland waterways: measurements and analyzes. *Report VNF, UTC, CETMEF*, 63 p.

Pham Van Bang, D.; Ouahsine, A.; Sergent, P.; Long, B.; Montreuil S.; Debaillon P. & Hissel F. (2008). Erosion of the sediments and vagueness generated by the passage of the boats: measurements and analyzes. *Proceedings of Xme Journées Nationales Génie Côtier-Génie-Civil*, pp. 605-614, ISBN 2-9505787-9-9, Sophia Antipolis, October 2008, Centre Français du Littoral, France

Van Rijn, L.C. (1986). Mathematical modeling off suspended sediment in non-uniform flows. *ASCE Journal of Hydraulic Engineering*, Vol. 12, No 6, June 1986, 1613-1641.

Verney R.; Deloffre, J.; Brown-Cottan J.C.; Lafite R.(2007). Intertidal The effect off wave-induced turbulence one mudflats: impact off boat traffic and wind. *Continental Shelf Reseach*, Vol 27, No 5, March 2007, 594-612.

4

Hybrid Model Approaches to Predict Multiscale and Multiphysics Coastal Hydrodynamic and Sediment Transport Processes

H. S. Tang[1,*] and Timothy Keen[2]
[1]City College, City University of New York
[2]Naval Research Laboratory
USA

1. Introduction

Coastal ocean processes are complicated and they happen as various phenomena that span a vast range of spatial and temporal scales. For instance, general circulations of oceans occur at global scales (Wunsch & Ferrari, 2004). Tropical waves that eventually impact coastal waters propagate with wavelengths of one thousand kilometers and periods of one month (Legeckis et al., 1983). Langmuir cells, which are pairs of vortices hanging below water surfaces, have spatial sizes ranging from one to hundreds of meters (Weller et al., 1985). Scour near coastal structures is under influence of large-scale current processes but occurs in a relatively small size (Sumer & Whitehouse, 2001). Flows around enormous numbers of swimming microorganisms can occur at scales of micrometers (Pedley, 1992). Here, the scales are either observation scales such as characteristic length and time or process scales such as those in wavelet analysis (Chui, 1992; Kumar and Foufoula-Georgiou, 1997).
Since a few decades ago, various geophysical fluid dynamics (GFD) models have been developed for individual coastal ocean phenomena at specific scales. The Princeton Ocean Model (POM), Finite-Volume Coastal Ocean Model (FVCOM), and HYbrid Coordinate Ocean Model (HYCOM) were developed to predict current velocity, sea level, salinity, and temperature at regional scales (Blumberg and Mellor, 1987; Chen et al., 2003; Halliwell, 2004). The WAVEWATCH and SWAN (Simulating Waves Nearshore) models were designed to simulate surface wave propagation at global to coastal scales (Tolman, 1991; Booij et al., 1999). Models have also been proposed to predict sediment transport and seabed morphology for near-coastal regions (e.g., Tonnon et al., 2007; Papanicolaou et al., 2008). In recent years, computational fluid dynamics (CFD), which can accurately model small-scale and detailed flow structures, has been applied to coastal engineering flows (e.g., Young et al., 2001). In view of the multiscale and multiphysics nature of coastal ocean processes, there is a great challenge to simulate them accurately and, until now, the efforts using numerical simulation have been successful merely for individual phenomena and scales. The challenge comes from model restrictions, numerical techniques, and computer capabilities. For

---

[*] Changsha University of science and Technology, China

instance, a deep-ocean model has difficulty in dealing with the vertical mesh at sudden bathymetry changes as well as the smaller scales of nearshore flows (Song & Hou, 2006; Heimusund & Berntsen, 2004). Limitations such as hydrostatic assumptions and/or two-dimensionality of GFD models are inherent restrictions that prohibit accurate simulations of many important phenomena such as vertical motions of Langmuir cells. It is frequently reported that coastal models become unstable at small time steps and grid spacing (e.g., Heimusund & Berntsen, 2004; Keen et al., 2003), which is not a surprise since they are designed for large-scale flow phenomena. It is anticipated that the solutions of the models do not necessarily converge to those of Navier-Stokes equations as grid spacing tends to zero. Therefore, it is not realistic to achieve accurate simulation of coastal phenomena, especially the small-scale processes, by just using existing coastal ocean models and simply reducing time steps and grid spacing. Although in principle CFD approaches have fewer limitations and can capture flow phenomena at much broader spectra of scales, they are computationally expensive and not applicable in simulating a complete actual coastal ocean flow. It should be noted that, although both GFD and CFD are based on the Navier-Stokes equations, they are different approaches with respect to numerical technique, turbulence closure, and parameterization for small scales.

It is now becoming a trend in prediction of coastal ocean flows to adapt to multiphysics/multiscale approaches (Fringer et al., 2006). Although computer modeling has reached the point where the simulation of individual flow phenomena over relatively narrow ranges of scales has become mature, a single, comprehensive model capable of dealing with multiphysics/multiscale problems is unlikely to be available in the near future. The hybrid method (HM) couples different models to each other, and the domain decomposition method (DDM) divides a flow domain into many subdomains, each of which is assigned to an individual model. Combining HM and DDM is one of the most promising currently available techniques to bridge the scales and overcome difficulties in multiphysics/multiscale modeling (Benek et al., 1983; Harten, 1993; Dolbow et al., 2004).

Since erosion and transport of erodible seabed sediment is coupled to various hydrodynamic forces, it is imperative to analyze them at different scales in correspondence to multiscale hydrodynamic processes (Chiew, 1991). It has been common to assume that seabed scour is a local process, which occurs within a few tens of diameters of a structure resting on the seafloor (Zang et al., 2009). This is true for event-scale erosion but it is unlikely for systematic or catastrophic scour and/or burial processes that operate at months or yearly time scales. Experimental results and parametric methods for scour around structures on the seafloor have been supplemented by numerical models that focus on the finest scales around the obstacles (Alam & Cheng, 2010). These high-resolution models typically focus on steady flow but recent laboratory studies have examined wave-induced scour (Xu et al., 2009) and numerical methods have been applied to shoaling waves (Myrhaug et al., 2008). These studies use simplified hydrodynamic forcing because of the disparity in scales between the external (ocean currents) and internal flows (around structures).

CFD approaches for the fluid flow around a structure are not easily implemented for sediments because of the interaction of turbulence with discrete particles. Thus, scour models have traditionally used parametric approaches (Myrhaug et al., 2008) and discrete particle models (Zamankhan & Doolatshahi, 2008). Lattice Boltzmann methods are being explored as well for simple geometries (Alam & Cheng, 2010). These approaches depend on a given flow field that is typically simplified from the ocean environment. The constitutive equations for these models are derived from observations in flumes and the seabed, and thus are specifically formulated for near-field flow. The other extreme is to use a numerical model of regional flow to calculate a large-scale mean scour that can be used as an indicator of variations in potential scour within a selected region (Keen & Glenn, 2002). These models tend to use the same macroscopic approach represented by the Navier-Stokes equations. There is a lack of knowledge of the impact of these external hydrodynamic processes on the local scour and burial problem. In order to use localized scour models to investigate the impact of external factors on seabed scour, it is necessary to implement them on much larger domains. This approach is problematic with current computational resources. The problem of applying a macroscopic model (i.e., those based on Navier-Stokes equations) for the external problem and a scour model for the local scour is prohibitive because of fundamental differences in their numerical formulations. This problem can be more easily examined using macroscopic approaches, which have common numerical formulations and parameterizations.

This chapter summarizes our recent work in modeling of multiscale and multiphysics hydrodynamics phenomena using HM and DDM. We also discuss related sediment transport, with emphasis on localized scour and erosion processes. First, a hybrid approach that couples the FVCOM and a CFD model is described, and results of multiscale simulation for an effluent thermal discharge from a diffuser at the ocean bottom is presented. Second, an analysis is made of the effects of local-scale hydrodynamics on sediment transport around the diffuser. Third, as a multiphysics modeling of interaction between different phenomena, simulation of flow over sand dunes under action of surface wind is presented to illustrate the interaction between surface waves, currents, and morphology. These examples demonstrate the multiscale/multiphysics methodology as applied to problems that cannot be simulated as either local or external phenomena. They also indicate that multiscale and multiphysics simulations of hydrodynamics are more advanced than conventional modeling because of the complex interaction between the flow and discrete particles in the ocean.

2. Hybrid CFD and FVCOM for simulation of thermal effluent into coastal flows

A well-tested CFD model is employed in this work (Lin & Sotiropoulos, 1997; Tang et al., 2003; Tang et al., 2008). The governing equations for hydrodynamic processes of the CFD model are the three-dimensional (3D) continuity and Navier-Stokes equations that, in general curvilinear coordinates, are expressed as follows:

$$\Gamma \frac{\partial Q}{\partial t} + J \frac{\partial}{\partial \xi^k}(F^k - F_v^k) + I = 0, \tag{1}$$

where

$$\Gamma = diag(0,1,1,1), \qquad Q = (p,u,v,w)^T,$$

$$F^k = \frac{1}{J}(U^k, uU^k + p\xi_x^k, vU^k + p\xi_y^k, wU^k + p\xi_z^k)^T, \tag{2}$$

$$F_v^k = \frac{1}{J}\left(\frac{1}{\mathrm{Re}} + \nu_t\right)(0, g^{lk}\frac{\partial u}{\partial \xi^l}, g^{lk}\frac{\partial v}{\partial \xi^l}, g^{lk}\frac{\partial w}{\partial \xi^l})^T,$$

$$I = -\frac{T}{Fr^2}e,$$

where: t is time; $J = |\partial x_i / \partial \xi^j|$ is the Jacobian of the geometric transformation from Cartesian coordinates, x_i (i = 1, 2, 3 for x, y, z axes, respectively) to curvilinear coordinates ξ^j (j = 1, 2, 3); p is the static pressure divided by the density; u, v, and w are the velocities in x, y, and z directions, respectively; T is the temperature. Furthermore, $U^k = u_i \xi^i_{x_k}$ are the contravariant velocities in ξ^k directions (Note that u_i are, respectively, u, v, and w, $\xi^i_{x_k}$ are the metrics of the geometric transformation, and repeated indices imply summation); Re is the Reynolds number; ν_t is turbulence eddy viscosity; g^{ij} ($g^{ij} = \xi^i_{x_k}\xi^j_{x_k}$) is the contravariant metric tensor; Fr is Froude number; e is the unit in the gravity direction. The governing equation for heat transfer reads as

$$\frac{1}{J}\frac{\partial T}{\partial t} - \frac{\partial}{\partial \xi^k}\left\{\frac{1}{J}\left(\frac{1}{\Pr \mathrm{Re}} + \frac{\nu_t}{\Pr_t}\right) g^{lk}\frac{\partial T}{\partial \xi^l}\right\} = 0. \tag{3}$$

where: Pr is the molecular Prandtl number; Pr_t is the turbulent Prandtl number; l = 1, 2, 3. The standard mixing length model is used in this work (Mason, 1989). The governing equations are discretized using a second-order accurate, implicit, finite-volume method on non-staggered grids, and they are solved using a dual time-stepping artificial compressibility method. A fourth-difference artificial dissipation method and V-cycle multigrid method are used. A domain decomposition method in conjunction with the Schwartz alternative iteration is employed to deal with complicated geometry. In order to achieve seamless transition of solutions between subdomains, an effective mass conservation algorithm is proposed. The CFD model has been tested and applied in various problems from academe as well as industry, such as vortex breakdown and flow past bridge piers (Sotiropoulos and Ventikos, 1998; Ge and Sotiropoulos, 2005). For details of the CFD model, readers are referred to Lin & Sotiropoulos (1997), Tang et al. (2003; 2008), and Xu & Sun (2009).

In the FVCOM, the governing equations are the continuity and momentum equations. The governing equations for the external mode in the model are vertically averaged two-dimensional (2D) continuity and momentum equations. The 2D continuity equation is (Chen et al., 2006):

$$\frac{\partial \eta}{\partial t} + \frac{\partial HU_l}{\partial x_l} = 0, \tag{4}$$

where: η is the water surface elevation; H is water depth; and U_l are depth-average current velocities in x_l directions. The momentum equations are:

$$\begin{aligned}\frac{\partial U_i H}{\partial t} + \frac{\partial U_i U_l H}{\partial x_l} = -gH\frac{\partial \eta}{\partial x_i} - \frac{gH}{\rho_0}\left(\int_{-1}^{0}\frac{\partial}{\partial x_i}\left(H\int_{\sigma}^{0}\rho d\sigma'\right)d\sigma + \frac{\partial H}{\partial x_i}\int_{-1}^{0}\sigma\rho d\sigma\right) \\ +(-1)^i fU_j H + \frac{\tau_{sx_i} - \tau_{bx_i}}{\rho_0} + D\tilde{F}_i + G_i,\end{aligned} \tag{5}$$

$$G_i = \frac{\partial U_i U_l H}{\partial x_l} - D\tilde{F}_i - \left(\frac{\partial \overline{u_i u_l}H}{\partial x_l} - D\overline{F}_i\right),$$

$$D\tilde{F}_i \approx \frac{\partial}{\partial x_l}\left(\bar{A}_m D\left(\frac{\partial U_i}{\partial x_l}+\frac{\partial U_l}{\partial x_i}\right)\right),\ D\bar{F}_i \approx \frac{\partial}{\partial x_l}\overline{\left(A_m D\left(\frac{\partial u_i}{\partial x_l}+\frac{\partial u_l}{\partial x_i}\right)\right)}, \tag{6}$$

where: the overbar denotes the vertical integration; g is the gravity constant; ρ_0 is the reference density of seawater; σ is vertical coordinate; f is the Coriolis parameter; τ_{sx_i} are the surface friction stresses, and τ_{bx_i} are the bottom friction stresses in x_i directions; D is the mean water depth; and A_m is the horizontal eddy viscosity (Smagorinsky, 1963). The governing equations of the internal mode in the model are 3D continuity and momentum equations with x and y as horizontal coordinates and σ as vertical coordinate:

$$\frac{\partial \eta}{\partial t}+\frac{\partial H u_l}{\partial x_l}+\frac{\partial \omega}{\partial \sigma}=0, \tag{7}$$

$$\begin{aligned}\frac{\partial u_i H}{\partial t}+\frac{\partial u_i u_l H}{\partial x_l}+\frac{\partial u_i \omega}{\partial \sigma} &= -gH\frac{\partial \eta}{\partial x_i}-\frac{gH}{\rho_0}\left(\frac{\partial}{\partial x_i}\left(H\int_\sigma^0 \rho d\sigma'\right)+\sigma\rho\frac{\partial H}{\partial x_i}\right)\\ &\quad +(-1)^i f u_j H+\frac{1}{H}\frac{\partial}{\partial \sigma}\left(K_m\frac{\partial u_i}{\partial \sigma}\right)+DF_i,\end{aligned} \tag{8}$$

$$DF_i \approx \frac{\partial}{\partial x_l}\left(A_m D\left(\frac{\partial u_i}{\partial x_l}+\frac{\partial u_l}{\partial x_i}\right)\right), \tag{9}$$

where: u_i are the layer horizontal velocity components; ω is the vertical velocity in σ coordinate; and K_m is the vertical eddy viscosity, which is determined using the Mellor and Yamada level-2.5 turbulent closure (Mellor & Yamada, 1982; Chen et al., 2003). In the FVCOM: i, j, l=1, 2 ($i \neq j$). The governing equations are discretized using a finite-volume method with triangle meshes on horizontal planes and layer meshes in the vertical direction. Second-order accurate upwind schemes are used to discretize the advection terms, and Runge-Kutta methods are employed to march in time. The external and internal modes may have different time steps. Mainly because it uses unstructured meshes, the FVCOM is becoming popular in coastal ocean modeling. Details for the FVCOM can be found in Chen et al. (2003; 2006).

In order to simulate multiscale and multiphysics coastal hydrodynamic processes, we proposed HM and DDM approaches (Tang & Wu, 2010; Wu & Tang 2010). In particular, the CFD model is coupled with the FVCOM; the CFD model is employed to resolve small-scale flow phenomena, and the FVCOM is used to model background circulation. The domains of the CFD model and FVCOM overlap over a region (Fig. 1). As a coupling strategy, the 3D CFD model is coupled to the 3D internal mode of the FVCOM, and the two models exchange solutions for the velocity distributions at grid interfaces between them. The strategy is based on the assumption that the obstacle covered by the CFD model only alters local velocity distribution but does not affect horizontal average velocities and water surface elevation, which are determined by the external mode. This assumption is consistent with the assumption in the FVCOM (Chen et al., 2003). Since Chimera grids (grids overlapping arbitrarily with each other) provide the best possible flexibility in connecting different models, we employed them to connect the CFD model and the FVCOM, as shown in Fig. 2. In the figure, a-b-c and d-e-f are grid interfaces between CFD model and FVCOM, and linear interpolations are made to provide solutions for velocities at the nodes and elements on the interfaces. The details of the coupling methods can be found in Tang & Wu (2010) and Wu & Tang (2010).

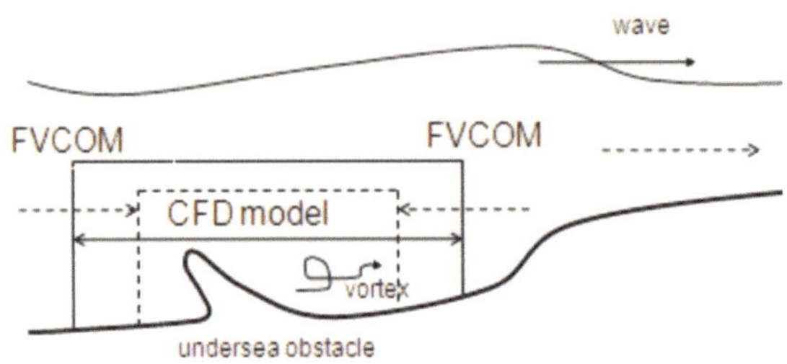

Fig. 1. Schematic representation of CFD model and FVCOM coupling.

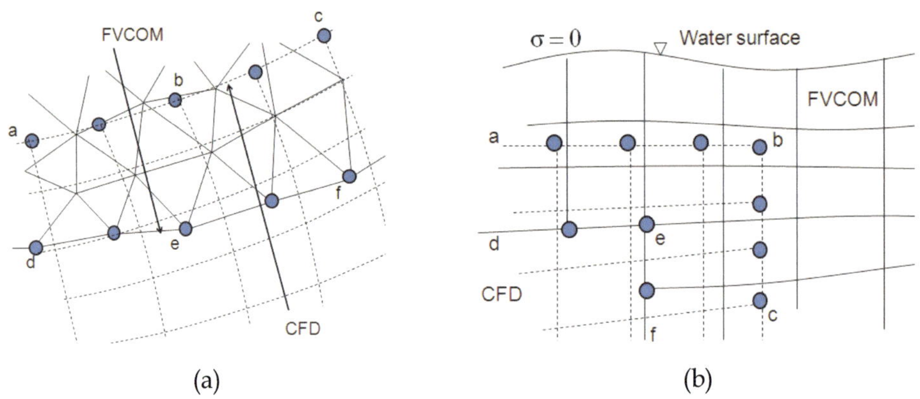

Fig. 2. Layout of grids of CFD and FVCOM. Solid lines - FVCOM grid, dash lines - CFD grid. a) Horizontal plane grids. b) Vertical direction grids.

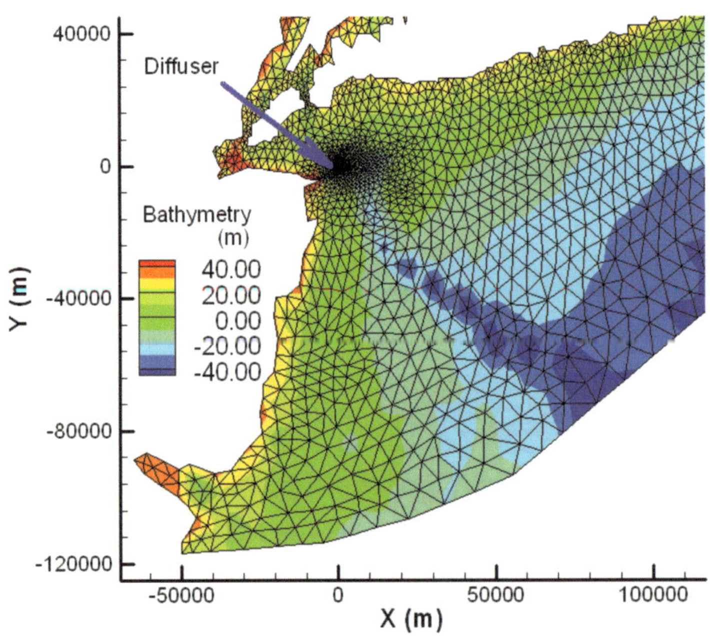

Fig. 3. Region of study, bathymetry, FVCOM mesh, and diffuser location.

Consider an effluent discharge from a diffuser in the setting of the New York and New Jersey coastal region under action of tides (Fig. 3). The x axis is in the east direction, and y axis is in the north direction. The diffuser is located in the New York Bight. The diffuser consists of a pipe with a diameter of 1.32 m lying on the ocean bottom, and ten ports with diameters of 0.175 m (Fig. 4a). All of the ten ports have an angle of 110° with respect to the x axis and upward angles with respect to the x-y plane ranging evenly from 45° to 18° from the first port, located near the origin of the plane, to the last port. Hot water at 32.0 °C is discharged at speed 3.92 ms^{-1} from the ports into the coastal water at 20.5 °C. The only driving force of the flow is tides; wind and other factors are ignored. This is a multiscale flow; the thermal effluent happens at scales of the discharge ports and the pipe, while the unsteady tides occur at scales of the tides.

The hybrid approach is used to simulate the flow; the CFD model captures the flow around the effluent from the diffuser, and the FVCOM describes the background large-scale flows. The mesh of the FVCOM is shown in Fig. 3, and those of the CFD model for the diffuser are shown in Fig. 4b. The CFD model and modeling of such thermal discharges have been intensively tested and calibrated using other models and measurement data (Tang et al., 2008). The coupling approach has been also validated (Tang & Wu, 2010; Wu & Tang, 2010).

The computed solutions are shown in Figs. 5, 6, 7, and 8. Fig. 5 presents the large-scale coastal flows at ebb and flood tides obtained with the FVCOM. It is seen from the figure that there are many total velocity patches, in red and blue, at scales ranging from 10^4 to 10^5 m. The thermal discharge, located near x, y = 0, is at the edges of a relative high velocity patch. No solution details for flows around the diffuser are available from simulations of the FVCOM; however, the CFD model provides the details. For example, Fig. 6 presents velocity distribution at a plane 6 m above the diffuser. Due to the presence of the diffuser and thermal effluent, the flow field during both ebb and flood tides is greatly altered; there are several velocity patches at scales of 10 m and larger, together with a low-velocity region at downstream side of the diffuser, which runs from x, y = 0 to the northwest direction (Fig. 6). Especially, there is a large vortex right behind the diffuser in case of flood tides.

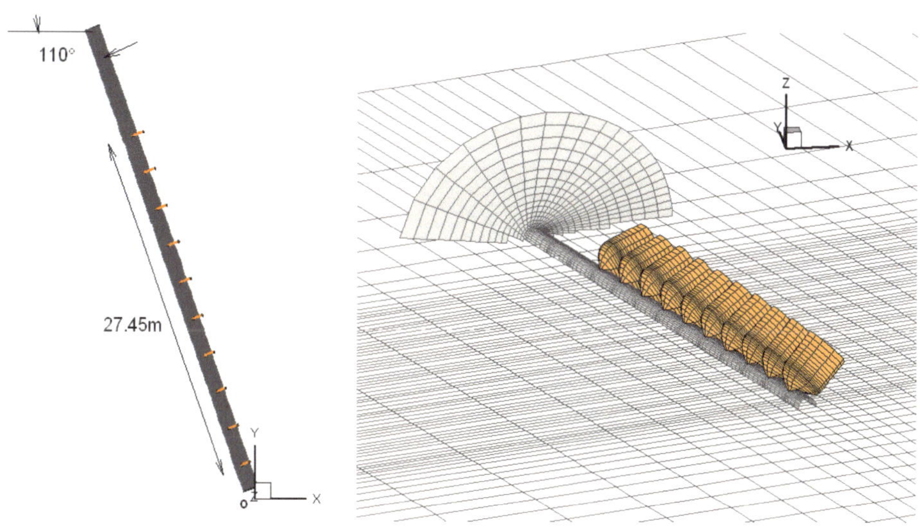

(a) (b)

Fig. 4. (a) Configurations of the diffuser. (b) CFD Meshes for the diffuser

As seen in Fig. 4a, the discharge ports point to the northeastern direction, and thus the hot water jets should be in the same direction. Fig. 7 indicates that during ebb tides, the jets are initially towards northeast but shortly later, due to the northwestern ambient tide currents, they turn towards the northwestern direction. While during flood tides, the jets and the currents are in about same directions, the thermal plume runs to a far downstream location. The CFD model has a high resolution at the mouth of the ports, with 10 grid nodes across the port diameter of order 1 cm resolution. As demonstrated in our previous studies (Tang et al., 2008), the CFD model accurately resolves velocities and temperature at the mouths of all ports using this mesh resolution.

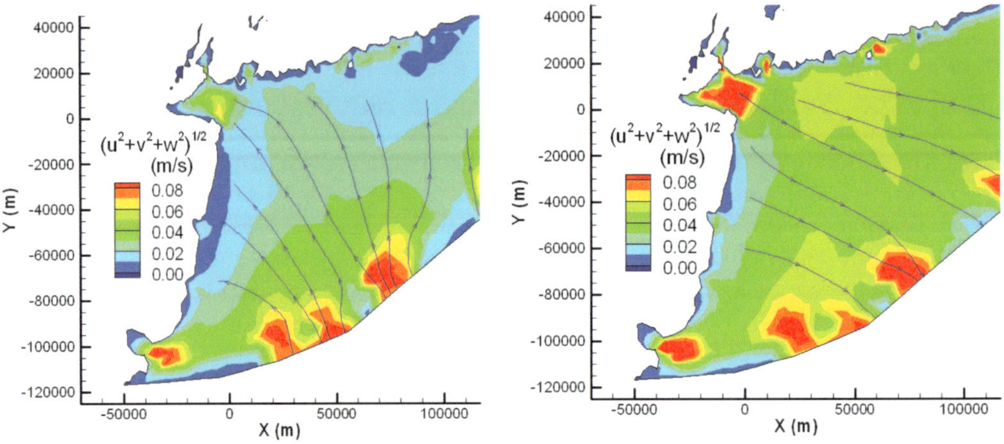

(a) (b)

Fig. 5. Large-scale surface velocity field at (a) flood tides and (b) ebb tides.

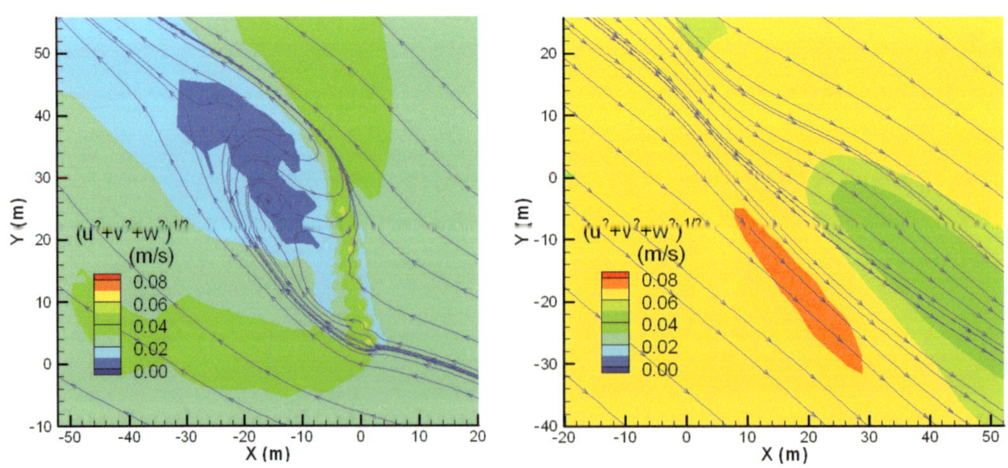

(a) (b)

Fig. 6. Local velocity on a plane 6 m above the diffuser at (a) flood and (b) ebb tide.

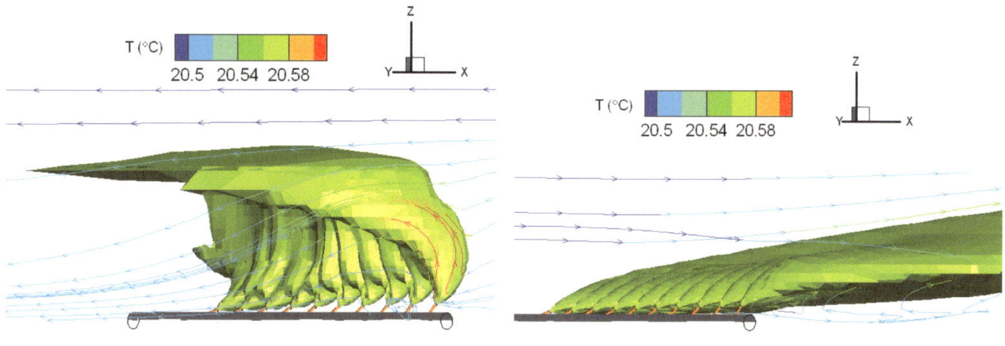

(a) (b)

Fig. 7. 3D temperature plume at (a) flood and (b) ebb tide.

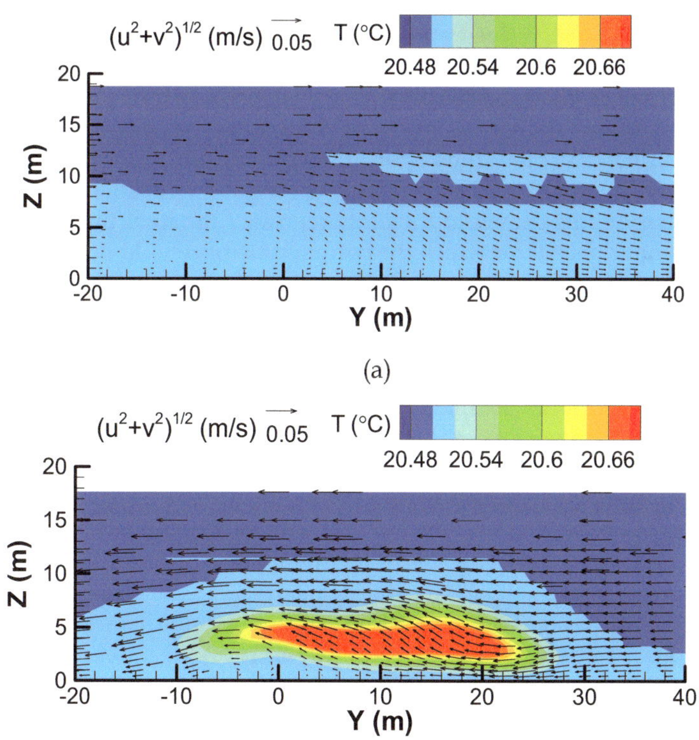

(b)

Fig. 8. Cross-section velocity and temperature at 10 m downstream from the diffuser in the x direction. (a) Flood tide. (b) Ebb tide.

Figure 8 shows the temperature and velocity field at a vertical cross-section, which is at the east side of diffuser. In case of the flood tides, the currents and thermal jets from all ports are in opposite directions and the thermal plume cannot reach this section whereas, for ebb tides, they are in the same directions and the plume is still strong on the section. In the latter case, the jets from individual ports already have merged with each other and no individual plumes are observed, and this is the result of intensive mixing in the cross-section (Fig. 8b). As shown in Fig. 8, there are fast lateral currents along the cross-section. Due to interaction of the effluent flows and the cross-section currents, there are vertical vortices next to the head of the diffuser, located near x, $y = 0$. It should be noted that all these small structures of the flow and temperature fields are only captured by the CFD model, and they are not available from the FVCOM.

3. Scour around a diffuser at the seabed

This section discusses erosion and scour around the diffuser based on the modeling results for the multiscale hydrodynamic processes in the previous section. Including potential scour computations on all of these FVCOM and CFD grids is beyond the scope of the current work. Instead, the results of the flow field at 1 mab (meter above bed) from the overlapping grids are averaged for 1 m^2 boxes. The area of interest is centered on the diffuser and represented by a grid of 40 x 70 cells that are 1 m along each axis. The output from the different grids is available at a range of heights above the seafloor; we have chosen 1 m as the representative height for the shear stress computations. All of the grids are included in the average, which produces fields for the flood and the ebb tides.
No wave data are used in these hydrodynamic simulations, which aim only to show the influence of tidal flow. The sedimentation computations do not include advection and no sediment is transported; the plots thus show the equilibrium resuspension depth H_R as defined by Keen and Glenn (1998). These depths are a good indication of scour at shorter time scales. The scour depths shown in the plots represent the total that would occur over an approximately 1-hour period for which the bottom currents would be representative.
The mean currents during flood tide are very low but the distribution has additional structure not seen in the FVCOM output (Fig. 9a). A very similar pattern is predicted during ebb tide (Fig. 9b). This area is likely to be the focus for erosion. The bottom sediment at this general location is expected to be variable because of its proximity to the Hudson River Canyon and the resulting large sediment outflow from the Hudson River. The available sediment samples indicate available sediment from clay to sand (USGS 2010). We are using a representative sediment distribution of 20 classes ranging from 4 microns to 2000 microns (clay to gravel). The uniform sediment is represented by a mean of 100 microns (very fine sand) with a standard deviation of 40 microns. This specific sediment population includes 55% clay and 44% very fine sand, with < 1 % fine sand. The computed critical shear stresses are 0.465 and 0.4388 Pa, respectively for the dominant sizes.
Detailed simulations of erosion would require unavailable field data for the area. We are using the erosion rate formulation from Ariathurai et al (1983):

$$E_i = M\,(\tau_b \,/\, \tau_{c,i} - 1) \tag{10}$$

where: M is the entrainment rate; τ_b is the bottom shear stress; $\tau_{c,i}$ is the critical shear stress for entrainment of size class i. This formula is useful for adjusting entrainment for unknown areas. The value of M used in these simulations is 0.001 kg m^{-2} s^{-1}. The bottom shear stress is estimated from the common formulation:

$$\tau_b = \rho_w C_d (u^2+v^2) \tag{11}$$

where: ρ_w is the water density, which is set to1025 kg m^{-3}, and C_d is the coefficient of drag, which is set to 1. The large drag coefficient is necessary because the currents from the model were too low to exceed τ_c for the sediment used. Including wave effects would improve this calculation. The resulting field of τ_b shows the effect of the diffuser (Fig. 10). Note that the bed stress is not symmetrical. Larger shear stresses are generated during the ebb tide and the pattern is slightly different from the flood. The persistent high shear stress is caused by the increased flow around the diffuser. This is partly attributable to the consistent seaward flow near x = -8 m during both flood and ebb tides, which is caused by a diffuser outflow of ~3.92 ms^{-1}. The diffuser flow is offshore, which appears to reinforce the ebb tide and make it more localized. This results in the concentration of higher stress around the outlets.

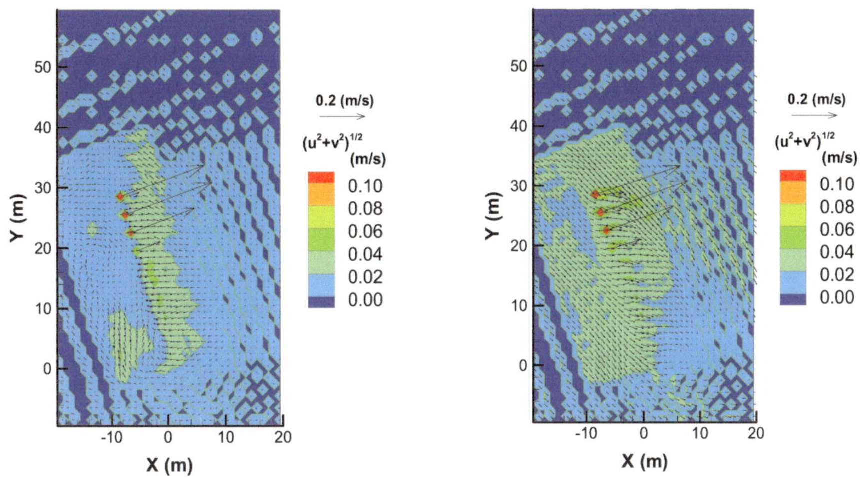

(a) (b)

Fig. 9. Averaged currents from all CFD grids at 1 mab. (a) Flood tide; maximum velocity is 0.40223 m s^{-1}. (b) Ebb tide; maximum velocity is 0.40676 m s^{-1}.

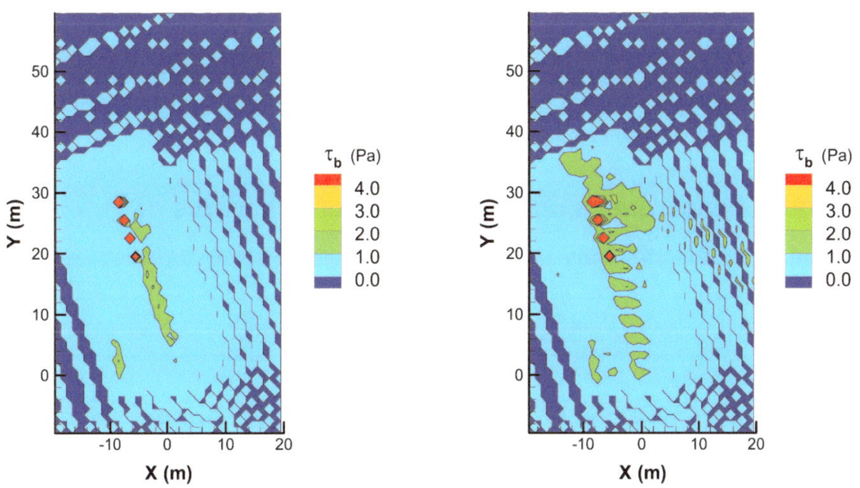

(a) (b)

Fig. 10. Bottom drag calculated from Eq. (11). (a) Flooding tide; maximum shear stress is 165.83 Pa. (b) Ebb tide; maximum shear stress is 169.59 Pa.

The pattern of potential scour for the flood (Fig. 11a) and ebb (Fig. 11b) tides is reflective of the shear stresses. Note the greater scour seaward of the northern end of the diffuser during the ebb tide. This pattern is due to the fact that, as shown in Fig. 4a, the ports near the end have lower angles in comparison with the ports at the other end, and this leads to a stronger horizontal velocity and more intensive shearing at the end. These calculations of potential scour demonstrate the importance of calculating the detailed flow around the obstruction. No calculations were performed for the FVCOM flow fields because it provided uniform near-bottom flow with a small magnitude. These simulations suggest that there could be substantial scour and possible damage to the structure if not considered in its design. The scour depths would be greatly increased by the impact of storm waves 1-4 m high, which occur frequently during the fall and winter in the Middle Atlantic Bight (Keen & Glenn, 1995).

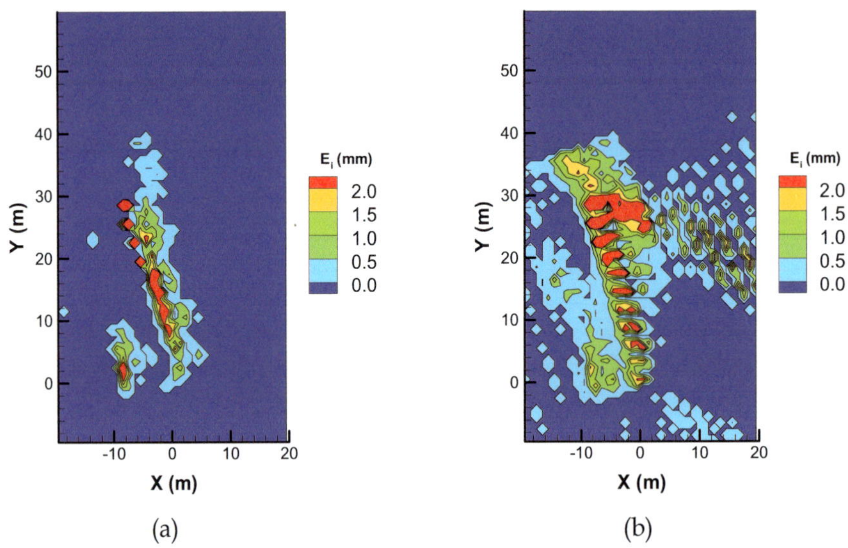

Fig. 11. Potential scour calculated from Eq. (10). (a) Flood tide; maximum erosion is 370.35 mm. (b) Ebb tide; maximum erosion is 378.76 mm.

4. Evolution of sand dunes under action of surface waves

The governing equation for wave action is given in a conservation form as (e.g., Mei, 1983):

$$\frac{\partial N}{\partial t}+\frac{\partial C_{gx_l}N}{\partial x_l}+\frac{\partial C_{\sigma}N}{\partial \sigma}+\frac{\partial C_{\theta}N}{\partial \theta}=\frac{S}{\sigma}, \tag{12}$$

where: t is the time; l =1, 2, and x_l correspond to x and y directions; σ is the frequency; and θ is the angle of the wave propagation direction; N is the wave action; C_{gx_l} are the wave speed in x_l direction in the physical space (x, y); C_{σ} and C_{θ} are, respectively, wave speed in σ and θ direction in the spectrum space (σ, θ); S is a source term that represents the combined effects of wind and other processes. In the current study, short waves are considered, which leads to a simple expression for C_{gx_l} (Mei, 1983).

The shallow water equations consist of the equation of mass conservation (e.g., Mei, 1983)

$$\frac{\partial H}{\partial t}+\frac{\partial HU_l}{\partial x_l}=0, \tag{13}$$

and the equations of momentum conservation:

$$\frac{\partial HU_i}{\partial t}+\frac{\partial HU_iU_l}{\partial x_l}=-gH\frac{\partial \eta}{\partial x_i}+\frac{\partial}{\partial x_i}\left(\nu_t H\left(\frac{\partial U_l}{\partial x_i}+\frac{\partial U_i}{\partial x_l}\right)\right)+\frac{\tau_{sx_i}-\tau_{bx_i}}{\rho}-\frac{1}{\rho}\frac{\partial S_{x_ix_l}}{\partial x_l}, \tag{14}$$

where: i,l =1, 2; ν_t is the turbulence eddy viscosity; ρ is the density; η is the water surface elevation; $S_{x_ix_l}$ are the radiation stresses resulting from waves.

The seabed morphology is controlled by the Exner equation (e.g., Henderson, 1966; Sleath, 1984):

$$\frac{\partial H_b}{\partial t}+\frac{\partial BU_l(U_1{}^2+U_2{}^2)^q}{\partial x_l}=0. \tag{15}$$

Here H_b is the elevation of the seabed, and B and q are constants. Typically, B depends on flow velocity, water depth, sediment grain sizes, and other factors, and q is usually in the range of $0.5 \le q \le 1.5$. In this paper, q = 1 is used (Hudson & Sweby, 2005; Kubatko et al., 2006).

The interactions among waves, currents, and morphology can be seen in the governing equations. In Eq. (12), the wave action is related to wave speed C_{gx_l}; these are coupled with the current through the velocity U_l in Eq. (12) (Mei, 1983). As indicated in Eq. (14), the current is affected by the wave field through the radiation $S_{x_ix_l}$, and morphology evolution through bottom elevation H_b ($\eta = H + H_b$). Actually, the current is also affected by wind through the bottom stresses τ_{bx_i}. Morphology is directly related to the current through the velocity field as shown in Eq. (15). Details for the governing equations can be found in Tang et al. (2009).

Equations (12) through (15) comprise a coupled non-homogeneous system of conservation laws. Each component in the system reproduces the framework of well-known models. For instance, the wave action equation (12) is employed in SWAN (Booij et al., 1999), the shallow water equations (13) and (14) are used for SHORECIRC (Luettich and Westerink, 2004), and the morphology evolution equation (15) is also widely employed in engineering (e.g., Wu, 2004; Kubatko et al., 2006). In order to solve the system, an extension of the Lax-Friedrich scheme (Lax, 1954) is applied to discretize the wave action equation (12). Second, the MacCormack scheme (MacCormack, 1969) is employed to solve the shallow water equations. Finally, the fourth Euler scheme and central difference operator are used to solve the morphology equation. The coupled system and the code have been validated and calibrated using a series of problems. For details of the discretization, validation, and calibration, the readers are referred to Tang & et al. (2009).

The above coupled wave, current, and morphology system is applied to study evolution of sand dunes under action of surface waves on the horizontal plane. The initial conditions for the wave energy, the velocity field, and the bottom shape are, respectively

$$N=0, \tag{16a}$$

$$U_1=0,\ U_2=0,\ H=2-\sum_{i=1}^{n}\exp\left\{-0.01((x-0.5i)^2+y^2)\right\}, \tag{16b}$$

$$H_b=\sum_{m=1}^{n}\exp\left\{-0.01((x-0.5m)^2+y^2)\right\}, \tag{16c}$$

where n is the number of sand dunes. In this study, n = 1, 2, 3. The upstream boundary condition is

$$H=2\text{m},\quad x=-15\text{m}. \tag{17}$$

The wind effect is introduced as a source term in Eq. (12) at the upstream end:

$$S=0.15(1-\tanh(20+x-0.01t)), \tag{18}$$

which generates surface waves starting at the upstream end and propagating in the x direction. In the simulation, Δx, $\Delta y=0.4$ m, $\Delta\sigma=0.1$ s^{-1}, and $\Delta\theta$=0.76 radians. Extrapolation is used to update solutions at the boundaries. Other related parameters are the same as those presented in our previous study on a single sand dune case (Tang et al., 2009).

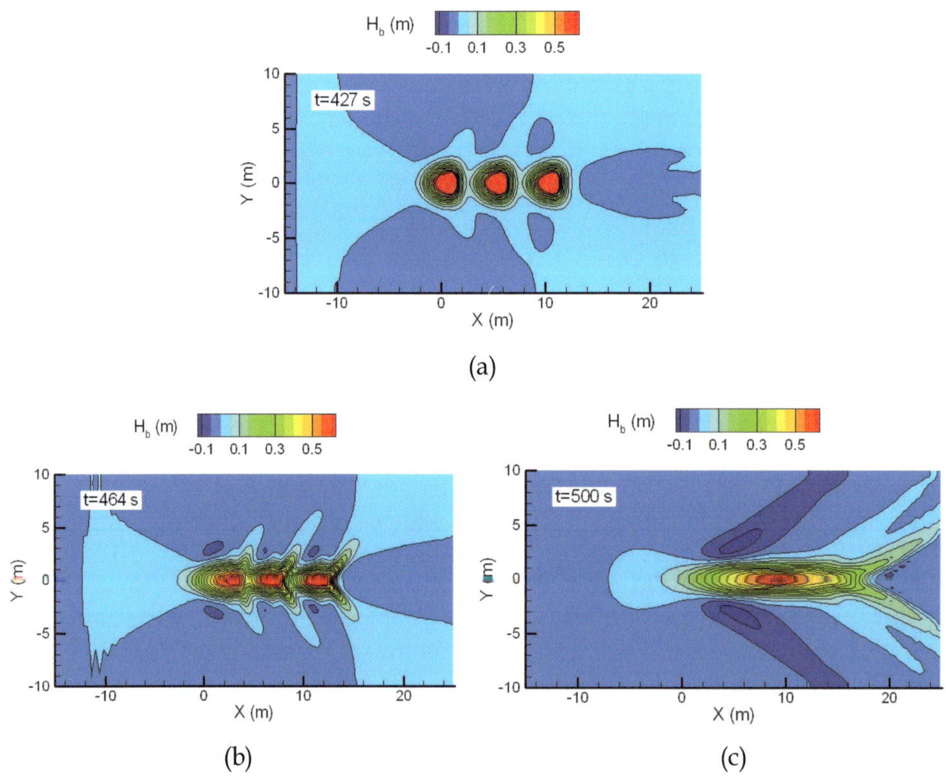

Fig. 12. Evolution of sand dunes (seabed elevation)

As seen in Eq. (18), the driving force S is a wave propagating in the x direction. The driving force generates a train of surface waves traveling in the same direction. The computed instantaneous evolution of sand dune elevation, surface height, and wave action, are presented in Figs. 12 through 14 for three sand dunes (n=3). As shown in Fig. 12, as the wind blows from left to the right, all three sand dunes move in the x direction, and gradually they merge with each other. During this process, each sand dune is changing from a circle into a triangle, a steroidal, and then a strip, which are the three typical shapes of sand dunes in their evolution. At the same time, low-elevation regions are forming at the front, lateral sides, and immediately behind the three sand dunes. In Fig. 13, it is seen that the water surface has a bump in front and a dip behind each of the three dunes. In comparison with sand dune configurations in Fig. 12, it is known from Fig. 14 that wave action reaches minimums above all sand dunes and maximums right behind them. From Figs. 12 and 13 it is seen that, during the evolution of the sand dunes, both the wave field and water surface elevation evolve congruently with the sand dunes. Figure 14 reveals clear traces of bed morphology, which indicates its strong effect on the wave field.

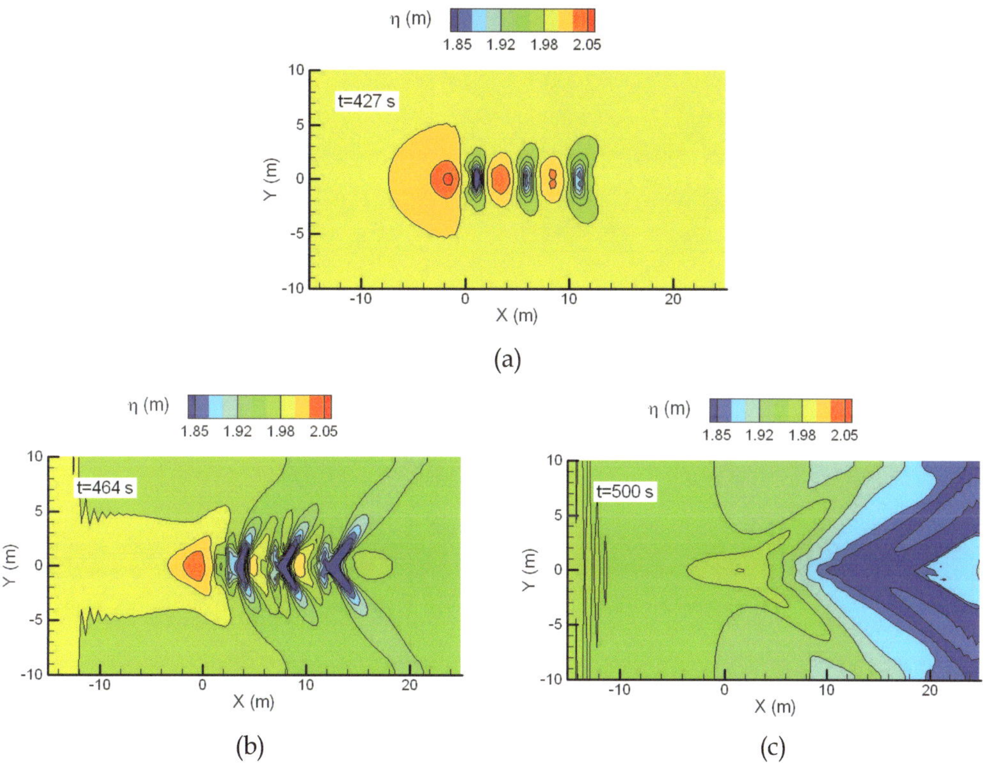

Fig. 13. Evolution of water surface elevation

Figure 15 shows the evolution of the heights and locations of three sand dunes as the wave is propagating towards the right. An interesting note is that the height of the first dune, the leading dune that is most upstream, decreases faster at the beginning (t < 470 s) in comparison with the heights of the other two dunes. As seen in Fig. 15a, however, at about t = 470 s the second and the third disappear or they merge with the first one, whereas the first still exists. After the disappearance of the second and third sand dunes, the height of the first dune remains about the same for a while. It is seen in Fig. 15b that, as all three dunes are about to merge with each other at t = 470 s, the first sand dune moves faster than the other two, with a large dx/dt, which accelerates the merging process.

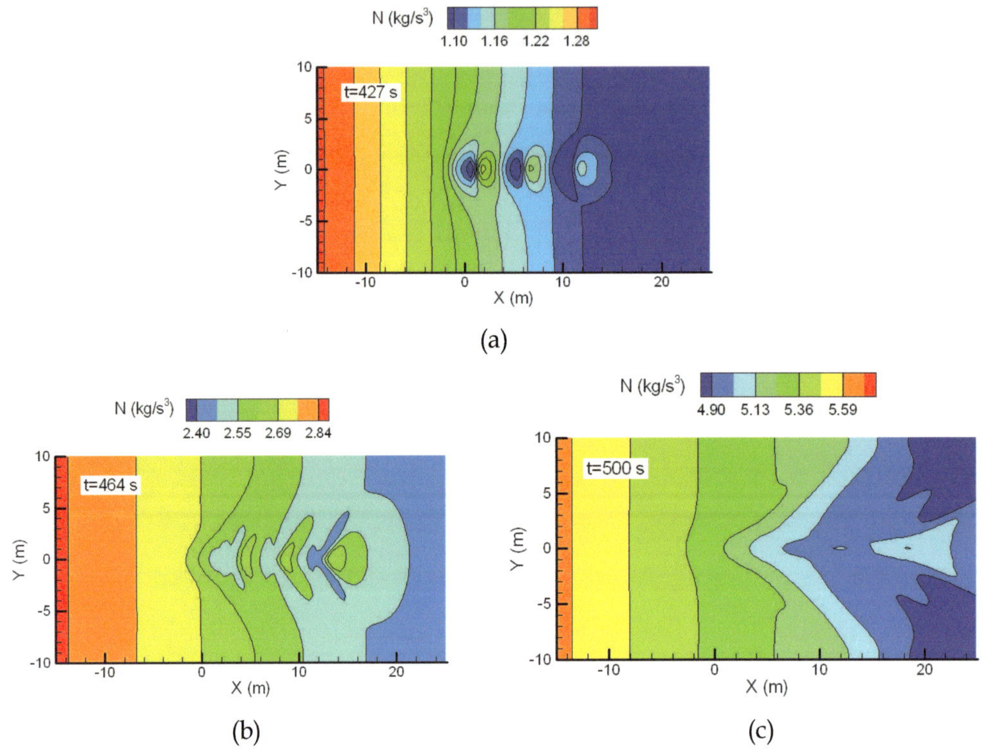

Fig. 14. Evolution of wave action

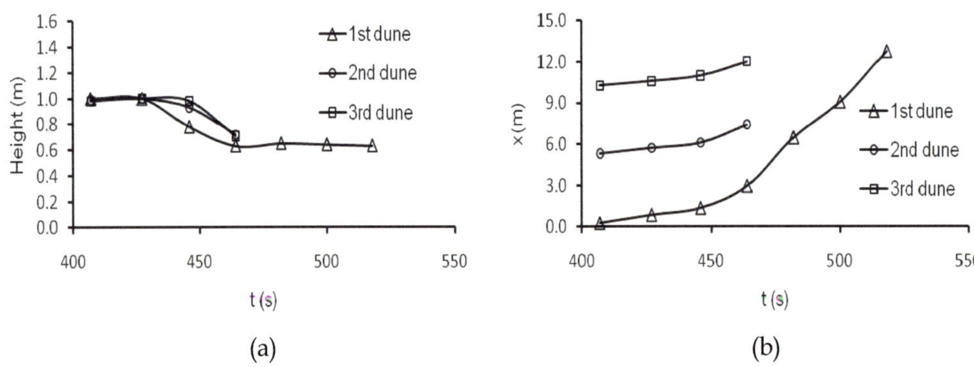

Fig. 15. Evolution of 3-dune. a) Heights of sand dunes. b) Location of sand dunes

Comparison of Figs. 12 through 14 with those obtained from modeling of single sand dunes (not shown) suggests that the evolution of a sand dune in the case of multiple sand dunes is consistent with a single sand dune in certain aspects, such as the shape of sand dunes (e.g., Hudson & Sweby, 2005; Tang et al., 2009). Nevertheless, sand dune development in situations of multiple sand dunes indeed behaves differently. In Fig. 16, the simulated results for situations of 1, 2, and 3 sand dunes (n = 1, 2, and 3 in Eq. 16) are presented, which shows the heights and locations of the leading or the most upstream sand dunes. The most interesting finding in this figure is that, in case of multiple sand dunes, the heights of the leading sand dunes decrease slower but they move downstream faster. The larger the number of the sand dunes, the more pronounced this trend. It is also seen that the evolutions of heights and locations of the leading sand dunes in case of 2 and 3 sand dunes are not much different, but they are obviously distinct from that of an isolated sand dune (Fig. 16). This is a clear indication of the interaction between individual sand dunes and their resulting hydrodynamics.

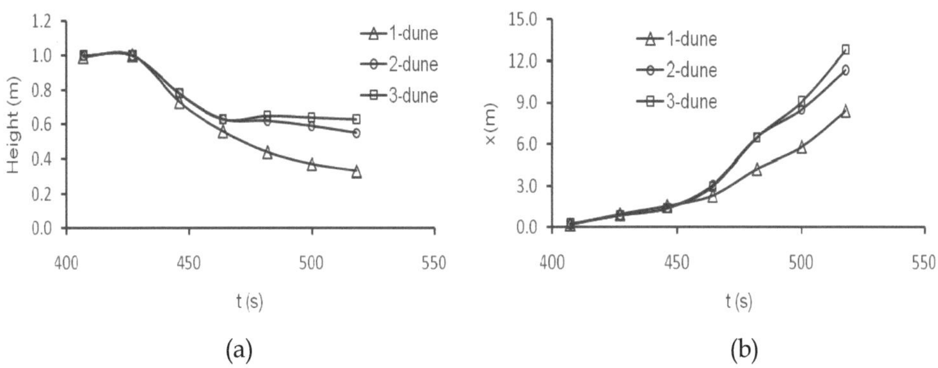

Fig. 16. Evolutions of the leading sand dunes in case of 1-, 2-, and 3-dune. a) Heights of sand dunes. b) Locations of sand dunes

5. Concluding remarks

This chapter presents the needs to simulate multiscale and multiphysics coastal ocean hydrodynamics and the necessity to include its effects in estimation of seabed sediment scour and morphology evolution. It describes the multiscale and multiphysics method the authors proposed as a prominent approach, which is a hybrid method in conjunction with the domain decomposition method. The feasibility and potential of the approach in resolving multiscale and multiphysics processes is demonstrated using example computations.

The potential scour for the seabed diffuser indicates two important results: (1) the additional detail of the flow field computed by the CFD model is critical in capturing the areas of scour by tides; and (2) it is necessary to include waves for both mean and storm conditions to estimate the cumulative potential scour around the diffuser. The local scour problem can be computed using a high-resolution model but this would not permit an examination of variations within the seafloor area of interest. This is especially true for waves, which are sensitive to slight changes in water depth. The modeling of sand dunes illustrates the strong interaction between surface waves, currents, and morphology. The simulation presents interesting features of sand dunes with respect to their heights and locations, and it clearly indicates that evolution of multiple sand dunes is distinct from that of single sand dunes.

6. Acknowledgement

The first author was supported by PSC-CUNY Research Award. The second author was funded by the Office of Naval Research through the Naval Research Laboratory 6.1 Core program.

7. References

Alam, M. S., & Cheng, L. (2010). A parallel three-dimensional scour model to predict flow and scour below a submarine pipeline. *Central European Journal of Physics*, Vol. 8, pp. 604-619.

Ariathurai, R., Golden, J., Jr., McNally, W. H., Jr., Stout, G. A., & Neiheisel, J. (1983). Shoaling processes in navigable waters. *J. Waterway, Port, Coastal and Ocean Engineering*, Vol. 109, 199-221.

Benek, J. A., Steger, J. L., & Dougherly., F. C. (1983). A flexible grid embedding technique with application to the Euler equations. AIAA Paper 83-1944.

Blumberg, A. F., & Mellor, G. L. (1987). A description of a three-dimensional coastal ocean circulation model. In: Heaps, N. (Ed.), *Three-Dimensional Coastal Ocean Models*, American Geophysical Union, 1-16.

Booij, N., Ris, R. C., & Hoithuijsen, L. H. (1999). A third-generation wave model for coastal regions. I. Model description and validation. *J. Geophysical Research*, Vol. 104, C4, pp. 7649-7666.

Chen, C. S., Beardsley, R. C., & Geoffrey, C. (2006). An unstructured grid, finite-volume coastal ocean model FVCOM user manual. SMAST/UMASSD-06-0602.

Chen, C., Liu, H., & Beardsley, R. C. (2003). An unstructured, finite-volume, three-dimensional, primitive equation ocean model: application to coastal ocean and estuaries. *J. Atmospheric and Oceanic Technology*, Vol. 20, pp. 159-186.

Chiew, Y. M. (1991). Prediction of maximum scour depth at submarine pipelines. *J. Hydraulic Engineering*, Vol. 117, pp. 452-466.

Chui, C. K. (1992). *An Introduction to Wavelet*, Academic Press, Harcourt Brace Jovanovich, New York.

Dolbow, J., Khaleel, M. A., & Mitchell, J. (2004). *Multiscale Mathematics -- Initiative: A Roadmap*, PNNL-14966.

Fringer, O.B., Gerritsen, M., & Street, R.L. (2006). An unstructured-grid, finite-volume, nonhydrostatic, parallel coastal ocean simulator. *Ocean Modelling*, Vol. 14, pp. 139–173.

Ge, L., & Sotiropoulos, F. (2005). 3D unsteady RANS modeling of complex hydraulic engineering flows. Part I: Numerical model. *J. Hydraulic Engineering*, Vol. 131, pp. 800–808.

Halliwell, G. R. (2004). Evaluation of vertical coordinate and vertical mixing algorithms in the HYbird-Coordinate Ocean Model (HYCOM), Ocean Modeling, 7, 285-322.

Harten, A. (1993). Discrete multi-resolution analysis and generalized wavelets. *Applied Numerical Mathematics*, Vol. 12, pp. 153-192.

Heimsund, B. -O., & Berntsen, J. (2004). On a class of ocean model instability that may occur when applying small time steps, implicit methods, and low viscosities. *Ocean Modeling*, Vol. 7, pp. 135-144.

Henderson, F. M. (1966). Open channel flow, MacMillan Publishing Company, NY.

Hudson, J., & Sweby, P. K. (2005). A high-resolution scheme for the equations governing 2D bed-load sediment transport. *J. Numerical Methods Fluids*, Vol. 47, pp. 1085-1091.

Lax, P. D. (1954). Weak solutions of nonlinear hyperbolic equations and their numerical computation. *Communications On Pure And Applied Mathematics*, Vol. 7, pp. 159-193.

Lin, F. B., & Sotiropoulos, F. (1997). Assessment of artificial dissipation models for three-dimensional incompressible flows. *ASME J. Fluids Engineering Transaction*, 119, 331–40.

Keen, T. R., & Glenn, S. M. (1995). A coupled hydrodynamic bottom boundary-layer model of storm and tidal flow in the middle Atlantic Bight of North-America. *J. Physical Oceanography*, Vol. 25, pp. 391-406.

Keen, T. R., & Glenn, S. M. (1998). Resuspension and advection of sediment during Hurricane Andrew on the Louisiana continental shelf. *Proc. Fifth Int. Conf. Estuarine and Coastal Modeling*, Alexandria, Virgina, October 1997.

Keen, T. R., & Glenn, S. M. (2002). Predicting bed scour on the continental shelf during Hurricane Andrew. *J. Waterway Port Coastal and Ocean Engineering*, Vol. 128, pp. 249-257.

Keen, T. R., Stone, G., Kaihatu, J., & Hsu, Y. L. (2003). Barrier island erosion during a winter cold front in Mississippi Sound. *Proc. Coastal Sediments '03*, New Orleans, Louisiana, March 2002.

Kubatko, E. J. ,Westerink, J. J., & Dawson, C. (2006). An unstructured grid mophodynamics model with a discontinuous Galerkin method for bed evolution. *Ocean Modelling*, Vol. 15, pp. 71-89.

Kumar, P., & Foufoula-Georgiou, E. (1997). Wavelet analysis for geophysical applications, *Reviews of Geophysics*, Vol. 35, pp. 385–412.

Legeckis, R., Pichel, W., & Nesterczuk, G. (1983). Equatorial long waves in geostationary satellite observations and in a multichannel sea surface temperature analysis. *Bull. American Meteorological Society* , Vol. 64, pp. 133-39.

Luettich, R., & Westerink, J. (2004). Formulation and numerical implementation of the 2D/3D ADCIRC, Finite Element Model Version 44.XX. .

MacCormack, R.W. (1969). The effect of viscosity in hypervelocity impact cratering, AlAA Paper 69-354.

Mason, P. J. (1989). Large-eddy simulation of the convective atmospheric boundary layer. *J. Atmospheric Sciences*, Vol. 46, pp. 1491–1516.

Mei, C. C. (1983). *Applied Dynamics of Ocean Waves*. John Wiley & Sons.

Mellor, G. L., & Yamada, T. (1974). A hierarchy of turbulence closure models for planetary boundary layers. *J. Atmospheric Sciences*, Vol. 31, pp. 1791–1806.

Myrhaug, D., Ong, M. C., & Gjengedal, C. (2008). Scour below marine pipelines in shoaling conditions for random waves. *Coastal Engineering*, Vol. 55, pp. 1219-1223.

Papanicolaou, A. N., Elhakeem, M., Krallis, G., Prakash, S., & Edinger, J. (2008). The morphodynamic modeling of tidal sand waves on the shoreface. *J. Hydraulic Engineering*,Vol. 134, pp. 1–14.

Pedley, T. J. (1992). Hydrodynamic phenomena in suspensions of swimming microorganisms. *Annual Review of Fluid Mechanics*, Vol. 24, pp. 313-58.

Sleath, J. F. A. (1984). Sea bed mechanics, John Wiley & Sons, New York.

Smagorinsky, J. (1963). General circulation experiments with the primitive equations, I. The basic experiment, *Monthly Weather Review*, Vol. 91, 99-164.

Song, Y. T., & Hou, Y. T. (2006). Parametric vertical coordinate formulation for multiscale, Boussinesq, and non-Boussinesq ocean modeling. *Ocean Modeling*, Vol. 11, pp. 298-332.

Sotiropoulos, F., & Ventikos Y. (1998). Transition from bubble-type vortex breakdown to columnar vortex in a confined swirling flow. *Int. J. Heat Fluid Flow*, Vol. 19, pp. 446–58.

Sumer, B. M., Whitehouse, R. J. S., & Torum, A. (2001). Scour around coastal structures: A summary of recent research. *Coastal Engineering*, Vol. 44, pp. 153-190.

Tang, H. S., Jones, C., & Sotiropoulos, F. (2003). An overset grid method for 3D unsteady incompressible flows, *J. Computational Physics*, Vol. 191, pp. 567-600.

Tang, H. S., Keen, T. R., & Khanbilvardi, R. (2009). A model-coupling framework for nearshore waves, currents, sediment transport, and seabed morphology, Comm. *Nonlinear Sciences & Numerical Simulations*, Vol. 14, pp. 2935-2947.

Tang, H. S. Paik, J. Sotiropoulos, F., & Khangaokar, T. (2008). Three-dimensional CFD modeling of thermal discharge from multports. *J. Hydraulic Engineering*, Vol. 134, pp. 1210-1224.

Tang, H. S., & Wu, X. G. (2010). Multiscale coastal flow simulation using coupled CFD and GFD models, *Proc. Int. Congress Environ. Modeling & Simulation Software*, July 5-8, 2010, Ottawa, Canada.

Tolman, H. L. (1991). A third-generation model for wind waves on slowly varying, unsteady and inhomogeneous depths and currents. *J. Physical Oceanography*, Vol. 21, pp. 782–797.

Tonnon, P. K., van Rijn, L. C., & Walstra, D. J. R. (2007). The morphodynamic modeling of tidal sand waves on the shoreface. *Coastal Engineering*, Vol. 54, 279–296.

USGS (2010). http://pubs.usgs.gov/of/2000/of00-358/graphics/chapter2/midatlan.jpg.

Weller, R. A., Dean, J. P., Marra, J., Price, J. F., Francis, E. A., & Boardman, D. C. (1985). Three-dimensional flow in the upper ocean. *Sciences*, Vol. 227, pp. 1552-1556.

Wu, W. (2004). Depth-averaged two-dimensional numerical modeling of unsteady flow and nonuniform sediment transport in open channels, *J. Hydraulic Engineering*, Vol. 130, pp. 1013-1024.

Wu, X. G., & Tang, H. S. (2010). Coupling of CFD model and FVCOM to predict small-scale coastal flows, *J. Hydrodynamics*, vol.22, pp. 284-289.

Wunsch C., & Ferrari R. (2004). Vertical mixing, energy, and the general circulation of the oceans. *Annual Review of Fluid Mechanics*, Vol. 36, pp. 281–314.

Xu, J. S., Li, G. X., Dong, P., & Shi, J. H. (2010). Bedform evolution around a submarine pipeline and its effects on wave-induced forces under regular waves. *Ocean Engineering*, Vol. 37, pp. 304-313.

Xu, K. L, & Sun, G. (2009). Assessment of an interface conservative algorithm MFBI in a Chimera grid flow solver for multi-element airfoils. *Proceedings of the World Congress on Engineering. Vol II.* WCE 2009, July 1 - 3, London, U.K.

Younis, B. A., Teigen, P., & Przulj, V.P. (2001). Estimating the hydrodynamic forces on a mini TLP with computational fluid dynamics and design-code techniques. *Ocean Engineering*, Vol. 28, pp. 585–602.

Zamankhan, P., & Doolatshahi, A. A., (2008). Analysis of submarine pipeline scour using large-eddy simulation of dense particle-liquid flows. *Pipeline and Riser Technology; Ocean space utilization*, Vol. 3, pp 545-562.

Zang, Z., Cheng, L., Zhao, M., Liang, D., & Teng, B. (2009) A numerical model for onset of scour below offshore pipelines. *Coastal Engineering*, Vol. 56, pp. 458-466.

5

Cohesive Sediment Flocculation and the Application to Settling Flux Modelling

A. J. Manning[1,2], J. V. Baugh[1],
R. L. Soulsby[1], J. R. Spearman[1] and R. J. S. Whitehouse[1]
[1]*HR Wallingford, Howbery Park, Wallingford, Oxfordshire, OX10 8BA,*
[2]*Marine Physics Research Group, School of Marine Science & Engineering, University of Plymouth, Portland Square Building (A410), Plymouth, Devon, PL4 8AA*
UK

1. Introduction

The majority of the particulate matter which accumulates within an estuary is commonly referred to as mud. Mud is typically composed of mineral grains which originate from both fluvial and marine sources, together with biological matter - both living and in various stages of decomposition. It is the combination of these features that make estuarine muds sticky in nature and for this reason these sediment types are referred to generically as cohesive sediment (Whitehouse et al., 2000).

The primary mineral component of cohesive muds are clay minerals. Clays have a plate-like structure, and generally have a diameter of less than 2 μm. Cohesion arises through the combined efforts of both the electrostatic charging of the clay minerals as they pass through brackish to highly saline water, and various biogenic long-chain polymer molecules which adhere to individual particle surfaces, such as sticky mucopolysaccharides. Edzwald and O'Melia (1975) conducted experiments with pure kaolinite, and found that their flocculation efficiency was less than 10%. Where as Kranck (1984) found that the flocculation of mineral particles which contained some organic matter, greatly enhanced the settling velocity of the aggregates (Fig. 1). The efficiency with which the particles coagulate is a reflection of the stability of the suspension (van Leussen, 1994). The statistical occurrence of collisions further increases as the abundance of particles in suspension rise. A suspension is classified as unstable when it becomes fully flocculated, and is stable when all particles remain as individual entities.

From a water quality perspective, cohesive sediments have the propensity to adsorb contaminants (Ackroyd et al., 1986; Stewart and Thomson, 1997). This in turn has a direct effect on water quality and related environmental issues (e.g. Uncles et al., 1998). Accurately predicting the movement of muddy sediments in an estuary therefore is highly desirable. In contrast to non-cohesive sandy sediments, muddy sediments can flocculate (Winterwerp and van Kesteren, 2004) and this poses a serious complication to modellers of estuarine sediment dynamics.

This chapter provides the following:

- An outline of the flocculation process
- Flocculation measurement methods and the importance of data
- Floc settling behaviour
- Examples of cohesive sediment depositional model approaches, including ways of parameterising flocculation

Fig. 1. Flocculation and destabilisation by adsorbed polymers (after Gregory, 1978).

2. Flocculation overview

Cohesive sediments have the potential to flocculate into larger aggregates termed flocs (Winterwerp and van Kesteren, 2004; see example in Fig. 2). Floc sizes (D) can range over four orders of magnitude, from individual clay particles of 1 μm to stringer-type floc structures several centimetres in length. An individual floc may comprise up to 10^6 individual particulates, and as flocs grow in size their effective densities (i.e. bulk density minus the water density), ρ_e, generally decrease (Tambo and Watanabe, 1979; Klimpel and Hogg, 1986; Droppo et al., 2000), but their settling speeds (Ws) rise due to a Stokes' Law relationship (Dyer and Manning, 1998). The general trend exhibited by floc effective densities by a number of authors is shown in Fig 3. A typical floc size vs. settling velocity distribution from the Tamar Estuary (UK) is illustrated in Fig 4. One can immediately see that for a constant Ws, there is a wide range in D and ρ_e. Similarly, for a constant D, there is a large spread in both Ws and ρ_e. Like the floc size, the settling velocities can also typically range over four orders of magnitude, from 0.01 mm s^{-1} up to several centimetres per second (Lick, 1994).

Kranck and Milligan (1992) observed that under the majority of estuarine conditions, most suspended particulate matter (SPM) within an estuary occurs in the form of flocs. Of the various physical processes which occur during a tidal cycle, flocculation of the sediment is regarded as one of the primary mechanisms which can affect the deposition, erosion and consolidation rates. The flocculation process is dynamically active which is directly affected by its environmental conditions, primarily being dependent on a complex set of interactions between sediment, fluid and flow within which the particles aggregation plays a major role (Manning, 2004a). Flocculation is therefore a principle mechanism which controls how fine sediments are transported throughout an estuary.

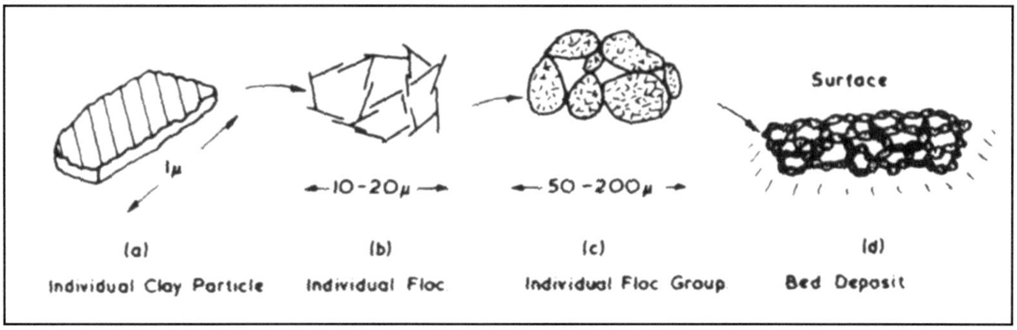

Fig. 2. Sizes of clay particles, flocs and floc groups (after McDowell and O'Connor, 1977)

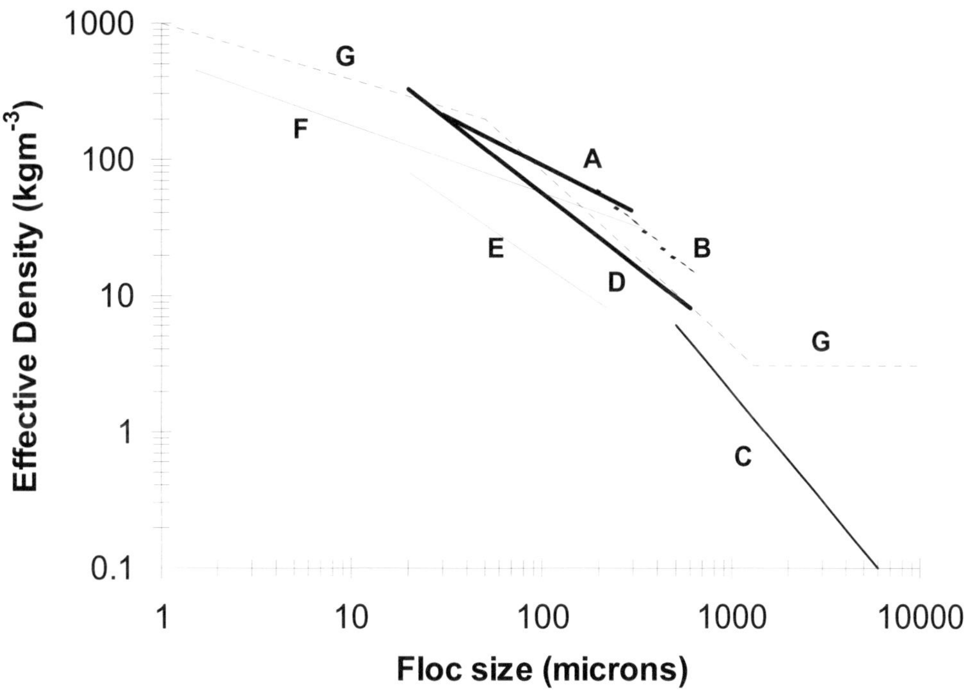

Fig. 3. Comparative results of effective density against floc size. A = Manning and Dyer (1999), B = Al Ani et al. (1991), C = Alldredge and Gotschalk (1988), D = Fennessy et al. (1994b), E = Gibbs (1985), F = McCave (1975), and G = McCave (1984) (from Manning and Dyer, 1999).

The degree of flocculation is highly dependent upon both the SPM concentration and turbulent shear (e.g. Krone, 1962; Parker et al., 1972; McCave, 1984; Burban et al. 1989; van Leussen, 1994; Winterwerp, 1998; Manning, 2004a), and both of these parameters can vary spatially and temporally throughout an estuary. A conceptual model which attempts to explain the linkage between floc structure and floc behaviour in an aquatic environment is provided by Droppo (2001). As a result of dynamic inter-particle collisions, floc growth implies large variations in the sediment settling flux with direct implications on the vertical distribution of sediment loading.

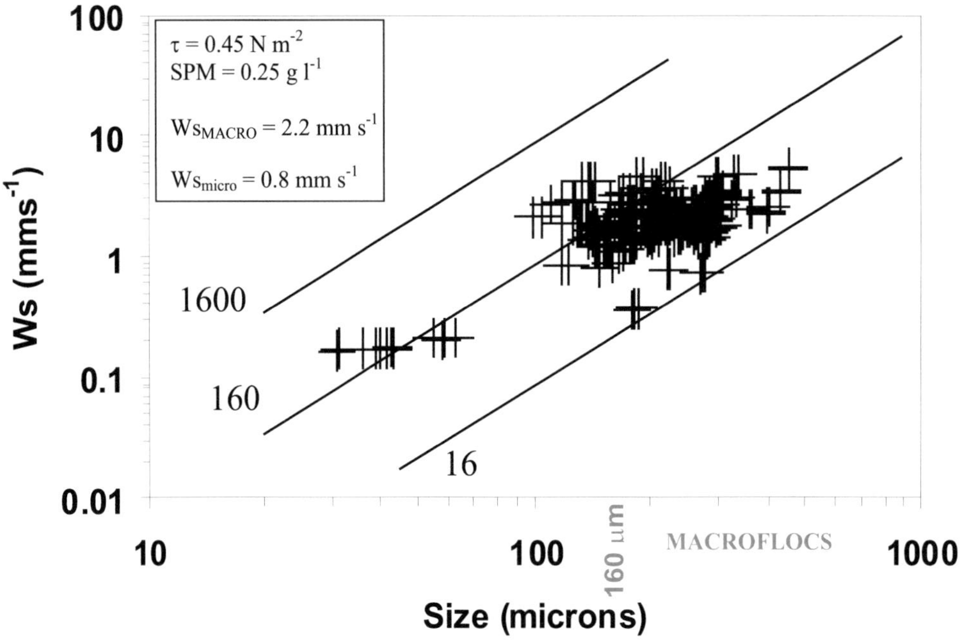

Fig. 4. An example of an INSSEV measured floc population from the Tamar Estuary (UK) illustrating the relationships between floc size and settling velocities of individual flocs during neap tide conditions. The diagonal lines represent values of constant effective density (kg m^{-3}). Macrofloc: microfloc segregation is indicated by the dotted line at 160 μm (*from Manning, 2004c*).

It has been generalised that there are two distinct component groups of flocs: *macroflocs* and *microflocs* (Eisma, 1986; Manning, 2001). Many floc suspensions exist as bi-modally distributed populations (e.g. Manning and Dyer, 2002a; Lee et al., 2010). These two floc fractions form part of Krone's (1963) classic order of aggregation.

Macroflocs (Fig. 5) are large, highly porous (> 90%), fast settling aggregates which are typically the same size as the turbulent Kolmogorov (1941) microscale. Macroflocs (D > 160 μm) are recognised as the most important sub-group of flocs, as their fast settling velocities tend to have the most influence on the mass settling flux (Mehta and Lott, 1987). Their fragile, low density structure means they are sensitive to physical disruption during sampling. Macroflocs are progressively broken down as they pass through regions of higher turbulent shear stress, and reduced again to their component microfloc sub structure (Glasgow and Lucke, 1980). They rapidly attain equilibrium with the local turbulent environment.

The smaller microflocs (Fig. 5; D < 160 μm) are generally considered to be the building blocks from which the macroflocs are composed. Microflocs are much more resistant to break-up by turbulent shear. Generally they tend to have slower settling velocities, but exhibit a much wider range in effective densities than the larger macroflocs (e.g. McCave, 1975; Alldredge and Gotschalk, 1988; Fennessy et al., 1994a).

In order for flocculation to occur, suspended particles must come into contact with each other. Van Leussen (1988) theoretically assessed the comparative influence of the three main collision mechanisms: Brownian motion, turbulent shear and differential settling (see Fig. 6),

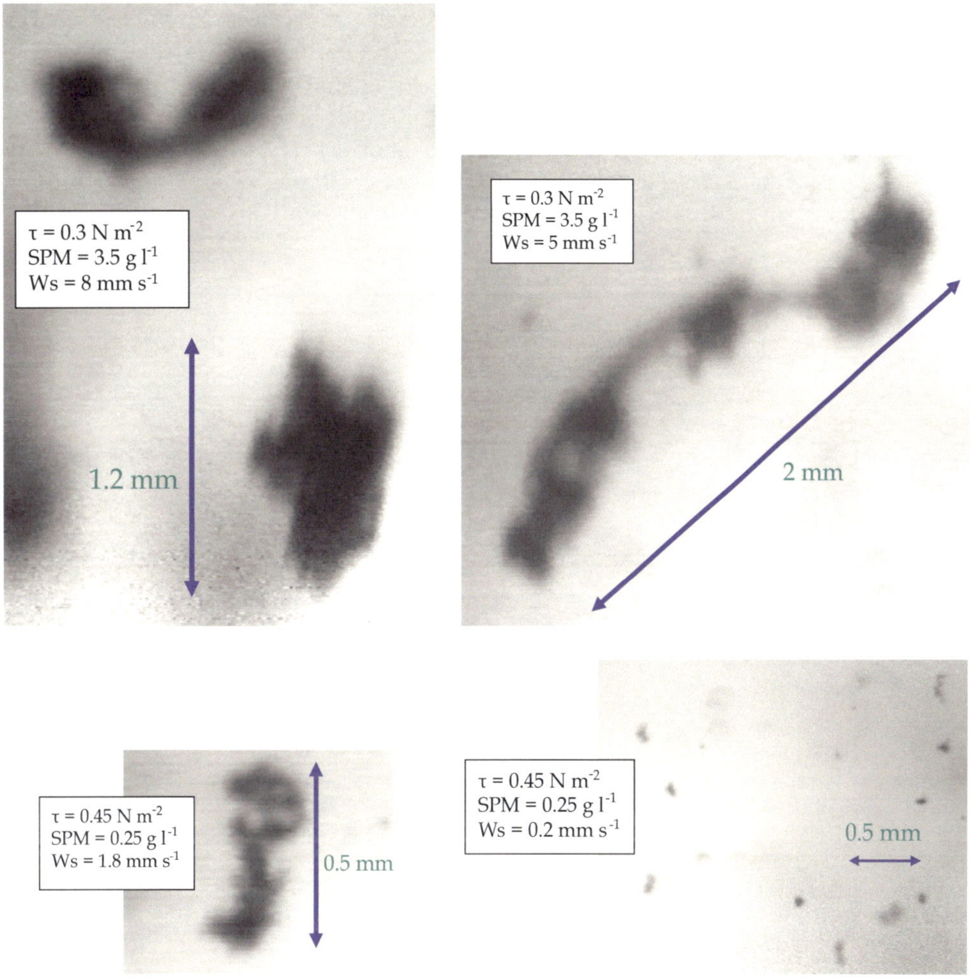

Fig. 5. Illustrative examples of real estuarine floc images. Ambient shear stress, concentration and settling velocity values are provided (*from Manning and Dyer, 2002*).

and deduced that turbulent shear stresses ranging between 0.03-0.8 Pa, provided the dominant flocculation collision mechanism. Turbulent shear stress can impose a maximum floc size restriction on a floc population in tidal waters (McCave, 1984). Tambo and Hozumi (1979) showed that when the floc diameter was larger than the length-scale of the energy dissipating eddies, the aggregate would break-up. Similarly, Eisma (1986) observed a general agreement between the maximum floc size and the smallest turbulent eddies as categorised by Kolmogorov (1941). Both Puls et al., (1988), and Kranck and Milligan (1992) have hypothesised that both SPM concentration and turbulence are thought to have an effect on the maximum floc size, and the resulting spectra. As SPM concentration increases, the influence of particle collisions can also act as a floc break-up mechanism. Floc break-up by three-particle collisions tends to be the most effective mechanism (Burban et al., 1989).

Settling velocity is regarded as the basic parameter used in determining suspended sediment deposition rates in either still or flowing water. Much has been documented on non-cohesive sediments (coarse silts and larger), and it is possible to calculate the settling velocity of low concentrations in suspension, using well defined expressions (e.g. Stokes' Law), from the relative density, size and shape of the particles, since the only forces involved are gravity and the flow resistance of the particle. However, the settling velocity of flocculated, cohesive sediments in estuaries are significantly greater than the constituent particles. Based on the research of Stolzenbach and Elimelich (1994) and Gregory (1978), Winterwerp and Van Kesteren (2004) concluded that although flocs are porous in composition, they can be treated as impermeable entities when considering their settling speeds.

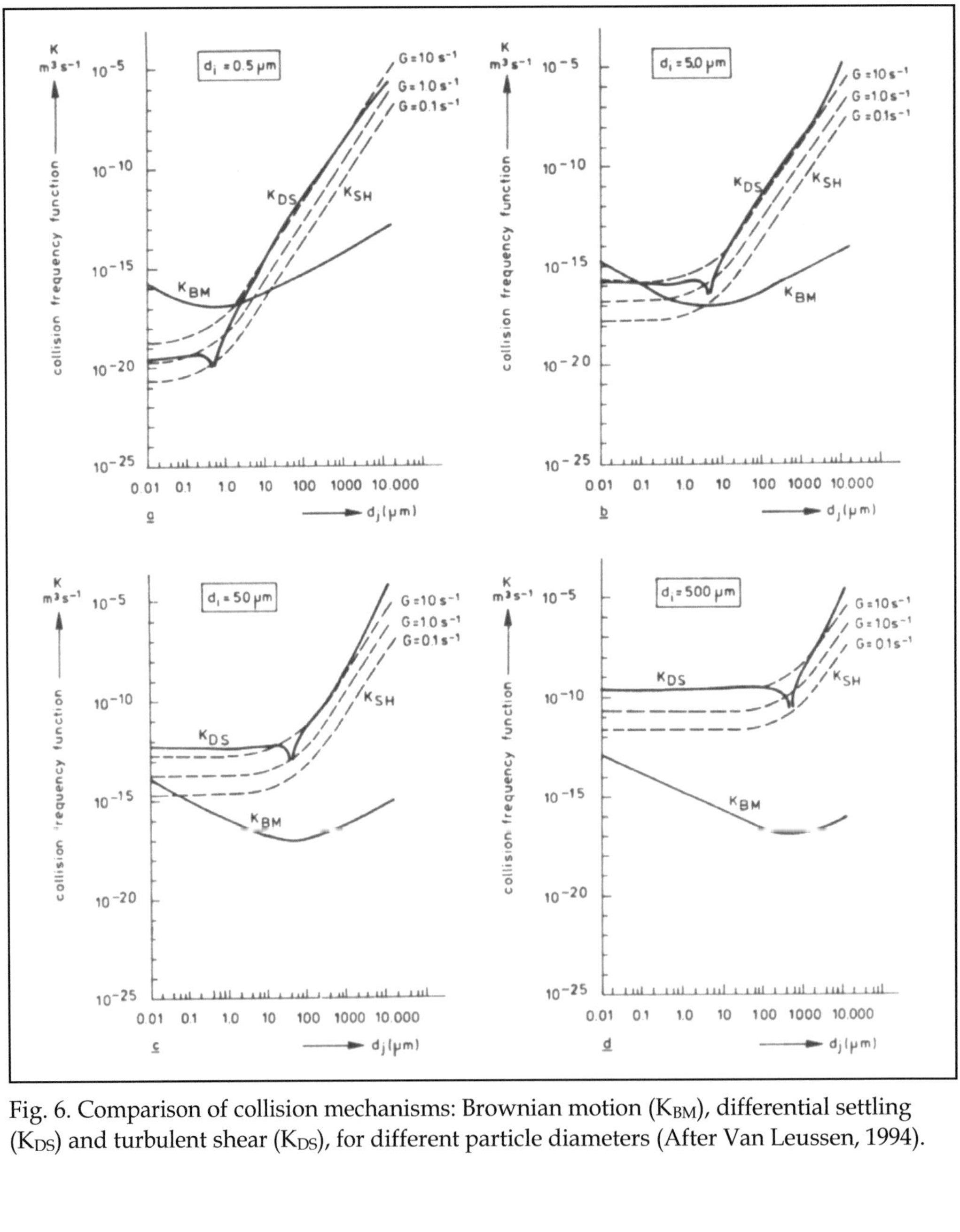

Fig. 6. Comparison of collision mechanisms: Brownian motion (K_{BM}), differential settling (K_{DS}) and turbulent shear (K_{DS}), for different particle diameters (After Van Leussen, 1994).

A knowledge of floc effective density is also important in the calculation of vertical settling fluxes since the majority of the suspended mass is contained in the low density, high settling velocity, large flocs (Mehta and Lott, 1987). Furthermore, the rheological properties of suspended particulate matter are governed by volume concentrations, as opposed to mass concentrations (Dyer, 1989).

3. Flocculation measurement methods and the importance of data

Hayter and Mehta (1982) and Whitehouse et al. (2000) both indicated that many parameters need to be determined in order to fully describe a cohesive sediment type and physical behaviour. Flocs are multi-component, being composed of varying proportions and types of inorganic and organic particles, and the packing (i.e. density) of these grains within a floc can significantly affect their resultant size and settling velocity. It is this complexity that means it is not a simple task to mathematically describe the mud flocculation process on a fundamental basis (Milligan and Hill, 1998; Mikkelsen and Pejrup, 1998). The principle reason for such a poor understanding of cohesive sediment settling fluxes and deposition rates, has been principally due to a lack of reliable floc data, although the situation is now improving. The influence of floc density variations are required for accurate settling flux determination. Therefore a key to rectifying this problem is to use a floc sampling system which directly measures (in-situ) both the simultaneous size and settling velocity of the larger and more fragile flocs.

It is difficult to accurately quantify the influence and occurrence of flocculation, as well as floc break-up, on in-situ estuarine floc distributions. The fragility of large, fastest settling macroflocs, which are easily broken-up upon sampling (Gibbs and Konwar, 1983), has tended to preclude the direct measurement of floc settling and mass characteristics due to instrumentation limitations (Eisma et al., 1997). Floc disruptive devices include field settling tubes (FST), such as the Owen tube (Owen, 1976). These instruments are the original devices used to determine the in situ settling properties of flocculated mud. The Owen tube is the most universally known of all FSTs. It was developed during the 1960's at Hydraulics Research Station Wallingford (now HR Wallingford Ltd) by M.W. Owen (1971, 1976). Collected water samples are extracted from the bottom of the tube at pre-selected time intervals and the settling velocity is inferred from gravimetric analysis (Vanoni, 1975), and tends to significantly under-estimated Ws.

Floc breakage occurs in response to the additional shear created during acquisition (Eisma et al., 1997). The presence of large estuarine macroflocs was initially observed *in-situ* by underwater photography (Eisma et al, 1990). To overcome this problem less invasive techniques for measuring floc size and settling velocity in situ have been developed, for example VIS (van Leussen and Cornelisse, 1994), INSSECT (Mikkelsen et al., 2004), INSSEV (Fennessy et al., 1994b; see Fig. 7), LabSFLOC (Manning, 2006; see Fig. 8), and the HoloCam (Graham and Nimmo Smith, 2010). Unlike particle sizers (e.g. Agrawal and Pottsmith, 2000; Benson and French, 2007; Law et al., 1997), these instruments can provide direct simultaneous measurements of floc size and settling velocity, in-situ, and permit an estimate of individual floc effective density by applying a modified Stokes' Law. These types of measurements make possible the computation of the floc mass distribution across a range of sizes (Fennessy et al., 1997).

Optical devices to measure concentration profiles by Spinrad et al. (1989), Kineke et al. (1989), and McCave and Gross (1991) have sought to quantify the rate of water clearance,

Fig. 7. a) Side view of INSSEV instrument mounted on a metal deployment frame. b) Front view of the INSSEV instrument (*right*), together with optical backscatter (OBS) sensors and an acoustic Doppler velocimeter (ADV) positioned on a vertical pole (*left*). The ADV provides high frequency turbulence data which can be directly related to the floc populations.

but they are unable, like all earlier instrumentation, to measure particle size and settling velocity spectra directly. Whereas sampling devices which directly observe D and Ws can provide an insight into the interaction of flocs with both turbulent eddies and SPM concentration variations during a tidal cycle, particularly within the lower layers of the flow where the turbulent shearing is at its greatest (Mehta and Partheniades, 1975). Deploying floc samplers in conjunction with high frequency velocimeters provides scientists a means of accurately acquiring time series of both the spectral distribution of the floc dry mass and settling velocities, together with information on the turbulence fluctuations, directly from within a turbulent estuarine water column. Such site-specific information of floc settling velocity spectra is a prerequisite for accurate physical process parameterisation, especially for the implementation into sediment transport modelling applications (Manning, 2004c).

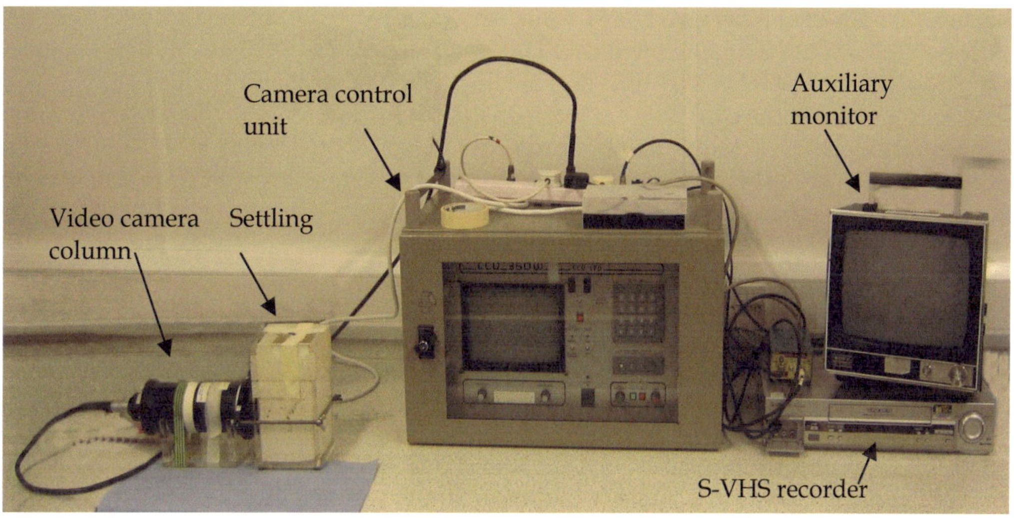

Fig. 8. LabSFLOC set-up (from Manning, 2006).

4. Floc settling behaviour

For non-cohesive sediment, the settling velocity can be regarded as being proportional to the particle size. However, as discussed earlier in this chapter, many estuarine locations tend to be dominated by muddy flocculated sediments and an accurate representation of the vertical sediment settling fluxes for cohesive sediments is problematic. As a result, the sizes and settling velocities of flocs are key parameters in the modelling of cohesive sediment transport in near-shore waters (e.g. Geyer et al., 2000; Cheviet et al., 2002).

Throughout a tidal cycle there are slack water periods (usually around times of high and low water), when the current flow which transports suspended matter in an estuary decreases quite significantly. It is at these times of slack water in the tidal cycle, there tend to be an absence of a significant amount of vertical exchange and allows suspended flocculated matter to deposit to the bed. Stringer flocs have been observed in many European estuaries by underwater cameras (Manning and Dyer, 2002a; Fennessy et al., 1994b) and it has been speculated that they are the result of particle scavenging through differential settling. Particle interaction by differential settling is where larger particles have larger settling velocities, and therefore fall onto relatively smaller particles. Stolzenbach and Elimelich

(1994) have suggested that differential settling is not an important contact mechanism, and may only occur within quiescent waters, particularly slack tide periods.

In still water, the rate of deposition of flocculated sediments is described by the rate of change of sediment mass (m) per unit area, or in other words the gradient *dm/dt*, where t is time. The depositional rate is equal to the product of the SPM concentration and settling velocity (at any point in time and space), which is known as the mass settling flux (MSF) towards the bed. Where detailed settling information is not available, it is quite common for a median settling velocity, Ws_{50}, to be used, although this can produce a misleading representation of the actual settling behaviour of individual floc populations.

Whitehouse et al (2000) note that it is highly probable that a concentration gradient will develop in an estuarine water column in very turbid and tidally active estuaries, e.g. Severn Estuary in the UK (Manning et al., 2009). This means that the near-bed concentration will be greater than the depth-averaged SPM value. As more flocs are deposited to the bed, the near-bed concentration gradient will rise. Thus the amount of SPM higher in water column will progressively decrease with time as more flocculated matter settles to the bed. This will lead to a gradually decrease in the depositional rate. For example, in the Severn Estuary, Whitehouse et al. (2000) found settling fluxes rose to a peak of about 60 x 10^{-3} kg $m^{-2}s^{-1}$ at a SPM of 25 kg m^{-3}. The MSF then rapidly decreases with rising concentration, due to hindered settling effects.

In tidal estuaries, the ambient hydrodynamics very rarely produce perfectly quiescent water column conditions. The mechanisms of the deposition of cohesive sediment in flowing water was originally studied through the use of laboratory flume experiments by Krone (1962), and Einstein and Krone (1962). Further flume experiments were conducted (see Partheniades, 1962; Postma, 1962; Mehta, 1988; Burt and Game, 1985; Delo, 1988). The role of turbulence on settling flocs was also examined by Owen (1971) and Wolanski et al. (1992). Furthermore, field measurements of the deposition of muddy sediments in estuaries during single tides have been undertaken by HR Wallingford (Diserens et al., 1991).

Classically deposition of the cohesive sediment to the bed occurs only when the bottom shear stress falls below some critical value, with the deposition rate proportional to the deficit of shear stress below that critical value. Erosion of cohesive sediment particles from near the estuary bed usually only occur when the bed shear stress rises above some critical value, with the erosion rate depending in some way on the excess of stress above that value. The critical stress for erosion is greater than or equal to the critical stress for deposition, such that an intermediate range of bottom shear stresses can exist for which neither deposition or erosion can occur.

Traditionally the rate of deposition of cohesive sediment from a suspension in flowing water has been modelled using the near-bed SPM concentration (C_b), median settling velocity Ws_{50}, the bed shear stress exerted by the flowing water τ_b and a critical bed shear stress for deposition τ_d (Whitehouse et al., 2000). τ_d is defined as the bed shear stress above which there is no deposition of suspended sediment. Lau and Krishnappan (1991) state that classically floc deposition can be identified in terms of a depositional shear stress τ_d, whereby this shear stress provides a threshold indicating that when the ambient shear stress falls below this stress level (i.e. τ_d), the flow is unable to support the matter in suspension and it is deposited.

In reality the whole sediment transport cycle in an estuary is particularly complicated, because it is not necessarily closed, and processes are interdependent. For instance, settling does not always lead to deposition, i.e. when entrainment dominates, and the sediment then remains in suspension. Sediment tends to respond behind the hydrodynamics of the ambient flow (Dyer, 1995). This lag is created by tidal asymmetry and a phase difference between the SPM concentration and flow velocity; this can produce a residual flux of sediment. For mud deposition, settling lag effects are the most relevant. On the decreasing tide, mud flocs will start to settle once the turbulence in the flow is incapable of maintaining them in suspension. As the flocs settle, they are transported along on the waning current flow, so that they eventually reach the estuary bed some distance up or downstream (depending upon the point in the tidal cycle) from the point at which settling was initiated. This effect is settling lag, and a qualitative model outlining these effects was developed by Postma (1961). A settling lag will sort out the flocs according to their threshold characteristics and settling velocity (Dyer, 1995).

Most previous research indicated that a paradigm of cohesive sediment transport research in that deposition (Dep) and erosion are mutually exclusive (e.g. Ariathurai and Krone, 1976; Officer, 1981; Dyer, 1986; Mehta, 1986, 1988; Partheniades, 1986, 1993; Sheng, 1986; Odd, 1988; Mehta et al., 1989; Uncles and Stephens, 1989). Sanford and Halka (1993) analysed a series of field measurements under tidal conditions in Chesapeake Bay. They observed that the suspended sediment concentration started to decrease when the flow velocity started to decrease. This behaviour could not be simulated by Krone's (1962) depositional formula, when used in conjunction with an erosion formula. Krone's (1962) formula predicted that the SPM concentration could not decrease before the ambient flow velocity, and thus the bed shear stress, fell below its τ_d ($\tau_b < \tau_d$). However, Sanford and Halka (1993) were able to model the observed concentration pattern only when they applied a continuous depositional formula. This is for the case where Dep = Ws . C_b. Sanford and Halka's (1993) research concluded that the paradigm of mutually exclusive deposition and erosion of cohesive sediment, is not valid for real estuarine scenarios, but is valid under laboratory conditions.

Winterwerp (2007) suggested that four elements comprise the deposition of cohesive sediments, which are: i) simultaneous erosion and deposition, ii) erosion rate, iii) bed shear stress, and iv) flocculation. Using this four-point framework, Winterwerp (2007) re-analysed Krone's (1962) original flume results and advocates that a critical shear stress for deposition does not exist. Instead this τ_d stress represents a critical shear stress for erosion of freshly deposited cohesive sediment, i.e. resuspension. Winterwerp's (2007) concludes that the use of Krone's deposition formula and Partheniades erosion formula for the modelling of the water-bed exchange processes (e.g. Ariathuria and Arulanandan, 1978), does not correctly represent the physics. Thus Winterwerp recommended that in order to model the sedimentation flux for applications at low mud SPM concentrations, only the depositional flux itself is required: Dep = Ws . Cb.

This approach concurs with the findings of Sanford and Halka (1993). For higher concentration suspensions, the flocculation process becomes more important. Similarly sediment-turbulence interaction can become significant through the formation of CBS or fluid mud, which can affect the bed shear stress through turbulence damping and drag reduction. However, both Le Hir et al. (2001) and Winterwerp (2002) indicate that the sedimentation flux is still correctly described by Dep = Ws . Cb, even at higher concentrations. The implication of Winterwerp's approach to deposition was reported by Spearman and Manning (2008).

In contrast, Maa et al. (2008) maintain that both τ_b (a hydrodynamic parameter) and τ_d (a sediment parameter) are the main controlling parameters for determining cohesive sediment deposition. Therefore further research is required in this area.

5. Examples of the application of different floc depositional models

Computer simulation models are commonly the chosen tools with which estuarine management groups attempt to predict sediment transport rates for tasks such as routine maintenance dredging, through to estimating the potential impacts new port related construction would have on an existing hydrodynamical regime. In order for these numerical models to provide sufficiently meaningful results, they require a good scientific understanding of the phenomena under consideration, and these processes need to be adequately described (i.e. parameterised) by the model coding.

Of particular importance, a quantitative understanding of the dynamics of the vertical structure of cohesive sediments in suspensions is essential for an accurate estuarine sedimentation model (Kirby, 1986; van Leussen, 1991). This requires an understanding of the physical processes related to the entrainment, advection and deposition of muddy sediments. One physical process which has caused particular difficulty is the modelling and mathematical description of the vertical mass settling flux of sediment, which becomes the depositional flux near to slack water. The MSF is the product of the concentration and the settling velocity. Manning and Bass (2006) have found that mass settling fluxes can vary over four or five orders of magnitude during a tidal cycle in meso- and macrotidal (Davies, 1964) estuaries, therefore a realistic representation of flux variations is crucial to an accurate depositional model.

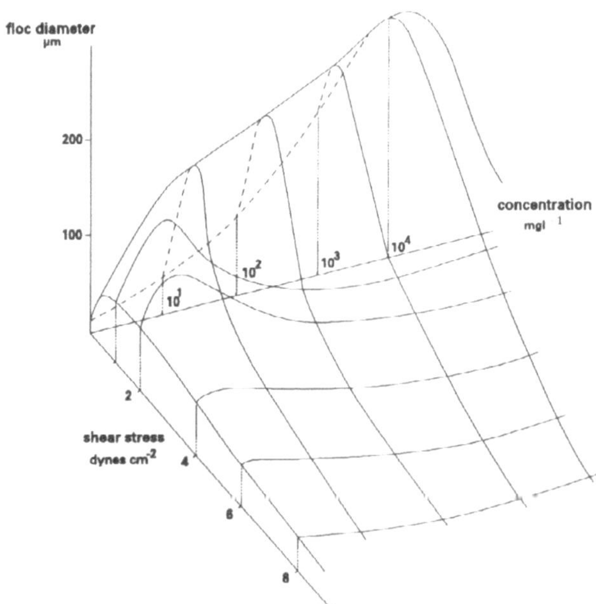

Fig. 9. Conceptual diagram showing the relationship between floc modal diameter, suspended sediment concentration and shear stress (Dyer, 1989).

The specification of the flocculation term within numerical models depends upon the sophistication of the model. Dyer (1989) proposed a conceptual relationship between D (Ws), SPM and τ (Fig. 9), but until recently was largely unproven. Therefore the simplest parameterisation is a settling velocity value which remains constant in both time and space.

Constant Ws of 0.5-1 mm s^{-1} have historically been used to represent mud settling, although these are now known to significantly under-estimates of macrofloc fall rates. Peterson et al. (2002) in contrast employed a constant Ws of 5 mm s^{-1} for the Tamar Estuary (UK), which tended to over-predict depositional rates. These fixed settling values are typically selected on an arbitrary basis and adjusted by model calibration. The next step has been to use gravimetric data provided by field settling tube experiments to relate flocculation to SPM concentration. Empirical results have shown a general exponential relationship between either the mean or median floc settling velocity (Ws_{50}) and SPM for concentrations ranging up to 10 g l^{-1} (Fig. 10). However, both of these parameterisation techniques do not include the important and influential effects of turbulence (Manning, 2004a). Beyond 10 g l^{-1}, the settling of flocs becomes hindered and their terminal velocities progressively slow with rising turbidity (see Fig. 11).

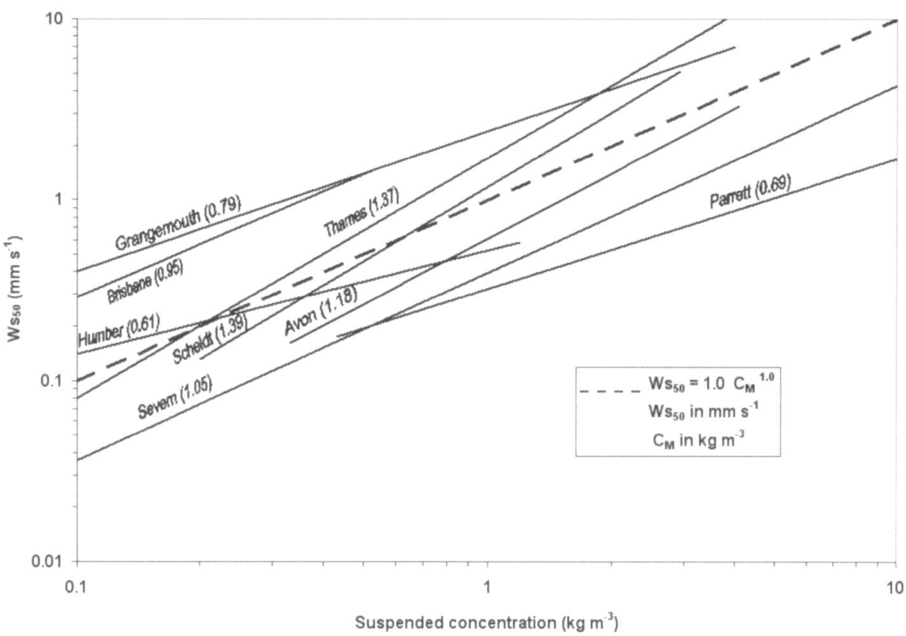

Fig. 10. Owen tube determined median settling velocity as a function of suspended sediment concentration (C_M) for diferent estuaries. The dotted line represents an exponent of unity (*Redrawn from Delo and Ockenden, 1992*).

More recently, a number of authors have proposed simple theoretical formulae inter-relating a number of floc characteristics which can then be calibrated by empirical study. Such an approach has been used by van Leussen (1994), who has utilised a formula which modifies the settling velocity in still water, by a growth factor due to turbulence and then divided by a turbulent disruption factor. This is a qualitative simplification of the Argaman and Kaufman (1970) model originally developed for the sanitation industry, with only a limited number of inter-related parameters, and hence does not provide a complete description of floc characteristics within a particular turbulent environment. Even so, Malcherek (1995) applied van Leussen's (1994) heuristic approach to the Wesser Estuary in Germany (Malcherek et al., 1996) with some degrees of success.

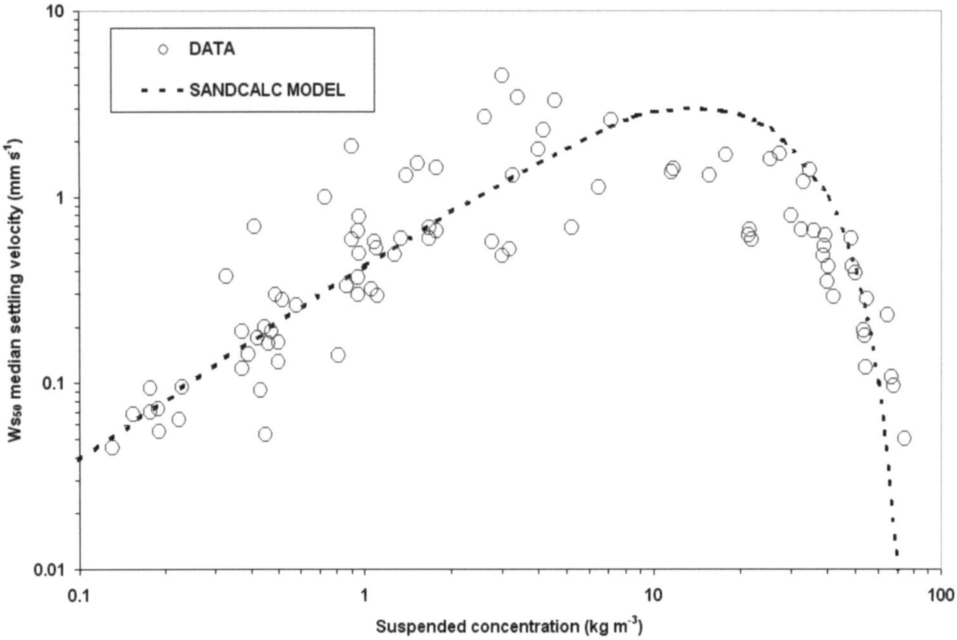

Fig. 11. Median settling velocity of Severn Estuary mud as a function of SPM concentration. The Owen tube data is from Odd and Roger (1986). The solid line represents the hindered settling effect based on the the *SandCalc* sediment transport computational software algorithm (*Redrawn from from Soulsby, 2000*).

A number of authors have attempted to observe how the floc diameter changes in turbulent environments. For example, Lick et al (1993) derived an empirical relationship based on laboratory measurements using a flocculator. They found the floc diameter varied as a function of the product of the SPM concentration and a turbulence parameter. However, this type of formulation says very little about the important floc settling or dry mass properties.

An approach which has recently gained much interest by mathematicians, is the fractal representation of flocs (e.g. Chen and Eisma, 1995; Winterwerp, 1999). Population balance approaches to flocculation modelling can also require floc fractal information (e.g. Maggi, 2005; Mietta et al., 2008; Verney et al., 2010). Fractal theory is dependent on the successive aggregation of self-similar flocs producing a structure that is independent of the scale considered. This is similar to Krone's (1963) order of aggregation. Winterwerp (1998) obtained a relationship, based on research by Kranenburg (1994), relating floc settling to the: floc size, primary particle diameter and the fractal dimension (nf). Fractal dimensions of 1.4 are representative of fragile aggregates, whilst values of 2.5 indicate strongly bonded estuarine flocs. However, in order to make a fractal based model solvable analytically within a numerical simulation, a mean nf of 2 is commonly assumed and this ignores important floc density variations. A less complex version of Winterwerp's (1998) original fractal flocculation model since been developed by Winterwerp et al. (2006) and has been incorporated into a Delft 3-D model to examine sediment transport in the Lower Scheldt Estuary.

Most floc settling velocity parameterisations do not include a component which represents floc density and hence floc mass flux variations. Also most floc parameterisations produce a single mean fall rate in time and space. However, a conclusion drawn from an

Intercomparison Experiment of various floc measuring devices conducted in the Elbe estuary (Dyer et al., 1996), was that a single mean or median settling velocity *did not* adequately represent an entire floc spectrum, especially in considerations of a flux to the bed. Dyer et al. (1996) recommended that the best approach for accurately representing the settling characteristics of a floc population was to split a floc distribution into two or more components, each with their own mean settling velocity. Both Eisma (1986) and Manning (2001) suggest a more realistic and accurate generalisation of floc patterns can be derived from the larger macrofloc and smaller microfloc sized fractions.

Significant advances into the modelling of flocculated cohesive sediment were made during the recent Defra funded EstProc (Estuary Processes Research) project (Estuary Process Consortium, 2005), where Manning (2004b; 2008) developed a series of algorithms. Manning's algorithms for settling velocity is based entirely on empirical observations made in situ using un-intrusive floc data collected with the INSSEV instrument (Fennessy et al., 1994; Manning and Dyer, 2002b) together with turbulence data, both acquired from a wide range of estuarine conditions. The Manning settling model includes aspects of floc mass representation and dual settling velocities, both of which vary in response to shear stress (τ) and SPM concentration changes.

The Manning algorithms were derived using a parametric multiple regression statistical analysis of key parameters which were generated from the raw spectral floc data (Manning, 2004b; 2008). The algorithms are based on the concept of macrofloc settling ($w_{s,macro}$) and the settling of the smaller microfloc size fraction ($w_{s,micro}$), and the ratio of macrofloc to microfloc mass present in each floc population termed the SPM_{ratio} (Manning, 2004c). The algorithms are illustrated in Fig. 12. The two fractions were demarcated at 160 μm (Manning, 2001). The representation of floc population settling characteristics by dividing distributions into bi-modal fraction, each with their own mean settling velocity, as advised by Dyer et al. (1996).

Since the development of the MFSV, Baugh and Manning (2007) have subsequently implemented the Manning algorithms into a Telemac 3-D numerical model of the Thames Estuary, and the parameterisation is now being used routinely for projects in the Coasts and Estuaries Group at HR Wallingford. Their findings of the 1-D case studies (see Figs 13 and 14.) found the MFSV model could reproduce 93% of the total mass settling flux observed over a spring tidal cycle. This increased to a near-perfect match within the turbidity maximum zone. A constant Ws of 0.5 mm s^{-1} only estimated 15% of the flux within the turbidity maximum zone (TMZ), whereas a fixed 5 mm s^{-1} settling rate over-predicted the TMZ mass flux by 47%. Both a power law Ws – SPM representation and van Leussen method did not fare much better, typically estimating less than half the observed flux during the various tidal and sub-tidal cycle periods. When the Manning settling flux model was applied to a highly saturated benthic suspension layer with SPM concentrations approaching 6 g l^{-1}, it calculated 96% of the observed flux. In contrast, the van Leussen approach only predicted a third of the total observed flux within a concentrated benthic suspension layer.

During 3-D model testing, Baugh and Manning (2007) reported the use of a constant settling velocity did not result in the representation of any of the observed structure of suspended concentrations (Fig. 15). The use of a linear settling velocity was a great improvement, firstly in that it introduced an element of vertical variation in the predicted suspended concentration. Secondly, the linear formulation also resulted in higher concentrations occurring on the inside of the river bend during the ebb tide and on the outside of the bend on the flood tide as observed. The use of the Manning algorithms further improved the

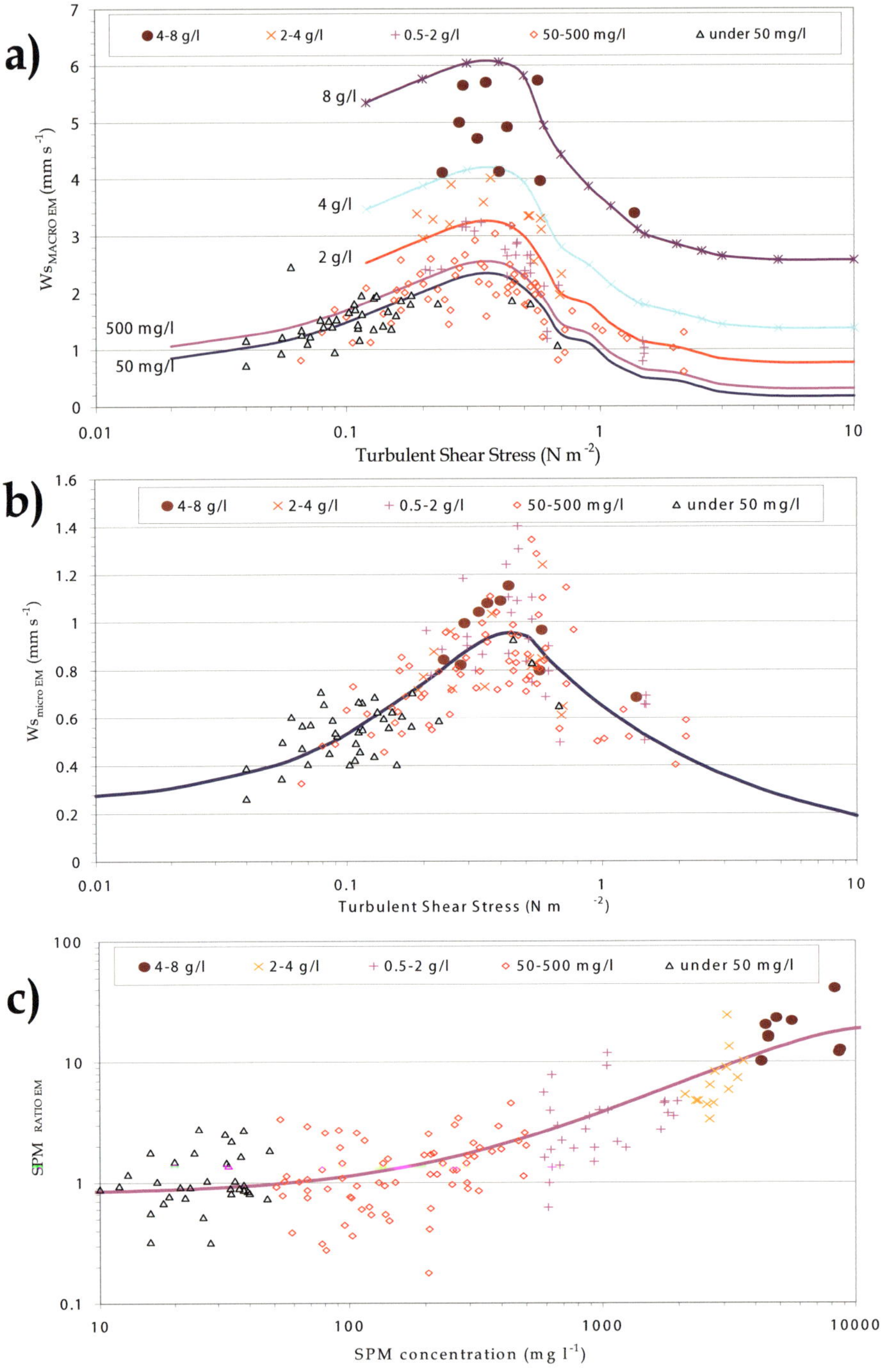

Fig. 12. Representative plots of the statistically generated regression curves, together with the experimental data points, illustrating the three contributing components for the empirical flocculation model: *a*) Ws_{macro} at various constant SPM values plotted against τ; *b*) Ws_{micro} plotted against τ; and *c*) SPMratio plotted against SPM (from Manning and Dyer, 2007).

representation of the observed distribution by increasing the level of vertical variation in suspended concentration. The Manning algorithm also introduced an area of higher concentration on the inside of the bend during the flood tide, as was observed. Similarly, Spearman (2004) reported significant improvements in a 3-D numerical mud transport simulation on tidal flats when Manning's settling model was included. Soulsby and Manning (see Soulsby et al., 2010 in prep.) have since derived a more physics-based flocculation model. This model is based on the key mass settling flux flocculation characteristics exhibited by the original Manning algorithms, but intends to provide a more generic, floc modelling solution, through a reduction in the number of coefficients used during calibration.

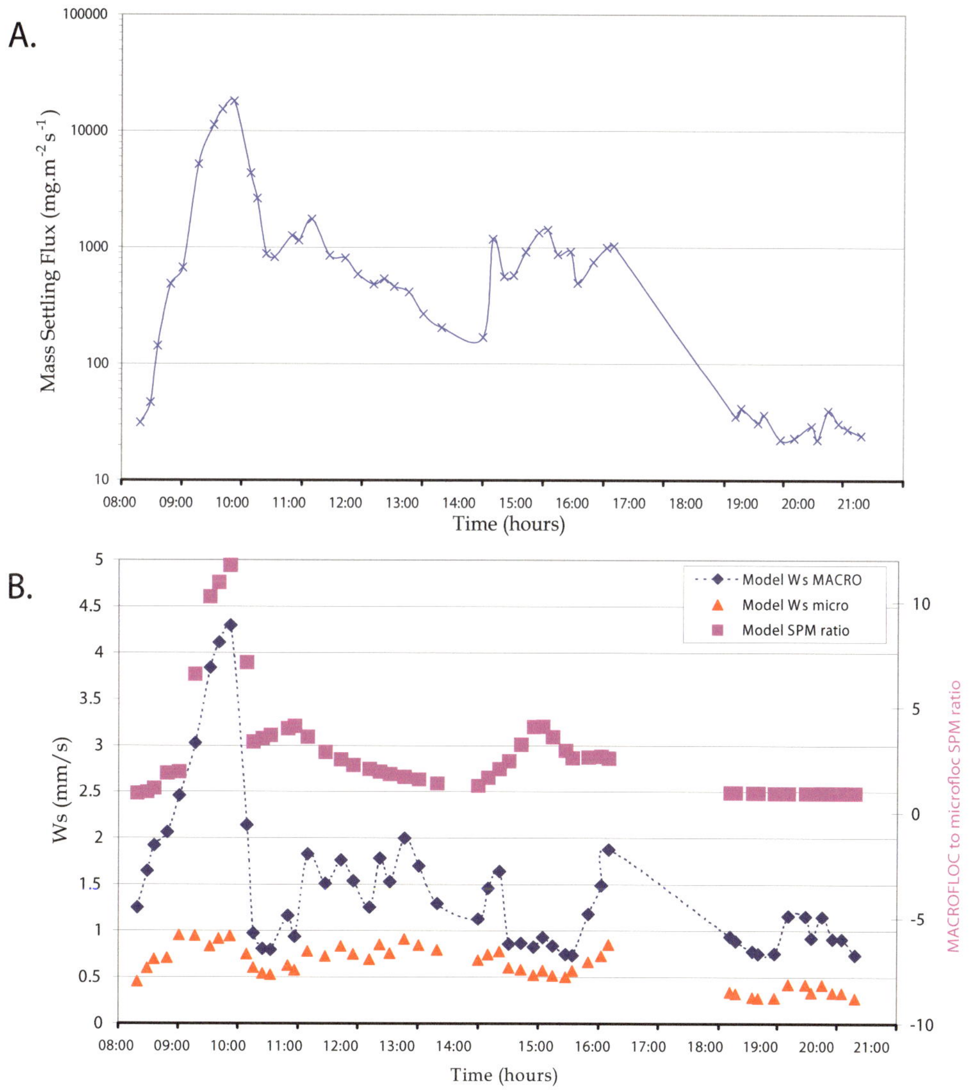

Fig. 13. A) Observed mass settling flux time series, B) Manning model component algorithm outputs for the tidal cycle (from Baugh and Manning, 2007).

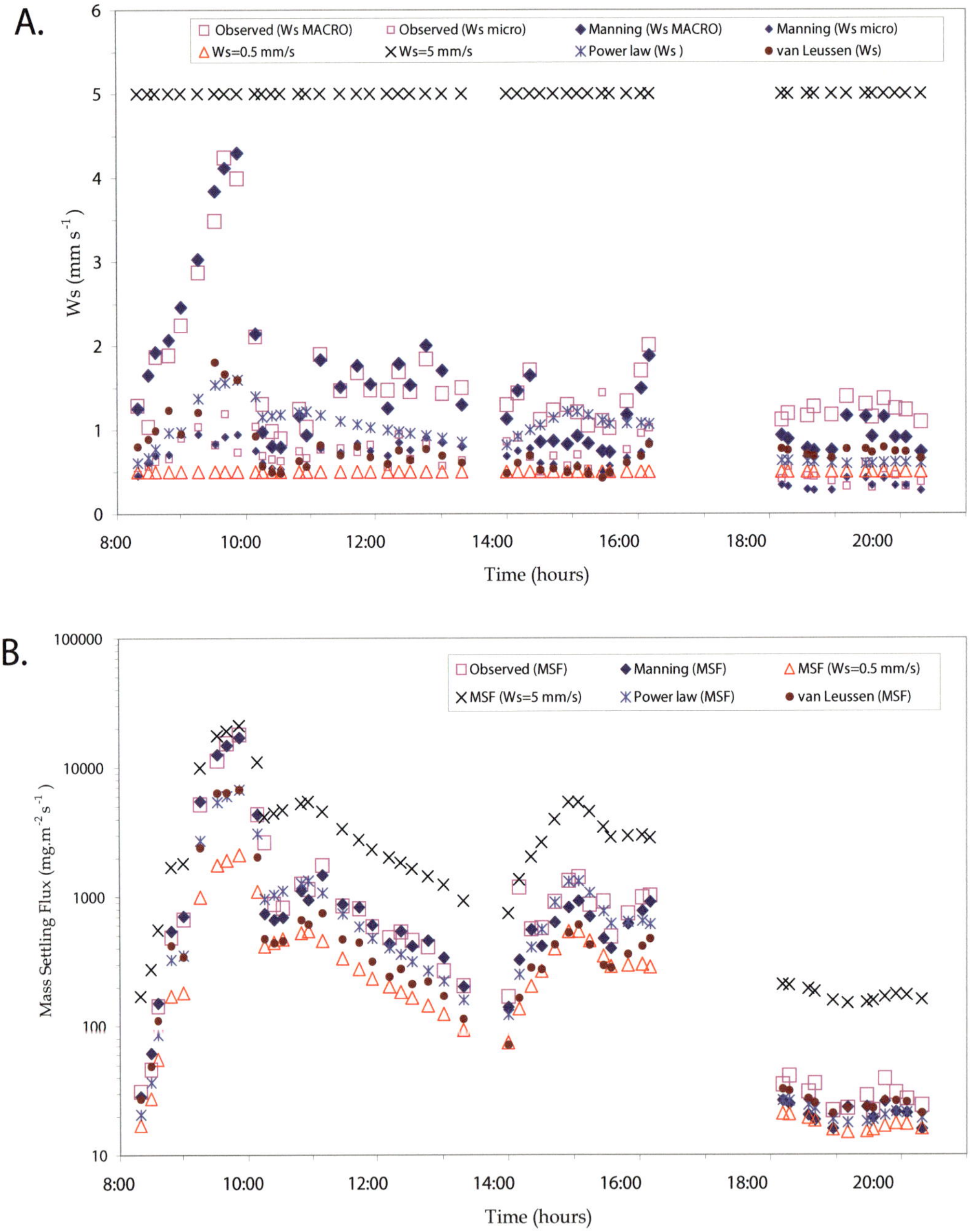

Fig. 14. Tidal cycle time series comparison of observed and predicted values of: A) settling velocity, B) mass settling flux (from Baugh and Manning, 2007).

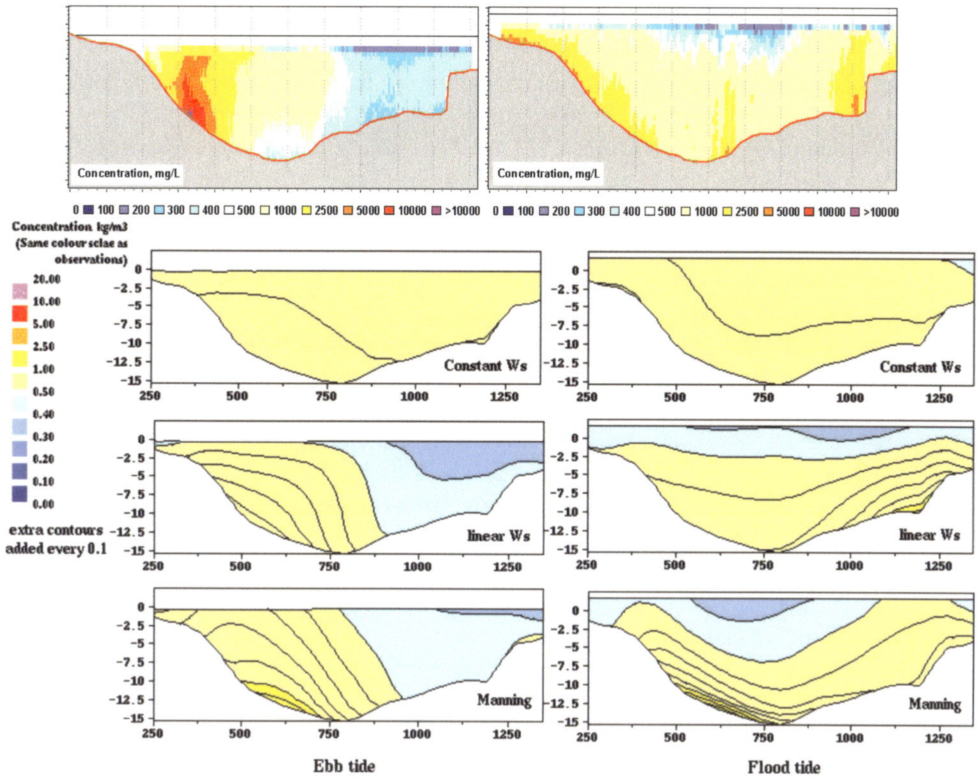

Fig. 15. Comparison of observed sediment concentrations across the river section (top) and predicted sediment concentrations using different assumptions about the settling velocity (from Baugh and Manning, 2007).

6. Acknowledgements

The preparation of this chapter was funded by the HR Wallingford Company Research project DDY0409.

7. References

Ackroyd, D.R., Bale, A.J., Howland, R.J.M., Knox, S., Millward, G.E. and Morris, A.W. (1986). Distributions and behaviour of Cu, Zn and Mn in the Tamar estuary. Estuarine Coastal and Shelf Science 23, 621-640.

Agrawal, Y.C. and Pottsmith, H.C. (2000). Instruments for particle size and settling velocity observations in sediment transport. Marine Geology, 168 (1-4): 89-114.

Al Ani, S., Dyer, K.R. and Huntley, D.A. (1991). Measurement of the influence of salinity on floc density and strength. Geo-Marine Letters, 11: 154-158.

Alldredge, A.L. and Gotschalk, C., (1988). In-situ settling behaviour of marine snow. Limnol. Oceanography, 33: 339-351.

Argaman, Y. and Kaufman, W.J. (1970). Turbulence and flocculation. J. Sanitary Eng., ASCE, 96, 223-241.

Ariathuria, C.R. and Arulanandan, K. (1978). Erosion rates of cohesive soils. Journal of the Hydraulics Division, ASCE, (104) 2, 279-282.

Ariathurai, R. and Krone, R.B. (1976). Finite element model for cohesive sediment transport. Journal of the Hydraulic Division, ASCE, vol. 102, No. HY3, pp. 323-338.

Baugh, J.V. and Manning, A.J. (2007). An Assessment of a New Settling Velocity Parameterisation for Cohesive Sediment Transport Modelling. Continental Shelf Research, doi:10.1016/j.csr.2007.03.003.

Benson T. French J. R. (2007). InSiPID: A new low-cost instrument for in situ particle size measurements in estuarine and coastal waters. J Sea Res 58:167-188.

Burban, P-Y., Lick, W. and Lick, J. (1989). The flocculation of fine-grained sediments in estuarine waters. Journal of Geophysical Research 94 (C6), 8323-8330.

Burt, T.N. and Game, A.C. (1985). Deposition of fine sediment from flowing water: an investigation of dependence on concentration. HR Wallingford Report SR 27.

Chen, S., Eisma, D. (1995). Fractal geometry of in-situ flocs in the estuarine and coastal environments. Netherlands Journal of Sea Research 32, 173-182.

Cheviet, C., Violeau, D. and Guesmia, M. (2002). Numerical simulation of cohesive sediment transport in the Loire estuary with a three-dimensional model including new parameterisations. In: Fine Sediment Dynamics in the Marine Environment - Proc. in Mar. Sci. 5., Amsterdam: Elsevier, J.C. Winterwerp and C. Kranenburg, (eds.), pp. 529-543, ISBN: 0-444-51136-9.

Davies, J.H. (1964). A morphogenetic approach to world shorelines. Z. Geomorphol., 8: 127-142.

Delo, E.A. (1988). Deposition of cohesive sediment from flowing water. HR Wallingford Report SR 152.

Delo, E.A. and Ockenden, M.C. (1992). Estuarine muds manual. HR Wallingford Report, SR 309.

Diserens, A.P., Delo, E.A. and Ockenden, M.C. (1991). Estuarine sediments - near-bed processes: field measurement of near-bed cohesive sediment processes. HR Wallingford Report SR 262.

Droppo, I.G. (2001). Rethinking what constitutes suspended sediments. Hydrological processes, 15, 1551-1564.

Droppo, I.G., Walling, D. and Ongley, E. (2000). The influence of floc size, density and porosity on sediment and contaminant transport. Journal of the National Centre for Scientific Research, 4, 141-147.

Dyer, K.R. (1986). Coastal and Estuarine Sediment Dynamics. Wiley & Sons, Chichester, 342p.

Dyer, K.R. (1989). Sediment processes in estuaries: future research requirements. Journal of Geophysical Research 94 (C10), 14327-14339.

Dyer, K.R. (1995). Sediment transport processes in estuaries. In: Geomorphology and Sedimentology of Estuaries (Ed. Perillo, G.M.E.), Elsevier Science, Oxford, 423-449.

Dyer, K.R., Cornelisse, J.M., Dearnaley, M., Jago, C., Kappenburg, J., McCave, I.N., Pejrup, M., Puls, W., van Leussen, W. and Wolfstein, K. (1996). A comparison of in-situ techniques for estuarine floc settling velocity measurements. Journal of Sea Research 36, 15-29.

Dyer, K.R. and Manning, A.J. (1998). Observation of the size, settling velocity and effective density of flocs, and their fractal dimensions. Journal of Sea Research 41, 87-95.

Edzwald, J.K. and O'Melia, C.R. (1975). Clay distributions in recent estuarine sediments. Clays and Clay Minerals, 23:39-44.

Einstein, H.A. and Krone, R.B. (1962). Experiments to determine modes of cohesive sediment transport in salt water. Journal of Geophysical Research, (67) 4, 1451-1461.

Eisma, D. (1986). Flocculation and de-flocculation of suspended matter in estuaries. Neth. J. Sea Res., 20 (2/3), 183-199.

Eisma, D., Dyer, K.R. and van Leussen, W. (1997). The in-situ determination of the settling velocities of suspended fine-grained sediment - a review. In: Burt, N., Parker, R., Watts, J. (Eds), Cohesive Sediments - Proc. of INTERCOH Conf. (Wallingford, England), Chichester: John Wiley & Son, pp. 17-44.

Eisma, D., Schuhmacher, T., Boekel, H., Van Heerwaarden, J., Franken, H., Lann, M., Vaars, A., Eijgenraam, F. and Kalf, J. (1990). A camera and image analysis system for in situ observation of flocs in natural waters. Neth. J. Sea Res., 27: 43-56.

Estuary Process Consortium (2005). Final Report of the Estuary Process Research Project (EstProc) - Algorithms and Scientific Information. Integrated Research Results on Hydrobiosedimentary Processes in Estuaries, R & D Technical Report prepared by the Estuary Process Consortium for the Fluvial, Estuarine and Coastal Processes Theme, co-funded by Defra & Environment Agency, Report No: FD1905/TR3, 140p.

Fennessy, M.J., Dyer, K.R. and Huntley, D.A. (1994a). Size and settling velocity distributions of flocs in the Tamar Estuary during a tidal cycle. Netherlands Journal of Aquatic Ecology, 28: 275-282.

Fennessy, M.J.; Dyer, K.R. and Huntley, D.A. (1994b). INSSEV: an instrument to measure the size and settling velocity of flocs in-situ. Marine Geology, 117, 107-117.

Fennessy, M.J., Dyer, K.R., Huntley, D.A. and Bale, A.J. (1997). Estimation of settling flux spectra in estuaries using INSSEV. In: N. Burt, R. Parker and J. Watts, (Eds.), Cohesive Sediments - Proc. of INTERCOH Conf. (Wallingford, England), Chichester: John Wiley & Son, pp. 87-104.

Geyer, W.R., Hill, P.S., Milligan, T.G. and Traykovski, P., (2000). The structure of the Eel River plume during floods. Con. Shelf Res., 20, 2067-2093.

Gibbs, R.J., (1985). Estuarine flocs: their size settling velocity and density. J. Geophys. Res., 90 (C2): 3249-3251.

Gibbs, R.J. and Konwar, L.N., (1983). Sampling of mineral flocs using Niskin bottles. Environ. Sci. Technol., 17 (6): 374-375.

Glasgow, L.A. and Lucke, R.H., (1980). Mechanisms of deaggregation for clay-polymer flocs in turbulent systems. Ind. Eng. Chem. Fundam., 19: 148-156.

Graham, G.W. and Nimmo Smith, W.A.M. (2010) The application of holography to the analysis of size and settling velocity of suspended cohesive sediments. Limnology and Oceanography: Methods 8, 1-15.

Gregory, J. (1978). Effects of polymers on colloid stability. In: The Scientific Basis of Flocculation (Ed. Ives, K.J.), Pub. Sijthoff & Noordhoff, Alphen aan den Rijn, The Netherlands: 89-99.

Hayter, E.J. and Mehta, A.J. (1982). Modelling of estuarial fine sediment transport for tracking pollutant movement. Final Report No. UFL/COEL-8L/009, University of Florida.

Kineke, G.C.; Sternberg, R.W. and Johnson, R. (1989). A new instrument for measuring settling velocity in-situ. Marine Geology, 90, 149-158.

Kirby, R. (1986). Suspended fine cohesive sediment in the Severn estuary and Inner Bristol channel. Rept. ESTU-STP-4042, Department of Atomic Energy, Harwell, U.K.

Klimpel R.C. and Hogg R. (1986). Effects of flocculation conditions on agglomerate structure. Journal of Colloid Interface Science 113, 121-131.

Kolmogorov, A.N. (1941). The local structure of turbulence in incompressible viscous fluid for very large Reynolds numbers. C. R. Acad. Sci. URSS, 30: 301; Dissipation of energy in locally iscotropic turbulence. C. R. Acad. Sci. URSS, 32: 16.

Kranck, K. (1984). The role of flocculation in the filtering of particulate matter in estuaries. In: The Estuary as a Filter (Ed. Kennedy, V.), Academic Press, Orlando Inc, 159-175.

Kranck, K. and Milligan, T.G. (1992). Characteristics of suspended particles at an 11-hour anchor station in San Francisco Bay, California. Journal of Geophysical Research 97, 11373-11382.

Kranenburg, C. (1994). The fractal structure of cohesive sediment aggregates. Estuarine Coastal Shelf Sci., Vol. 39, 451-460.

Krishnappan, B.G. (1991). Modelling of cohesive sediment transport. Preprints Int. Symp. Trans. Suspended Sediments & its Mathematical Modelling, IAHR, Florence, Italy, September 2-5 1991, 433-448.

Krone, R.B. (1962). Flume studies of the transport of sediment in estuarial shoaling process: Final report, Hydraulic Engineering Laboratory and Sanitary Engineering Research Laboratory, University of California, Berkeley, California, 110 p.

Krone, R. B. (1963). A study of rheological properties of estuarial sediments. Report No. 63-68, Hyd. Eng. Lab. and Sanitary Eng. Lab., University of California, Berkeley.

Lau, Y. L. and Krishnappan, B. G. (1991). Size distribution and settling velocity of cohesive sediments during settling.

Law, D.J., Bale, A.J. and Jones, S.E. (1997). Adaptation of focused beam reflectance measurement to in-situ particle sizing in estuaries and coastal waters. Marine Geology, 140: 47-59.

Le Hir, P., Bassoulet, P. and Jetsin, H. (2001). Application of the continuous modelling concept to simulate high-concentrated suspended sediment in a macro-tidal estuary. In: W.H. McAnally and A.J. Mehta (eds), Coastal and Estuarine Fine Sediment Processes, Amsterdam: Elsevier, pp. 229-248.

Lee, B.J., Toorman, E., Molz, F.J. and Wang, J. (2010). A Two-Class Population Balance Equation Yielding Bimodal Flocculation of Marine or Estuarine Sediments. Water Research, DOI: 10.1016/j.watres.2010.12.028.

Lick, W. (1994). Modelling the transport of sediment and hydrophobic contaminants in surface waters. In: U. S. / Israel Workshop on monitoring and modelling water quality, May 8 13, 1994, Haifa, Israel.

Lick, W., Huang, H. and Jepsen, R. (1993). Flocculation of fine-grained sediments due to differential settling. J. Geophs. Res., 98 (C6): 10,279-10,288.

Maa, J. P.-Y., Kwon, J.-I., Hwang, K.-N. and Ha, H.K. (2008). Critical bed shear stress for cohesive sediment deposition under steady flows. Journal of Hydraulic Engineering, 134(12), 1767-1771.

Maggi, F. (2005). Flocculation dynamics of cohesive sediments. PhD Thesis, TU Delft, 136p.

Malcherek, A. (1995). Mathematische modellierung von stromungen und stofftransportprozessen in Astuaren. Dissertation, Institut fur Stromungsmechanik und Elektronisch Rechen im Bauwessen der Universitat Hannover, Bericht Nr. 44/1995 (in German).

Malcherek, A., Markofsky, M., Zielke, W., Peltier, E., Le Normant, C., Teisson, C., Cornelisse, J., Molinaro, P., Corti, S. and Greco, G. (1996). Three dimensional numerical modelling of cohesive sediment transport in estuarine environments. Final report to the EC contract MAS2-CT92-0013.

Manning, A.J. (2001). A study of the effects of turbulence on the properties of flocculated mud. Ph.D. Thesis. Institute of Marine Studies, University of Plymouth, 282p.

Manning, A.J. (2004a). The observed effects of turbulence on estuarine flocculation. In: P. Ciavola and M. B. Collins (Eds), Sediment Transport in European Estuaries, Journal of Coastal Research, SI 41, 90-104.

Manning, A.J. (2004b). The development of new algorithms to parameterise the mass settling flux of flocculated estuarine sediments. HR Wallingford Ltd (UK) Technical Report No. TR 145, 26p.

Manning, A.J. (2004c). Observations of the properties of flocculated cohesive sediment in three western European estuaries. In: P. Ciavola and M. B. Collins (Eds), Sediment Transport in European Estuaries, Journal of Coastal Research, SI 41, 70-81.

Manning, A.J. (2006). LabSFLOC – A laboratory system to determine the spectral characteristics of flocculating cohesive sediments. HR Wallingford Technical Report, TR 156.

Manning, A.J. (2008). The development of algorithms to parameterise the mass settling flux of flocculated estuarine sediments. In: T. Kudusa, H. Yamanishi, J. Spearman and J.Z. Gailani, (eds.), Sediment and Ecohydraulics - Proc. in Marine Science 9, Amsterdam: Elsevier, pp. 193-210, ISBN: 978-0-444-53184-1.

Manning, A.J., and Bass, S.J. (2006). Variability in cohesive sediment settling fluxes: observations under different estuarine tidal conditions. Marine Geology, 235, 177-192.

Manning, A.J. and Dyer, K.R. (1999). A laboratory examination of floc characteristics with regard to turbulent shearing. Marine Geology, 160: 147-170.

Manning, A.J. and Dyer, K.R. (2002a). A comparison of floc properties observed during neap and spring tidal conditions. In: J.C. Winterwerp and C. Kranenburg, (eds.), Fine Sediment Dynamics in the Marine Environment - Proc. in Marine Science 5, Amsterdam: Elsevier, pp. 233-250, ISBN: 0-444-51136-9.

Manning, A.J. and Dyer, K.R. (2002b). The use of optics for the in-situ determination of flocculated mud characteristics. J. Optics A: Pure and Applied Optics, Institute of Physics Publishing, 4, S71-S81.

Manning, A.J., Dyer, K.R. (2007). Mass settling flux of fine sediments in Northern European estuaries: measurements and predictions. Marine Geology 245, 107-122.

Manning, A.J., Langston, W.J. and Jonas, P.J.C. (2009). A Review of Sediment Dynamics in the Severn Estuary: Influence of Flocculation. Marine Pollution Bulletin, 61, 37–51, doi:10.1016/j.marpolbul.2009.12.012.

McCave, I.N. (1975). Vertical flux of particles in the ocean. Deep-Sea Res., 22: 491-502.

McCave, I.N., (1984). Erosion, transport and deposition of fine-grained marine sediments. In: D.A.V., Stow, DJ.W., Piper (Eds), Fine-Grained Sediments: Deep Water Processes and Facies, pp. 35-69.

McCave, I.N. and Gross, T.F., (1991). In-situ measurements of particle settling velocity in the deep sea. Marine Geology, Vol. 99, 403-411.

McDowell, D.N. and O'Connor, B.A. (1977). Hydraulic behaviour of estuaries. MacMillan, London, 292p.

Mehta, A.J. (1986). Characterisation of cohesive sediment properties and transport processes in estuaries. In: A.J. Mehta (ed.), Lecture notes on Coastal and Estuarine Studies, Vol. 14, Estuarine Cohesive Sediment Dynamics, Proceedings of a Workshop on Cohesive Sediment Dynamics with Special Reference to Physical Processes in Estuaries, pp. 290-325.

Mehta, A.J. (1988). Laboratory studies on cohesive sediment deposition and erosion. In: J. Dronkers and W. van Leussen (eds), Physical Processes in Estuaries, Springer-Verlag, New York.

Mehta, A.J., Hayter, E.J., Parker, W.R., Krone, R.B. and Teeter, A.M. (1989). Cohesive sediment transport, 1: Process description. ASCE, J. Hyd. Eng., 115, 1076-1093.

Mehta, A.J. and Lott, J.W. (1987). Sorting of fine sediment during deposition. Proc. Specialty Conf. Advances in Understanding Coastal Sediment Processes. Am. Soc. Civ. Eng., New York, pp. 348-362.

Mehta, A.J. and Partheniades, E. (1975). An investigation of the depositional properties of flocculated fine sediment. Journal of Hydrological Research, 92 (C13), 361-381.

Mietta, F., Maggi, F. and Winterwerp, J.C., (2008). Sensitivity to break-up functions of a population balance equations for cohesive sediment. In: T. Kudusa, H. Yamanishi, J. Spearman and J.Z. Gailani, (eds.), Sediment and Ecohydraulics - Proc. in Marine Science 9, Amsterdam: Elsevier, pp. 275-286, ISBN: 978-0-444-53184-1.

Mikkelsen, O.A., Milligan, T., Hill, P., Moffatt, P. (2004). INSSECT - an instrumented platform for investigating floc properties close to the bottom boundary layer, Limnology and Oceanography: Methods 2, 226-236.

Mikkelsen, O.A., Pejrup, M. (1998). Comparison of flocculated and dispersed suspended sediment in the Dollard estuary. In: Black, K.S., Patterson, D.M., Cramp, A. (Eds), Sedimentary processes in the intertidal zone. Geological Society London, Special Publications, 139, pp. 199-209.

Milligan, T.G. and Hill, P.S. (1998). A laboratory assessment of the relative importance of turbulence, particle composition and concentration in limiting maximal floc size. J. Sea Res., 39, 227-241.

Odd, N.V.M. (1988). Mathematical modelling of mud transport in estuaries. In: Dronkers, J., van Leussen, W. (Eds), Physical Processes of Estuaries, Berlin: Springer, pp. 503-531.

Officer, C.B. (1981). Physical dynamics of estuarine suspended sediment. Mar. Geol., 40: 1-14.

Owen, M.W. (1971). The effects of turbulence on the settling velocity of silt flocs. Proc. 14th Cong. Int. Assoc. Hydraul. Res. (Paris), pp. D4-1 - D4-6.

Owen, M.W. (1976). Determination of the settling velocities of cohesive muds. Hydraulics Research, Wallingford, Report No. IT 161, 8p.

Parker, D.S., Kaufman, W.J. and Jenkins, D. (1972). Floc break-up in turbulent flocculation processes. J. Sanitary Eng. Div., Proc. Am. Soc. Civil Eng., 98 (SA1): 79-97.

Partheniades, E. (1962). A study of erosion and deposition of cohesive sediment in salt water. PhD Thesis, University of California.

Partheniades, E. (1986). The present state of knowledge and needs for future research on cohesive sediment dynamics. Proceedings of the 3rd International Symposium on River Sedimentation, pp. 3-25.

Partheniades, E. (1993). Turbulence, flocculation, and cohesive sediment dynamics. In: A.J. Mehta (ed.), Nearshore and estuarine cohesive sediment transport. AGU, 40-59.

Petersen, O., Vested, H.J., Manning, A.J., Christie, M.C. and Dyer, K.R. (2002). Numerical modelling of mud transport processes in the Tamar Estuary. In: J.C. Winterwerp and C. Kranenburg, (eds.), Fine Sediment Dynamics in the Marine Environment - Proc. in Marine Science 5, Amsterdam: Elsevier, pp. 643-654, ISBN: 0-444-51136-9.
Postma, H. (1961). Transport and accumulation of suspended matter in the Dutch Wadden Sea. Netherland Journal of Sea Research, 1, 148-190.
Postma, H. (1962). Sediment. In: Demerara coastal investigation: Delft, Netherlands. Hydrodynamics Laboratory, 105-164.
Puls, W., Kuehl, H. and Heymann, K. (1988). Settling velocity of mud flocs: results of field measurements in the Elbe and the Weser Estuary. In: J. Dronkers, and W. van Leussen, (eds), Physical Processes in Estuaries. Berlin: Springer-Verlag, pp. 404-424.
Sanford, L.P. and Halka, J.P. (1993). Assessing the paradigm of mutually exclusive erosion and deposition of mud, with examples from upper Chesapeake Bay. Marine Geology, 114, 37-57.
Sheng, P.Y. (1986). Modelling bottom boundary layer and cohesive sediment dynamics in estuarine and coastal waters. In: A.J. Mehta (ed.), Lecture Notes on Coastal and Estuarine Studies, 14: Estuarine Sediment Dynamics, Chapter XVII, pp. 360-400.
Soulsby, R.L., Manning, A.J., Whitehouse, R.J.S. and Spearman, J.R. (2010). Development of a generic physically-based formula for the settling flux of natural estuarine cohesive sediment. Final Report - summary, HR Wallingford company research project DDY0409, 1p.
Spearman, J. (2004). Note on the use of algorithms for modelling mud transport on tidal flats. HR Wallingford Ltd (UK) Technical Report No. TR 144, 14p.
Spinrad, R.W., Bartz, R. and Kitchen, J.C. (1989). In-situ measurements of marine particle settling velocity and size distributions using the remote optical settling tube. J. Geophys. Res., Vol. 94 (C1), 931-938.
Soulsby, R.L. (2000). Methods for predicting suspensions of mud. HR Wallingford Report, TR 104.
Spearman, J. and Manning, A.J. (2008). On the significance of mud transport algorithms for the modelling of intertidal flats. In: T. Kudusa, H. Yamanishi, J. Spearman and J.Z. Gailani, (eds.), Sediment and Ecohydraulics - Proc. in Marine Science 9, Amsterdam: Elsevier, pp. 411-430, ISBN: 978-0-444-53184-1.
Stewart, C. and Thomson, J.A.J. (1997). Vertical distribution of butyltin residues in sediments of British Columbia harbours. Environmental Technology 18, 1195-1202.
Stolzenbach, K.D. and Elimelich, M. (1994). The effect of density on collisions between sinking particles: implications for particle aggregation in the ocean. Journal of Deep Sea Research I, 41 (3):469-483.
Tambo, N. and Hozumi, H. (1979). Water Research, 13:409-419. Physical characteristics of flocs - II. Strength of flocs.
Tambo, N. and Watanabe, Y. (1979). Physical characteristics of flocs-I. The floc density function and aluminium floc. Water Research 13, 409-419.
Uncles, R.J. and Stephens, J.A. (1989). Distributions of suspended sediment at high water in a macrotidal estuary. J. Geophysical Res., 94, 14395-14405.
Uncles, R.J., Stephens, J.A. and Harris, C. (1998). Seasonal variability of subtidal and intertidal sediment distributions in a muddy, macrotidal estuary: the Humber-Ouse, UK. In: Sedimentary Processes in the Intertidal Zone, Black, K.S., Paterson,

D.M. and Cramp, A. (Eds), Geological Society, London, Special Publications, 139, 211-219.

van Leussen, W. (1988). Aggregation of particles, settling velocity of mud flocs: a review. In: Dronkers, J., van Leussen, W. (Eds), Physical Processes of Estuaries, Berlin: Springer, pp. 347-403.

van Leussen, W. (1991). Fine sediment transport under tidal action. Geo-Marine Letters, 11:119-126.

van Leussen, W. (1994). Estuarine macroflocs and their role in fine-grained sediment transport. Ph.D. Thesis, University of Utrecht, The Netherlands, 488p.

van Leussen, W. and Cornelisse, J.M. (1994). The determination of the sizes and settling velocities of estuarine flocs by an underwater video system. Journal of Sea Research 31, (3) 231-241.

Vanoni, V.A. (1975). Sedimentation engineering. ASCE Task Comm., New York, 745p.

Verney, R., Lafite, R., Brun-Cottan, J.C. and Le Hir, P. (2010). Behaviour of a floc population during a tidal cycle:Laboratory experiments and numerical modelling. Continental Shelf Research,doi:10.1016/j.csr.2010.02.005.

Whitehouse, R.J.S., Soulsby, R., Roberts, W. and Mitchener, H.J. (2000). Dynamics of Estuarine Muds. Thomas Telford Publications, London, 232p.

Winterwerp, J.C. (2002). On the flocculation and settling velocity of estuarine mud. Continental Shelf Research, 22, 1339-1360.

Winterwerp, J.C. (2007). On the sedimentation rate of cohesive sediment. In: J.P.-Y. Maa, L.P. Sanford and D.H. Schoellhamer (eds), Coastal and Estuarine Fine Sediment Processes - Proc. in Marine Science 8, Amsterdam: Elsevier, pp. 209-226, ISBN: 0-444-52238-7.

Winterwerp, J. C. (1998). A simple model for turbulence induced flocculation of cohesive sediment. J. Hyd. Eng., 36 (3), 309-326.

Winterwerp, J.C. (1999). On the dynamics of high-concentrated mud suspensions. Ph.D. Thesis, Delft University of Technology, Faculty of Civil Engineering and Geosciences, The Netherlands, 172p.

Winterwerp, J.C., Manning, A.J., Martens, C., de Mulder, T., and Vanlede, J. (2006). A heuristic formula for turbulence-induced flocculation of cohesive sediment. Estuarine, Coastal and Shelf Science, 68, 195-207.

Winterwerp, J.C. and van Kesteren, W.G.M. (2004). Introduction to the physics of cohesive sediment in the marine environment. Developments in Sedimentology, 56, van Loon, T. (Ed.), Amsterdam: Elsevier, 466p.

Wolanski, E., Gibbs, R., Mazda, Y., Mehta, A.J. and King, B. (1992). The role of buoyancy and turbulence in the settling of mud flocs. Journal of Coastal Research, 8, 35-46.

6

Determination of Longshore Sediment Transport and Modelling of Shoreline Change

H. Anıl Arı Güner, Yalçın Yüksel and Esin Özkan Çevik
Yıldız Technical University
Turkey

1. Introduction

1.1 Longshore sediment transport

Littoral transport is the movement of sedimentary material in the littoral zone, that is, the zone close to the shoreline. Littoral transport is classified as cross-shore transport or as alongshore transport. Littoral transport results from the interaction of winds, waves, currents, tides, sediments and other phenomena in the littoral zone. Transport can be described by the product of instantaneous concentration and the instantaneous velocity. Generally the sediment transport through a plane of unit width and height equal to the water depth is denoted by,

$$q_s = \frac{1}{t'} \int_0^{h+\eta} \int_0^{t'} c(z,t)\, u(z,t)\, dt\, dz \tag{1}$$

where q_s is the sediment transport rate (m^3/ms), t' is the integration period (s), h is the local water depth (m), η is the instantaneous water surface elevation (m), c(z,t) is the instantaneous concentration of material, u(z,t) is the instantaneous velocity component (m/s), z is the elevation above the bed level (m), t is the time (s). When waves approach a shoreline at an angle, alongshore sediment transport (also often called littoral transport in some literatures) takes places. Equation (1) for the alongshore sediment transport can therefore be reduced to a much more convenient equation;

$$q_s = \frac{1}{t'} \int_0^{h+\eta} u(z)\, \overline{c(z)}\, dz \tag{2}$$

where the velocity, u, is taken equal to the alongshore velocity, V. This velocity is practically independent of time, because the waves, causing the time dependent velocity component, are assumed to act almost perpendicularly to the coast and thus to the alongshore velocity direction. Because the velocity u(z,t) is almost independent of time $[u(z,t) \rightarrow u(z)]$ the time averaged concentration $\overline{c(z)}$ can be used instead of the instantaneous concentration $[c(z,t)]$. Equation (2) for longshore sediment transport is similar to the sediment transport formulas used for rivers. Principally both types of transport formulas are same, so it is possible to use these transport formulas also for a current alone situation. The methods used for calculating the velocity and the concentration, however, are different in each case. The velocity depends on the generating forces and on the bottom shear stress. The bottom shear stress is influenced by the wave action. The sediment concentration along the coast is much higher than in a river because wave action stirs up a lot of material from the bottom and the current transports it (van der Velden, 1989). Alongshore transport has an average net direction parallel to the shoreline. Sediment, moved by alongshore transport, will generally not return to the same area.

When waves break in the surf zone they release momentum, giving rise to a "radiation stress". The cross-shore component of the radiation stress forces water onshore and causes a set-up of the water level, which rise in the onshore direction above the still water level. The water surface slope that produces balances the cross-shore gradient of the shore-normal component of the radiation stress. For waves incident obliquely on the shoreline is also an alongshore component of the radiation stress, whose gradient gives rise to a "longshore current" within (and just outside) the surf zone which is balanced by friction with bed. This in turn drives sediment longshore as a longshore sediment transport.

There are several expressions to calculate the potential velocity of longshore current (current for an infinitely long, strait beach). One commonly used expression is based on Longuet-Higgins (1970).

$$V_L = 20.7m\sqrt{gH_b}\sin 2\alpha_b \tag{3}$$

where V_L is the longshore current velocity, m is the beach slope, H_b is the breaking wave height, α_b is the wave breaking angle.

Sediment transport due to the various incident waves from the one end of a beach can then be added up to yield a sediment transport rate to the other end ($q_{t(+)}$). Similarly the sediment transport rate can be moved to opposite side ($q_{t(-)}$). The sum of these two is called the gross sediment transport rate and the difference is the net sediment transport rate. This net rate has a direction and the terms updrift and down drift are relative to the direction of the net sediment transport (Kamphuis, 2010). This transport process will drive the form of the beaches.

In the past, many researchers have attempted to determine the longshore transport rate. Both u(z) and c(z) depend on various parameters (wave characteristics, H_0 and T, breaker index, γ, sediment characteristics, d_{50}, angle of wave incidence, θ_0, slope of beach profile, m, bed roughness, k_s). The difference between the various solutions is the way in which they take into account the mentioned parameters above. Furthermore the transport formulas are defined differently. Since most coastal sediments are sandy, formulas have generally been developed for sandy beaches. Some longshore sediment transport formulas are discussed in the present chapter which are mostly used longshore sediment transport calculation.

1.2 Longshore sediment transport formulas

Bijker formula (1971)

Bijker formula can calculate sediment transport and the transport distribution through the surf zone. Bijker (1971) developed a formula to calculate the sediment transport as a function of a given wave field and a given alongshore current irrespective of its origin (wave-induced current or tidal current). Bijker formula is based on that transport equals velocity times concentration. The Bijker transport formula is built up of two components, namely, a bed load transport component and a suspended load transport component. The bed load transport formula was adopted from the Kalinske-Frijlink formula (for bed load transport under river conditions). Bijker divided the Kalinske-Frijlink formula into a stirring parameter and a transport parameter. He introduced the influence of the waves via a modification of the bottom shear stress, τ_c, in the stirring parameter into τ_{cw}. The remaining part, the transport parameter was adapted simply by neglecting the ripple factor, μ. The bottom transport formula was written as;

$$q_b = \frac{5d_{50}V\sqrt{g}}{C}\exp\left[\frac{-0.27\Delta d_{50}\rho g}{\mu\overline{\tau_{cw}}}\right] \quad (4)$$

where q_b is the bed load transport (m^3/sm), d_{50} is the particle diameter, V is the mean current velocity, C is Chezy coefficient (=18log(12h/k_s)), k_s is the bottom roughness, g is the gravitational acceleration, Δ is the relative apparent density of bed material (=(ρ_s-ρ)/ρ), ρ_s is the mass density of bed material, ρ is the mass density of water, μ is the ripple factor (=$(C/C_{90})^{1.5}$), C_{90} is Chezy coefficient based on d_{90}, τ_{cw} is the bed shear stress due to waves and current (time averaged).

$$\overline{\tau_{cw}} = \frac{\rho g V^2}{C^2}\left[1 + 0.5\left(\xi\frac{\overline{u}_o}{V}\right)^2\right] \quad (5)$$

where ξ is Bijker parameter.

$$\xi = C\sqrt{\frac{f_w}{2g}} \quad (6)$$

where f_w is Jonsson's friction factor. Bijker assumed that the bottom transport occurred in a layer with a thickness equal to the bottom roughness, k_s, of the bed transport layer.

$$C_a = \frac{q_b}{6.34\ k_s u_*} \quad (7)$$

Bijker coupled the adapted bed load transport formula to the suspended load transport formula of Einstein. After a lot of algebra the suspended sediment can be shown to be directly proportional to the bed load transport.

$$q_s = 1.83\ Q q_b \quad (8)$$

where

$$Q = \left[I_1 \ln(\frac{33h}{k_s}) + I_2\right] \quad (9)$$

I_1 and I_2 are known as the Einstein integrals;

$$I_1 = R\int_A^1 \left\{\frac{1-\xi}{\xi}\right\}^b d\xi \quad (10)$$

$$I_2 = R\int_A^1 \left\{\frac{1-\xi}{\xi}\right\}^b \ln(\xi)d\xi \tag{11}$$

where ξ is the dimensionless height (z/h), A is the dimensionless roughness (k_s/h),

$$R = \frac{0.216A^{(b-1)}}{(1-A)^b} \tag{12}$$

$$b = \frac{w}{\kappa\, u_*} \tag{13}$$

u_* is the shear stress velocity (influence of waves),

$$u_* : u_{*cw} = \sqrt{\frac{\tau_{cw}}{\rho}} \tag{14}$$

values for the Einstein integral factor can be found in Table 1. When both the bed load and the suspended load are known, the total transport can be calculated. Additionally, since the suspended load transport is directly related to the bed load, a very simple relationship results may be found.

$$q_t = q_b + q_s = q_b\,(1+1.83\,Q) \tag{15}$$

CERC formula (USACE, 1984)

The CERC formula was developed from prototype and model measurements long before much of the alongshore current theory had been developed. Indeed, the CERC formula was developed soon after World War II by the Beach Erosion Board, the predecessor of the U.S. Army Coastal Engineering Research Center. Observations in both prototype and model, made in the decade following World War II, indicated a correlation between the longshore transport rate and the longshore component of energy flux at the outer edge of the surf zone.

Assuming a dense sand with ρ_s=1800 kg/m³ and porosity with n=0.32, the formula is given:

$$q_t = 2.9\times10^6 H_{sb}^{5/2} \sin\alpha_b \left(m^3/yr\right) \tag{16}$$

$$q_t = 330 H_{sb}^{5/2} \sin\alpha_b \left(m^3/hr\right) \tag{17}$$

Bailard and Inman formula (1981, 1984)

Bailard and Inman (1981) derived a formula for both the suspended and bed load transport based on the energetics approach by Bagnold (1966). Bagnold assumed that the work done in transporting the sediment is a fixed portion of the total energy dissipated by the flow. The formula takes into account bed load and suspended load and flow associated with waves and current. Assuming that a weak longshore current prevails, neglecting effects of the slope term on the total transport rate for near-normal incident waves, the local time averaged longshore sediment transport rate is (Bailard, 1984).

ks/h	b =0.00		b =0.20		b =0.40		b =0.60		b =0.80	
	Q	q_s/q_b	Q	q_s/q_b	Q	q_s/q_b	Q	q_s/q_b	Q	q_s/q_b
0.00001	303000	554000	32800	60000	3880	7100	527	964	88	161
0.00002	144000	263000	17900	32700	2430	4440	377	689	71.6	131
0.00005	53600	98000	7980	14600	1300	2370	239	438	53.6	98
0.0001	25300	46300	4320	7900	803	1470	169	310	42.7	78.2
0.0002	11900	21800	2330	4260	496	907	119	218	33.9	62
0.0005	4360	7980	1020	1870	260	475	74.3	136	24.6	45
0.001	2030	3720	545	998	158	290	51.2	93.7	19.1	34.9
0.002	940	1720	289	529	95.6	175	35.1	64.2	14.6	26.7
0.005	336	615	123	226	48.5	88.7	20.8	38.1	10.0	18.3
0.01	153	280	63.9	117	28.6	52.3	13.8	25.2	7.3	13.4
0.02	68.9	126	32.8	60.0	16.5	30.2	8.9	16.3	5.2	9.5
0.05	23.2	42.4	13.1	24.0	7.7	14.1	4.8	8.7	3.1	5.7
0.1	9.8	18.0	6.3	11.5	4.1	7.5	2.8	5.1	2.0	3.6
0.2	3.9	7.1	2.8	5.1	2.0	3.7	1.5	2.8	1.2	2.1
0.5	0.8	1.5	0.7	1.3	0.6	1.1	0.5	0.9	0.4	0.7
1	0	0	0	0	0	0	0	0	0	0

ks/h	b =1.00		b =1.50		b =2.00		b =3.00		b =4.00	
	Q	q_s/q_b	Q	q_s/q_b	Q	q_s/q_b	Q	q_s/q_b	Q	q_s/q_b
0.00001	20.0	36.6	2.33	4.26	0.973	1.78	0.432	0.790	0.276	0.505
0.00002	17.9	32.8	2.31	4.23	0.973	1.78	0.432	0.790	0.276	0.505
0.00005	14.4	28.2	2.28	4.17	0.967	1.77	0.432	0.790	0.276	0.505
0.0001	13.6	24.9	2.25	4.11	0.967	1.77	0.432	0.790	0.276	0.505
0.0002	11.9	21.8	2.21	4.04	0.967	1.77	0.431	0.789	0.275	0.504
0.0005	9.8	17.9	2.13	3.90	0.962	1.76	0.431	0.789	0.275	0.504
0.001	8.4	15.3	2.05	3.76	0.951	1.74	0.430	0.787	0.275	0.504
0.002	7.0	12.8	1.96	3.58	0.940	1.72	0.428	0.784	0.274	0.502
0.005	5.4	9.8	1.78	3.26	0.907	1.66	0.424	0.776	0.273	0.499
0.01	4.3	7.8	1.62	2.96	0.869	1.59	0.417	0.763	0.270	0.494
0.02	3.3	6.0	1.42	2.59	0.809	1.48	0.404	0.740	0.264	0.483
0.05	2.2	4.0	1.10	2.02	0.694	1.27	0.374	0.684	0.249	0.456
0.1	1.5	2.7	0.84	1.53	0.568	1.04	0.339	0.620	0.236	0.432
0.2	0.9	1.6	0.55	1.01	0.414	0.76	0.317	0.580		
0.5	0.3	0.6	0.17	0.32						
1	0	0	0	0						

Table 1. Values of Einstein integral factor Q and values of the ratio suspended load to bed load.

$$q = 0.5\rho f_w u_0^3 \frac{e_b}{(\rho_s - \rho) g \tan\gamma}\left(\frac{\delta_v}{2} + \delta_v^3\right) + 0.5\rho f_w u_0^4 \frac{e_s}{(\rho_s - \rho) g w_s}\left(\delta_v u_3^*\right) \tag{18}$$

where e_b and e_s are the efficiency factors and typically used values in calculations are 0.1, 0.02 respectively, although work has indicated that these are related to the bed shear stress and the particle diameter they are assumed to be constant, $\tan\gamma$ (=0.63) is a dynamic friction factor. In Equation (18) the first term is bed load and the second term is the suspended load component.

$$\delta_v = \frac{V}{u_0} \tag{19}$$

$$u_3^* = \frac{\left\langle \left| U_t' \right|^3 \right\rangle}{u_0} \tag{20}$$

in which U_t' is the instantaneous velocity vector near the bed (wave and current)

Van Rijn formula (1984, 1993)

Van Rijn (1984) presented comprehensive formulas for calculating the bed load and suspended load, and only a short description of the method is given in the following. For the bed load he adapted the approach of Bagnold assuming that sediment particles jumping under the influence of hydrodynamic fluid forces and gravity forces dominate the motion of the bed load particles.

$$q_b = 0.25\gamma\rho_s d_{50} D_*^{-0.3} \sqrt{\frac{\tau'_{b,wc}}{\rho}} \left[\frac{\tau'_{b,wc} - \tau_{b,cr}}{\tau_{b,cr}} \right]^{1.5} \tag{21}$$

in which D* is the dimensionless grain diameter, $\tau'_{b,wc}$ is the effective bed shear stress for waves and current combined /calculated according to Van Rijn's own method (not discussed here).

$$D_* = d_{50} \left[\frac{(s-1)g}{\nu^2} \right]^{1/3} \tag{22}$$

$$\gamma = 1 - \sqrt{\frac{H_s}{h}} \tag{23}$$

in which H_s is the significant wave height.

The depth integrated suspended load transport in the presence of current and waves is defined as the integration of the product of velocity and concentration from the edge of the bed-load layer to the water surface yields

$$q_s = c_a V h \frac{1}{h} \int_a^h \frac{v}{V} \frac{c}{c_a} dz = c_a V h F \tag{24}$$

where c is the concentration distribution, V is the mean longshore current, and

$$F = \frac{V_*}{\kappa V} \left(\frac{a}{h-a} \right)^Z \left(\int_{a/h}^{0.5} \left(\frac{h-z}{z} \right)^{Z'} \ln(z/z_0) d(z/h) + \int_{0.5}^{1} e^{-4Z'(z/h-0.5)} \ln(z/z_0) d(z/h) \right) \tag{25}$$

$$c_a = 0.015 \frac{d_{50}}{a} \frac{T^{1.5}}{D_*^{0.3}} \tag{26}$$

$$Z' = Z + \Psi \tag{27}$$

$$Z = \frac{w}{\beta \kappa V_*} \tag{28}$$

$$\Psi = 2.5\left(\frac{w}{V_*}\right)^{0.8}\left(\frac{c_a}{c_0}\right)^{0.4} \tag{29}$$

$$\beta = 1 + 2\left(\frac{w}{V}\right)^{2} \tag{30}$$

in which Z is the suspension parameter reflecting the ratio of the downward gravity forces and upward fluid forces acting on a suspended sediment particle in a current, ψ is an overall correction factor representing damping and reduction in particle fall speed due to turbulence and β is a coefficient quantifying the influence of the centrifugal forces on suspended particles.

Van Rijn (1984) calculated the concentration distribution c in three separate layers, namely;

1. from the reference level a to the end of a near-bed mixing layer (of thickness δ_s)
2. from the top of the δ_s-layer to half the water depth (h/2)
3. from (h/2) to h

Different exponential or power functions are employed in these regions with empirical expressions depending on the mixing characteristics in each layer.

Kamphuis formula (1991)

Kamphuis (1991) proposed a formula for the longshore sediment transport (LST) rate based on three-dimensional, mobile-bed hydraulic beach model experiments performed with both regular and irregular waves. After a detailed dimensional analysis he expressed the longshore sediment transport rate as a function of wave steepness, beach slope, relative grain size and wave angle. The scale effect for the sediment transport rate was obtained very small. The transport formula is expressed as;

$$Q = 2.27 H_{sb}^{2} T_p^{1.5} m_b^{0.75} d_{50}^{-0.25} \sin^{0.6}(2\alpha_b) \text{ (kg/s)} \tag{31}$$

where H_{sb} is the significant wave height at the breaking point, T_p is the peak wave period, m_b is the beach slope near the breaking point, d_{50} is the median grain size, α is the breaking wave angle. Equation (31) can be converted to

$$Q = 6.4.10^4 H_{sb}^{2} T_p^{1.5} m_b^{0.75} d_{50}^{-0.25} \sin^{0.6}(2\alpha_b) \text{ (m}^3\text{/yr)} \tag{32}$$

or

$$Q = 7.3 H_{sb}^{2} T_p^{1.5} m_b^{0.75} d_{50}^{-0.25} \sin^{0.6}(2\alpha_b) \text{ (m}^3\text{/hr)} \tag{33}$$

The formula is more sensitive to wave period than the earlier expressions. Kamphuis (2010) expressed that these expressions over-predicts transport for gravel beaches because they do not include a critical shear stress (they assume that particles move even for small wave conditions, which is true for sand but not for gravel).

Watanabe et al. formula (1986, 1992)

Watanabe et al. (1986) proposed a power-model type formula for local sediment transport rate under combined action of waves and currents as the summation of the transport rate due to mean currents and that due to the direct action of waves.

$$q(x) = (1-\varepsilon_v) s A_c \left[\frac{(\tau_{b,wc} - \tau_{b,cr}) V(x)}{\rho g}\right] \tag{34}$$

in which ε_v is the porosity, s (=ρ_s/ρ-1) is the immersed specific density of the sediment, A_c is a dimensionless coefficient, $\tau_{b,wc}$ is the maximum value of the periodical bottom friction in a coexistent wave-current field, $\tau_{b,cr}$ is the critical shear stress for the onset of general sand movement and V is the longshore current velocity. A value of 2.0 was adopted for the coefficient A_c. Total immersed-weight rate of the alongshore transport can be computed by the cross-shore integration of the Equation (34). The total volumetric transport rate Q is calculated by the below equation:

$$Q = \frac{\int_{x_0}^{\infty} q(x)d}{(1-\varepsilon_v)(\rho_s-\rho)g} \tag{35}$$

Watanabe (1992) studied the cross-shore distribution of the longshore transport rate. He expressed that the transport rate q(x) becomes maximum between the breaking point and the location of the maximum longshore current velocity.

Damgaard and Soulsby formula (1997)

Damgaard and Soulsby (1997) derived a physics-based formula for bed load alongshore sediment transport. It is intended primarily for use on shingle beaches, although it is also applicable to the bed load component on sandy beaches. The general principles are that the mean bed shear stress are calculated from the gradient of the radiation stress in the surf zone and the oscillatory bed shear stresses recalculated from the wave orbital velocities using a wave friction factor corresponding to the maximum of the values for rough turbulent flow and for sheet flow using an expression derived by Wilson (1989). The resulting formula is:

q_t=maximum of q_{t1} and q_{t2}

$$\begin{aligned} q_{t1} &= \frac{0.19(g\tan\beta)^{1/2}(\sin 2\alpha_b)^{3/2}H_b^{5/2}(1-\hat{\theta}_{cr})}{12(s-1)} \quad for \quad \hat{\theta}_{cr} < 1 \\ &= 0 \quad for \quad \hat{\theta}_{cr} \geq 1 \end{aligned} \tag{36}$$

$$\begin{aligned} q_{t2} &= \frac{0.24 f(\alpha_b) g^{3/8} d_{50}^{1/4} H_b^{19/8}}{12(s-1)T^{1/4}} \quad for \quad \theta_{wr} \geq \theta_{wsf} \\ q_{t2} &= \frac{0.046 f(\alpha_b) g^{2/5} H_b^{13/5}}{12(s-1)^{6/5}(\pi T)^{1/5}} \quad for \quad \theta_{wr} < \theta_{wsf} \end{aligned} \tag{37}$$

Subject to q_{t2}=0 for $\theta_{max} \leq \hat{\theta}_{cr}$

where

$$\hat{\theta}_{cr} = \frac{16.7\theta_{cr}(s-1)d_{50}}{H_b(\sin 2\alpha_b)(\tan\beta)} \tag{38}$$

$$f(\alpha_b) = (0.95 - 0.19\cos 2\alpha_b)(\sin 2\alpha_b) \tag{39}$$

$$\theta_{wr} = \frac{0.15 H_b^{3/4}}{(s-1)g^{1/4}(Td_{50})^{1/2}} \tag{40}$$

$$\theta_{wsf} = \frac{0.0040 H_b^{6/5}}{(s-1)^{7/5} g^{1/5} T^{2/5} d_{50}} \tag{41}$$

θ_w=maximum of θ_{wr} and θ_{wsf}

$$\theta_m = \frac{0.1 H_b (\sin 2\alpha_b)(\tan\beta)}{(s-1)d_{50}} \tag{42}$$

$$\theta_{\max} = \left[\left(\theta_m + \theta_w \sin\alpha_b\right)^2 + (\theta_w \cos\alpha_b)^2\right]^{1/2} \tag{43}$$

$\hat{\theta}_{cr} = \dfrac{0.30}{1+1.2D_*} + 0.055\left[1-\exp\left(-0.02D_*\right)\right]$ threshold Shields parameter

$$D_* = \left[\frac{g(s-1)}{\nu^2}\right]^{1/3} d \tag{44}$$

where H_b is the wave height at breaker line, T is the wave period, d_{50} is the median grain diameter, tanβ is the beach slope.

Bayram et al. formula (2007)

Bayram et al. (2007) presented a new formula for calculating the longshore sediment transport (LST) rate. This formula is derived based on an average concentration and alongshore current velocity for the surf zone, where the current may originate from breaking waves, wind, or tides. The new formula was validated with an extensive data set covering a wide range of conditions. They assumed that the presence of breaking waves cause to mobilize sediment, whereas any type of current (e.g., from breaking waves, wind and tide) can transport the sediment. This new formula consider that the flux of wave energy towards shore is $F(=EC_g)$ and a certain portion ε of this energy is used for the work W, that is, W=εF. This formula is given as follows;

$$Q_{lst} = \frac{\varepsilon}{(\rho_s-\rho)(1-a)gw_s} F\overline{V} \tag{45}$$

where $\overline{V}$ is the mean alongshore current velocity over the surf zone and a is the porosity, ε is the transport coefficient. The transport coefficient depends on wave and sediment conditions. They also indicated that the wave-energy flux is given by,

$$F_b = E_b C_{gb} \cos\theta_b \tag{46}$$

where E_b is the wave energy per unit crest width, C_{gb} is the group velocity and index b denotes incipient breaking. Neglecting energy dissipation seaward of the surf zone (ie., bottom friction), the wave-energy flux can be estimated at any depth since $F=F_b$. The transport coefficient is given by,

$$\varepsilon = (9.0 + 4.0\frac{H_{s,b}}{w_s T_p})10^{-5} \tag{47}$$

where $H_{s,b}$ is the significant wave height at breaking, T_p is the peak wave period, w_s is the particle settling velocity.

2. Modelling of shoreline change

2.1 Introduction

Determination of the shoreline change and coastal morphology accurately in coastal zones is a very important issue for the "Integrated Coastal Zone Management". For the assessment of the coastal morphology and shoreline change, the most important environmental parameters as wind, wave, current, water level, sediment transport phenomena and its supply should be well known. However, these parameters can not be detected accurately and properly due to the complicated behaviour of the nature. Prediction of the beach evolution for beach stabilization or shore protection is one of the main responsibilities in the field of coastal engineering. Despite of the difficulty of the problem, accurate estimations must be performed for the design and maintenance of the shore protection projects.

In the planning of projects located in the nearshore zone, prediction of beach evolution with numerical models has proven to be a powerful technique to assist in the selection of the most appropriate design. Models provide a framework for developing problem formulation and solution statements, for organizing the collection and analysis of data, and, importantly, for efficiently evaluating alternative designs and optimizing the selected design (Hanson and Kraus, 1991).

Various modelling approaches ranging from simple one-dimensional to sophisticate three-dimensional models for prediction and assessment of the evolution of coastal morphology have been developed over the last decades. Recently, these models have become useful in representing the governing physical processes in cross-shore and alongshore dimensions.

In numerical modelling of coastal processes there is a tendency to develop that attempt to simulate the physical processes in fine detail. This results in very specialized computationally extensive models, which are very difficult to handle. Given the complexity of the coastal processes, which can not be represented by deterministic formula and the high uncertainty in wave and sediment transport data (Kamphuis, 1999), numerous runs are required for sensitivity analysis and model calibration and verification which prohibit the practical application of such sophisticated model (Dabees, 2000).

Numerical models of beach evolution expand from simple one-dimensional models to complicated three dimensional models. These models can be classified as:

- Beach profile change models:
  - Descriptive models
  - Equilibrium profile models
  - Process based models
- Analytical shoreline change models
- Numerical shoreline change models
- Numerical three dimensional beach change models:
  - Deterministic three dimensional models
  - Schematic three dimensional models (N-line)

2.2 Numerical shoreline change models

Analytical and numerical models are developed based on the equations of sediment continuity (48) and longshore transport rate (49) which are the governing processes in the long-term response of shoreline evolution.

$$\frac{dy}{dt} = -\frac{1}{d_p}\left\{\frac{dQ}{dx} - q\right\} \tag{48}$$

where y is the shoreline position, x is the longshore coordinate, t is the time, Q is the volume rate of longshore sediment transport, q is the rate of cross-shore sediment transport or sediment sources along the coast, d_p is the profile depth which equals to the closure depth (d_c) plus the beach berm height (d_b).

Analytical solutions

The first one-line theory used for the determination of the shoreline change was introduced by Pelnard-Considere (1956). The analytical one-line theory was used for planform shoreline evolution by employing equilibrium profile. By assuming constant wave direction and a small incident wave angle, the longshore transport rate can be calculated with Equation (49).

$$Q=f(H_b, \alpha_b, \tan\beta, d_{50}) = Q_0 \sin 2\alpha_e = Q_0 \sin 2(\alpha_b - \frac{dy}{dx}) \tag{49}$$

where, Q_0 is the amplitude of longshore transport rate, Q is the volume rate of longshore sediment transport, α_b is the wave breaking angle with respect to the X-axis and α_e is the effective breaking angle.

The effective breaking angle is defined as:

$$\alpha_e = \alpha_b - \alpha_s \tag{50}$$

$\alpha_s = \tan^{-1}(dy/dx)$ is the angle the shoreline makes with the X-axis. Following the assumption that α_s is small such that $\alpha_s = (dy/dx)$. Equation (50) becomes:

$$\alpha_e = \alpha_b - \frac{dy}{dx} \tag{51}$$

Assuming small incident angle, Equation (49) is given by,

$$Q = 2\,Q_0\,(\alpha_b - \frac{dy}{dx}) \tag{52}$$

Differentiating Equation (52) to determine $\partial Q/\partial x$ and substituting in Equation (48) assuming that q=0, the governing equation is simplified in the form of diffusion equation for shoreline evolution below:

$$\frac{dy}{dt} - D\frac{d^2y}{dx^2} = 0 \tag{53}$$

where;

$$\underline{D} = \frac{2Q_0}{d_p} = \frac{Q}{\alpha_b d_p} \tag{54}$$

Many solutions to various initial and boundary conditions for Equation (53) have been presented (Carlslaw and Jaeger, 1959; Crank and Jaeger, 1975). Grijm (1961) derived the analytical solution for the shoreline evolution of horizontal bed response to the sediment discharges from a river. Le Mehaute and Soldate (1977) developed a mathematical model which incorporates different effects of variations of sea level, wave refraction, diffraction, rip currents, coastal structures and beach nourishment for long-term shoreline evolution. Larson et al. (1987) surveyed 25 analytical models for simulating the evolution of a sandy beach. A general solution using Laplace transformation techniques for beach evolution with and without structures was developed. Walton (1994) derived three one-line models in which two of them are linearised shoreline change models and the other one is non-linear model using different approaches. Dean (2002) has extended his previous research (Dean, 1984) on analytical solutions of shoreline change applicable to beach nourishment projects. Several solutions to the diffusion equation for shoreline evolution are presented for various geometries, different initial and boundary conditions.

The analytical models supply simple and economic solutions for the shoreline behaviour for various coastal problems. However analytical models with several constraints limit the practical applications in the field. In Table 2 assumptions and limitations of the analytical solutions are given.

	Assumption	Limitation
1	Constant profile shape	Unable to describe changes in profile shape in time and space.
2	No cross-shore transport q=0	Unable to account for storm induced offshore losses or seasonal variations.
3	Small breaking angle ($\sin 2\alpha_b = 2\alpha_b$)	Valid only for small breaking angles.
4	Small shoreline change angle $\left(\alpha_s = (dy/dx)\right)$	Valid only for straight shorelines. Provides inaccurate results near structures.
5	Constant wave condition in space	Unable to account for changes in wave heights alongshore due to refraction or diffraction. Exclusion of wave refraction provides inaccurate predictions of shoreline change for curved shorelines.
6	Constant wave condition in time	Valid only for one representative wave condition. Unable to provide time dependent simulations using actual wave climate.
7	Simple boundary conditions	Valid only for simple cases where Q is specified at the boundary. Unable to account for multiple coastal structures.

Table 2. Assumptions and limitations of the analytical solution of the one-line theory (Dabees, 2000)

Numerical solutions

In the numerical models of shoreline change, the governing equation (48) is converted to finite difference forms (55)-(57) using a typical staggered grid representation. In Figure 1, shoreline position (y), cross-shore supply (q) and longshore sediment transport rate (Q) are defined for the longshore position (x). i and prime (') denote the cell number and variables in the next time step, respectively.

$$\frac{y_i' - y_i}{\Delta t} = \lambda \left(\frac{\partial y}{\partial t}\right)_i + (1-\lambda)\left(\frac{\partial y}{\partial t}\right)_i' \tag{55}$$

$$\left(\frac{\partial y}{\partial t}\right)_i = -\frac{1}{d_p}\left(\frac{Q_{i+1} - Q_i}{\partial y} - q_i\right) \tag{56}$$

$$\left(\frac{\partial y}{\partial t}\right)_i' = -\frac{1}{d_p}\left(\frac{Q'_{i+1} - Q'_i}{\partial y} - q'_i\right) \tag{57}$$

where Δt is the time step and λ ($0 \le \lambda \le 1$) is a parameter for selected numerical schemes. For the shoreline change models, both of the finite difference equations (explicit or implicit) can be used depending on the selected λ. For explicit schemes, the shoreline can be directly computed using previous information (the solution corresponding to $\lambda = 1$). Implicit schemes are more complicated to program than explicit schemes. The further reading for the implicit and explicit finite difference schemes for shoreline change modelling can be found in Kamphuis (2010).

A considerable number of numerical one-line models have been explained in the literature with different formulations of initial and boundary conditions for shoreline evolutions in response to coastal structures and beach nourishments. There are a lot of numerical models apart from "in-house" numerical models that are all based on one-line theory and all widely known numerical modelling systems used in coastal engineering practice (for example GENESIS (Hanson and Kraus, 1991); UNIBEST (Delft, 1993); SAND94 (Szmytkiewicz et al., 2000); LITPACK (DHI, 2008)).

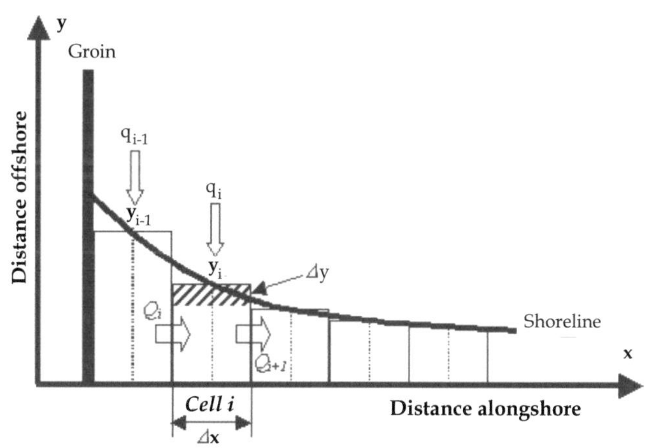

Fig. 1. Sketch of shoreline position and transport rates in finite difference form (Horikowa, 1988)

3. Case study

The site of the field study is Karaburun coastal village located near the south west coast of the Black Sea at 41°21'05"N and 28°41'01"E, which is Northwest of Istanbul (Figure 2). It has a WNW-ESE general orientation and an extension of approximately 4 km. The fishery harbor of the village is at the western end of the 4 km sandy beach. The harbor operations are effected by the sedimentation problem because of considerable rate of westward sediment transport towards the harbor entrance, thus the water depth shallows and the navigation to and from the harbor is prevented (Photo 1).

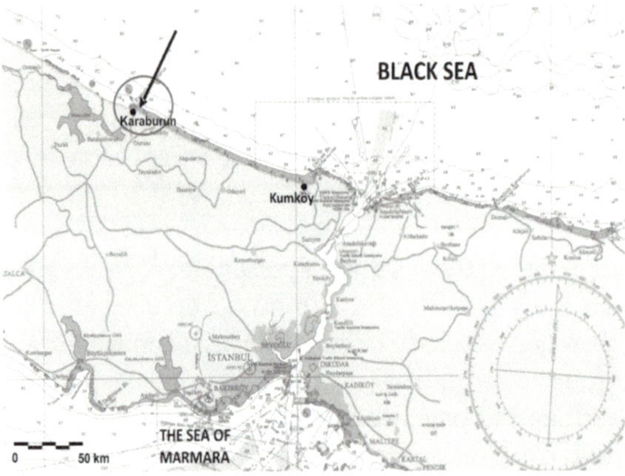

Fig. 2. Location of Karaburun coastal village

Photo 1. Karaburun Fishery Harbor and the nearby beach at east of Harbor (looking towards south east direction) (Ari et al., 2007)

3.1 Prediction of longshore sediment transport rate

In order to determine the LST rates in this region; the long-term observations of shoreline changes, sea bottom topography, sediment properties, wind, wave and current measurement campaigns were performed. The LST rates are obtained from three different methods which are:

- CERC (USACE,1984) formula
- Kamphuis (1991) formula
- Numerical model (LITPACK (DHI, 2008))

CERC (USACE,1984) and Kamphuis (1991) formulas are explained in the first part of this chapter in detail. The LST rates obtained from these models are given in Table 2.

The numerical model used in the current work is LITPACK package which is an integrated modelling system for LITtoral Processes and Coastline Kinetics developed by Danish Hydraulic Institute. LITPACK is used for modelling of non-cohesive sediment transport in waves and currents, littoral drift, coastline evolution and profile development. It is an integration and enhancement of deterministic numerical models: STP, LITDRIFT, LITLINE,

LITREN, LITPROF. For the current study, the LITPACK package (ver. 2008, the newest available for this study) licensed at the Hydraulic and Coastal Engineering Laboratory at Yildiz Technical University has been used. LITPACK uses a sophisticated approach involving wave radiation stress computations for the evaluation of longshore sediment flux. A detailed description of the approach can be found in Fredsøe and Deigaard (1992).

LITDRIFT module was used for the estimation of LST rates of the study field. The module includes important sediment transport mechanisms such as non-linear wave motion, turbulent bottom boundary layer, wave breaking and sediment grading. It is an essentially combination of a 1D wave model, a 1D hydrodynamic model and an intra-wave sediment transport model (STP). The model shoals, refracts and breaks the input wave from the toe of the profile to the shoreline.

The input data for the LITDRIFT module is the wave data from the wave modelling results calibrated with the locally measured wave data, wind data, bottom profiles at different coordinates, initial coastline and sediment characteristics. In reference to the site measurements the median grain diameter (d_{50}) and the sediment spreading ($\sigma_g = \sqrt{d_{84} / d_{16}}$) were chosen as 1.53 mm and 1.36, respectively.

According to the calculations for the research area, the gross and the net longshore sediment transport rates were obtained as shown in Table 2.

	Q_{net} (m^3/year)	Q_{gross} (m^3/year)
CERC (USACE, 1984)	443292	772350
Kamphuis (1991)	76448	186510
Numerical Model (LITDRIFT)	85210	152600

Table 2. The net and gross longshore sediment transport rates predicted by different methods

LST rates are estimated both by modelling and field measurements. However, both of these methods have major limitations and there is no independent argument of the accuracy of either. The limitations and/or the defects of the sediment transport models are criticised by several researchers. For this reason, LST rates predicted by these methods differ from each other by various reasons as shown in Table 2.

CERC formula used to be the most frequently applied expression which yields to the new models that include more coastal parameters nowadays. It is recommended to use CERC formula in storm conditions where the wave heights exceed 4 m (Wang et al, 1998). However, in Karaburun coastal region according to the wave climate studies, the average wave height is 0.9 m. Therefore, CERC equation overestimates LST rates obtained from Kamphuis (1991) and numerical model (LITDRIFT) by factors of 5.8 and 5.2, respectively for the net LST and by factors of 4.1 and 5.1, respectively for the gross LST.

It is recommended to use Kamphuis (1991) formula in low-wave energy conditions with breaker heights of less then 1 m (Wang et al., 1998). The recommended value corresponds to the study area's wave condition. The percentage differences with respect to CERC equation and numerical model (LITDRIFT) are 141.2% and 10.8%, respectively for the net LST, 122.2% and 20.0%, respectively for the gross LST.

The other method to predict the LST rate in the region is the numerical model (LITDRIFT). The percentage differences for the net LST with respect to CERC equation and Kamphuis (1991) formula are 135.5% and 10.8%, respectively, 134.0% and 20.0%, respectively for the gross LST.

3.2 Determination of shoreline change

The direction of the longshore sediment transport was determined as towards Northwest (towards the fishery harbor) from the field studies and measurements. Because of this transport, there becomes a considerable sediment deposition in and nearby the Karaburun fishery harbor.

Determination of the shoreline change evolution was performed with two methods. One of them is "in-house" numerical model and the other one is widely known numerical modelling system called LITLINE which is a module of LITPACK software package. Both of the models are based on one-line theory for shoreline change modelling.

In the first model (Ari et al., 2007), the shoreline change was investigated between the years of 1996 and 2005. Multitemporal and georeferenced IKONOS images (in 2003 and in 2005, multispectral) and IRS1C/D images (in 1996 and in 2000) were used in this study. The Karaburun shoreline was digitized manually by using ERDAS software for each image. The shorelines were superimposed and the changes of them had been measured in 50m interval.

Karaburun shoreline was modeled by using 71 grid-cells, each 50 m long for the distance of 3.55 km. In the first model the shoreline was idealized as shown in Figure 3. Both updrift and downdrift boundaries were assumed as complete barriers. The fishery harbor is located at downdrift boundary and there is a headland at the updrift boundary.

The first simulations were carried out for the period from 1996 to 1997 using a 3-hour time step. The initial shoreline obtained from the image in 1996 was used in the model. In order to calculate shoreline position in 2000, each model run considered the previous computed shoreline change for one year as an initial shore.

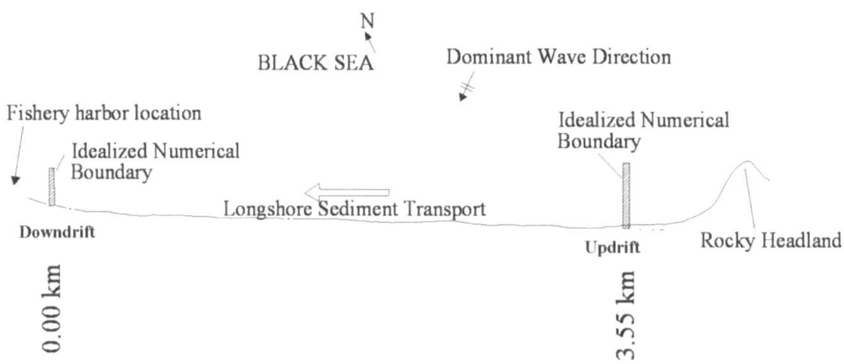

Fig. 3. Idealized numerical shoreline (Ari et al., 2007)

The results of the simulation of the present condition are seen in Figure 4. As shown from the figure, erosion occurs at the updrift side and deposition occurs at the downdrift side. The computed shoreline change was found nearly close the shoreline in 2000 from the image. However, sand fill was made in the eroded area in 2000. So erosion was reduced after 2000. The IKONOS images in 2003 and 2005 were agreed the result because shoreline change reached almost its equilibrium stage (Ari et al., 2007).

The second model used in the study is LITLINE module of LITPACK software package. LITLINE is the module that computes the changes of a shoreline over a period of time using spatially and temporally varying longshore transport. LITLINE calculates the coastline position based on input of the wave climate as a time series. The model is, with minor modifications, based on an one-line theory, in which the cross-shore profile is assumed to remain unchanged during erosion/accretion.

Karaburun shoreline was modeled by using 463 grid-cells, each 10 m long for the distance of 4.63 km (1.08 km longer than the first model). In the second model no idealized boundaries were used. The headland was used in the LITLINE module as it is. The simulation was carried out for the period from 2004 to 2005 using a 1-hour time step.

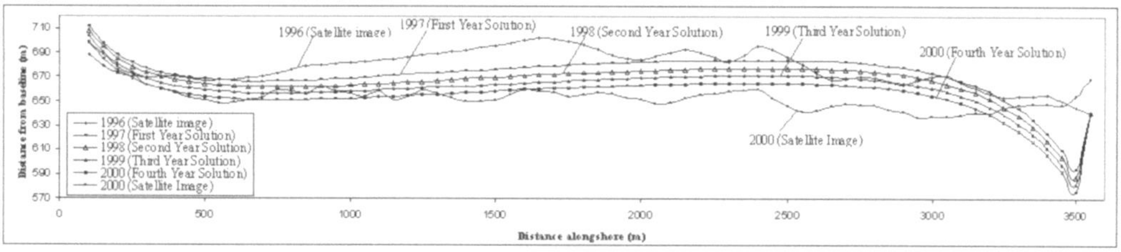

(a) Numerical solution (for four years) versus images in 1996 and 2000

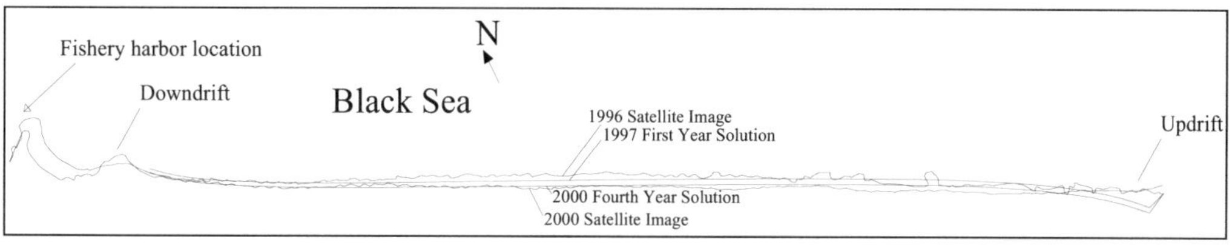

(b) Numerical solution in 1997 and 2000 versus images in 1996 and 2000

Fig. 4. Karaburun shoreline changes (Ari et al., 2007)

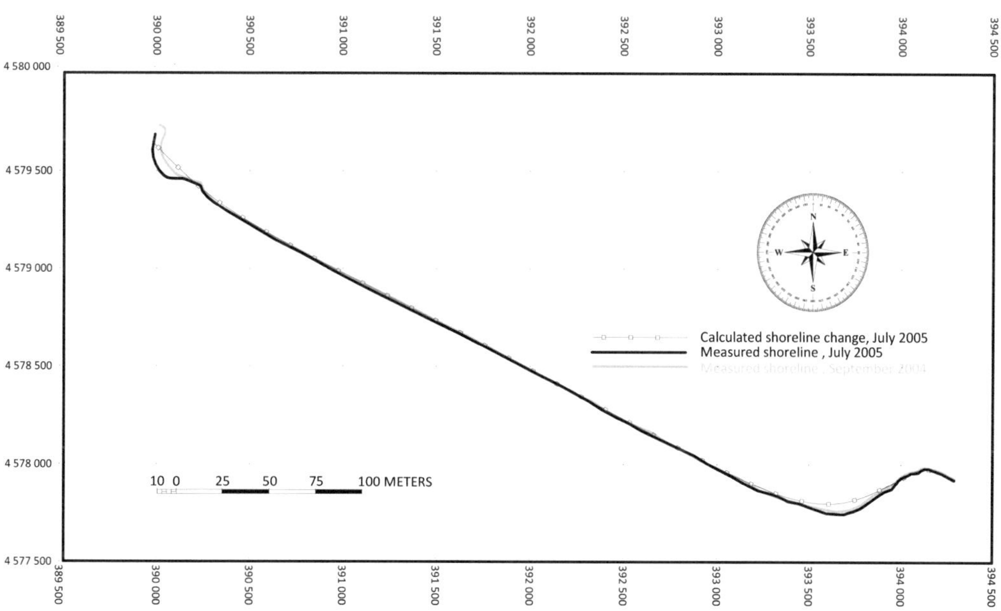

Fig. 5. Comparison of the coastline changes (measured vs. model output)

The study shows that the prediction of long term shoreline variation needs field work, analysis, observations and reliable and continuous wind, wave and current data. If there is

no time history of shoreline or original shoreline measurement in any site, the remote sensing technology helps to monitor the shoreline.
The validation of the LITLINE model was performed with the field measurements initially. Karaburun coastline was measured with RTK-GPS between the years of 2004-2006. The RTK-GPS measurements provide accuracy of 2-3 cm horizontally. The coastline change between the years of 2004-2005 was simulated with LITLINE module and compared with RTK-GPS measurements. The model output showed that the model gave very reasonable results compared with in-situ measurement (Figure 5). Secondly, the coastline change obtained from LITLINE module was compared with coastline extracted from satellite image. IKONOS satellite images used in this study have radiometric solution of 11 bite and resolution of 1 m. The compared results showed again that the simulated shoreline change with LITLINE module is very compatible with the extracted coastline using remote sensing technique (Figure 6).

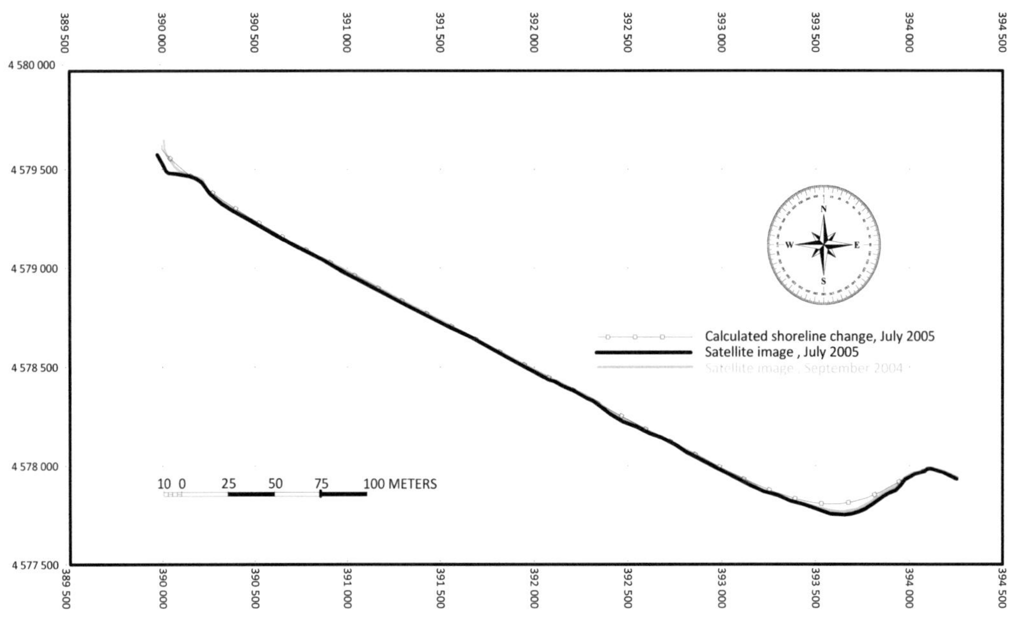

Fig. 6. Comparison of the coastline changes (extracted coastline from satellite image (IKONOS) vs. model output)

4. References

Ari, H.A., Yuksel, Y., Cevik, E.O., Guler, I., Yalciner, A.C. & Bayram, B. (2007). Determination and control of longshore sediment transport: A case study. *Ocean Engineering,* 34 (2), pp 219–233

Bagnold, R.A. (1966). An approach to the sediment transport problem from general physics, *Geological Survey Professional Papers,* 422-1, Washington, USA

Bailard, J.A. (1984). A simplified model for longshore sediment transport, *Proceedings of the 19th Coastal Engineering Conference*, pp.1454-1470

Bailard, J.A. & Inman, D.L., (1981). An energetic bedload model for plane sloping beach: local transport, *Journal of Geophysical Research*, 86(C3), pp. 2035-2043

Bijker, E.W. (1971). Alongshore transport computations. *Journal of Waterways, Harbors and Coastal Engineering Division, ASCE*, Vol. 97, ww4, pp 687-701

Bayram, A., Larson, M., Miller, H.C. & Kraus, N.C. (2001). Cross-shore distribution of longshore sediment transport: comparison between predictive formulas and field measurements, *Coastal Engineering*, 44, pp 79-99

Bayram, A., Larson, M. & Hanson, H. (2007). A new formula for the total longshore sediment transport rate, *Coastal Engineering*, 54, pp 700-710

Carslaw, H. & Jaeger, J. (1959). *Conduction of Heat in Solids*, Clarendon Press, Oxford, UK

Crank., J. & Jaeger, J. (1975). *The Mathematics of Diffusion*, 2nd Edition, Clarendon Press, Oxford, UK

Dabees, M.A. (2000) *Efficient Modelling of Beach Evolution*, Ph.D. Thesis, Queen's University, Kingston, Ontario, Canada

Damgaard, J.S. & Soulsby, R.L. (1997). Alongshore bed-load transport. *Proceedings of the. 25th Int. Conf. Coastal Eng.*, Orlando, 3, pp. 3614-3627. ASCE

Dean, R.G., (1984). *Principles of Beach Nourishment.* In: P.D. Komar (Editor), CRC Handbook of Coastal Processes and Erosion. CRC Press, Boca Raton, Fla, pp. 217-232

Dean, R.G., (2002). *Beach Nourishment: Theory and Practice*, 118. World Scientific

Delft, (1993). *UNIBEST User's Manual. Version 4.00.* Delft Hydraulics Laboratory. The Netherlands

DHI (2008). *LITPACK-An Integrated Modelling System for LITtoral Processes And Coasline Kinetics*, Short Introduction and Tutorial, DHI Water and Environment

Fredsøe, J. & Deigaard, R. (1992). *Mechanics of Coastal Sediment Transport: Advanced Series on Ocean Engineering*, World Science, Singapore

Grijm, W. (1961), Theoretical Forms of Shorelines. *Proceedings of 7th Coastal Engineering Conf.*, ASCE: 197-202

Hanson, H. & Kraus, N.C. (1991). *Genesis: Generelized Model for Simulating Shoreline Change*, CERC Report 89-19, Reprint, U. S. Corps of Eng., Vicksburg

Horikowa, K., (1988). *Nearshore Dynamics and Coastal Processes: Theory, Measurement, and Predictive Models.* University of Tokyo Press, Japan

Kamphuis, J.W. (1991). Alongshore sediment transport of sand, *Journal of Waterway, Port, Coastal and Ocean Engineering*, ASCE, Vol. 117 No.6, pp. 624-641

Kamphuis, J.W. (1999). Marketing Uncertainty, *Proceedings of 5th Int. Conf. On Coastal and Port Eng. In Developing Countries*, Capetown

Kamphuis, J. W. (2010). *Introduction to Coastal Engineering and Management, 2nd Edition,* World Scientific Publishing Co. Pte. Ltd.,ISBN-13 978-981-283-484-3, Singapore

Larson, M., Hanson, H. & Kraus, N.C. (1987), *Analytical Solutions of the One-Line Model of Shoreline Change*, CERC Report 87-15, US Corps of Engineers, Vicksburg

Le Mehaute, B. & Soldate, M. (1977). *Mathematical Modelling of Shoreline Evolution*, Misc. Rep. 77-10, US Corps of Engineers, Vicksburg

Longuet-Higgins, M.S., (1970). Longshore Currents Generated by Obliquely Incident Sea Waves, *Journal of Geophysics Res.*, Vol.75, pp. 6778-6801

Pelnard-Considere, R. (1956). Essai de Theorie de l'Evolition des Formes de Rivage en Plages de Sable et de Galets, *4-ieme Journees de l'Hydraulique*, Les engeries de la mer, Question III, Rapport No. 1

Soulsby, R. (1997). *Dynamics of Marine Sands,* Thomas Telford, London
Szmytkiewicz, M., Biegowski, J., Kaczmarek, L.M., Okrój, T., Ostrowski, R., Pruszak, Z., Rózynsky, G. & Skaja, M. (2000). Coastline Changes Nearby Harbour Structures: Comparative Analysis Of One-Line Models Versus Field Data, *Coastal Engineering,* Vol. 40., 199-139
USACE, (1984). *Shore Protection Manual.* Department of the Army, U.S. Corps of Engineers, Washington, DC 20314
Van der Velden, E.T.J.M. (1989). *Coastal Engineering.* vol.2, Delft TU
Van Rijn, L.C. (1984). Sediment transport: PartI: Bed load transport; Part II: Suspended load transport; Part III: Bed forms and alluvial roughness, *Journal of Hydraulic Division,* Vol.110, No.10, pp. 1431-1456; Vol.110, No.11, pp.1613-1641, Vol.110, No 12, pp.1733-1754
Van Rijn, L.C. (1993). *Principles of sediment transport in rivers, estuaries and coastal seas,* Aqua Publication, The Netherlands, Amsterdam
Walton, T.L., (1994). Shoreline Solution for Tapered Beach Fill. *Journal of Waterway Port and Ocean Engineering, ASCE,* 120(6): 651-655
Wang, P., Kraus, N.C. & Davis, R.A., Jr. (1998). Total Rate of Longshore Sediment Transport in the Surf Zone: Field Measurements and Empirical Predictions, *Journal of Coastal Research,* 14(1), 269-283
Watanabe, A, Maruyama, T., Shimizu, T. & Sakakiyama, T. (1986). Numerical prediction model of three-dimensional beach deformation around a structure, *Coastal Engineering in Japan,* JSCE, Vol.29, pp. 179-194
Watanabe, A. (1992). Total rate and distribution of longshore sand transport, *Proceeding of the 23th Coastal Engineering Conference,* pp.2528-2541
Wilson, K.C. (1989). Friction of wave induced sheet flow. Coastal *Engineering,* 12, 371-379

7

Three- Dimensional Numerical Simulation of Cohesive Sediment Transport in Natural Lakes

Xiaobo Chao and Yafei Jia
The University of Mississippi
USA

1. Introduction

Sediment has been identified as one of the leading nonpoint-source pollutants. Most sediments are transported into surface water bodies from agricultural lands and watersheds through runoff, and greatly affect the surface water quality. Sediment particles will change channel topography and play an important role in affecting water quality physically or chemically by the pollutants, nutrients and pesticides they carried. Shallow lakes in the Southern U.S. are often surrounded by agriculture lands. The suspended sediments in these shallow lakes are normally very fine, and they can be classified as cohesive sediments.

The basic processes involved in cohesive sediment transport, such as flocculation, deposition, erosion, consolidation, etc., have been studied by many scientists. Burban et al. (1990) presented a formula to calculate the settling velocity of flocs in fresh water based on laboratory experiments. Thorn (1981), Ziegler and Nisbet (1995), Li and Mehta (1998) established several empirical formulas for settling velocity of flocs by considering the effects of sediment size, sediment concentration, salinity, turbulence intensity, and bed shear stress. Krone (1962) and Mehta and Partheniades (1975) investigated deposition of cohesive sediment and proposed formulas to estimate deposition rates. Partheniades (1965) proposed a formula to calculate the erosion rate of cohesive sediment. Hamm and Migniot (1994) studied the consolidation of cohesive bed material using an approach of three stage process.

In recent decades, some researchers have studied the cohesive sediment transport in rivers, lakes, and coastal waters using numerical models. Willis and Krishnappan (2004) reviewed a number of numerical models and gave an overview of the knowledge base required for modeling cohesive sediment transport in river flow. Nicholson and O'Connor (1986) developed a 3D cohesive sediment transport model using a splitting method in conjunction with a characteristics technique and a mixed explicit-implicit finite difference approach. Ziegler and Nisbet (1995), Bailey and Hamilton (1997), and Wu and Wang (2004) developed several two-dimensional (2D) depth-averaged models to simulate cohesive sediment transport. Liu (2007) developed a vertical (laterally integrated) two-dimensional model to simulate the cohesive sediment transport in Danshuei River estuary by considering the effects of reservoir construction upstream of the river. Normant (2000), Jin and Ji (2004) proposed 3D layer models to simulate the cohesive sediment transport in estuaries and lakes, respectively.

Sediment erosion, transport and suspension often result from turbulent flows. In inland lakes, however, the water flow is often dominated by wind, and the wave induced by wind generates the most dynamic conditions along shorelines. Jin and Sun (2007) studied the flow circulation, wave dynamics and their impacts on sediment resuspension and vertical mixing in Lake Okeechobee based on field measurements. Their results show that wave action is the dominant factor in sediment re-suspension in that lake. Cozar et al. (2005) presented empirical correlations between total turbidity and wind speed based on the field observations. They also obtained an empirical formula to calculate the suspended sediment concentration using wind speed and water depth. Some researchers have shown that sediment resuspension in shallow lakes is primarily a result of wave action (Luettich et al 1990; Hawley and Lesht 1992). Field observations in Deep Hollow Lake, a shallow oxbow lake in Mississippi showed that during the period from October to December of 1999, the concentration of suspended sediment in the lake varied from 20 to 90 mg/l, even though there was no runoff discharged into the lake during this time (Rebich and Knight 2001). It was apparent that these levels of suspended sediment concentration reflected the influence of wind-induced currents and waves.

This paper presents a 3D model developed based on the mass transport equation for simulating the concentration distribution of cohesive sediment in shallow lakes. It was assumed that low concentration of sediment does not affect the motion of flow, therefore, the decoupled approach was used to calculate the flow field and sediment transport separately. Flow information such as velocity fields, free surface elevations, and eddy viscosity parameters were obtained from a three dimensional hydrodynamic model CCHE3D (Jia et al 2001), and processes of flocculation, deposition, erosion, etc., were considered in the cohesive sediment transport simulation. Because the wind-driven hydrodynamics are considered as important features for sediment transport and re-suspension in lakes, in this flow model, the bottom shear stresses induced by currents and waves were calculated.

This model was first verified by a simple mathematic solution consisting of the movement of a non-conservative tracer in a prismatic channel with uniform flow, and the numerical results agreed well with the analytical solutions. Then it was applied to Deep Hollow Lake, a shallow oxbow lake in Mississippi. Simulated sediment concentrations were compared with available field observations. The trend obtained from the numerical model was generally in good agreement with the observations. It was found that without considering the effect of wind-induced wave, a numerical model can never capture the observed suspended sediment distribution.

This paper presents detailed technical information on cohesive sediment transport in lakes. The background and objectives of this paper are introduced in Section 1. Section 2 describes the general cohesive sediment transport processes in a lake, including flocculation, deposition, erosion, consolidation, etc. Section 3 describes the numerical model for simulating the cohesive sediment transport. Section 4 describes the effect of wind-driven flow and wind-induced wave on sediment transport. Section 5 presents two cases for model verifications. Section 6 provides an application case of the model to a shallow lake in Mississippi. The discussions and conclusions are described in Section 7 and Section 8, respectively.

2. General cohesive sediment transport processes in a lake

The general processes governing cohesive sediment transport in lakes are flocculation, settling, deposition, erosion, consolidation, etc.

2.1 Flocculation

Due to the action of electrostatical forces, the individual fine sediment particles may move toward each other and form the so called flocs/aggregates when they collide. This process is called flocculation, and it is affected by sediment size, sediment concentration, turbulence intensity, temperature, organic matters, etc. (Thorn, 1981; Mehta, 1986; McConnachie, 1991). Flocculation becomes weaker when sediment size increases. In general, high sediment concentration would enhance flocculation as the collision is intensified. The turbulence of flow also affects flocculation. In the range of low shear stress, turbulence increases the chance of collision among sediment particles so that flocculation increases. When the shear stress exceeds the critical level, flocculation may reduce as turbulence increases. Temperature has a significant influence on flocculation. High temperature intensifies flocculation as the thermal motions of ions increase.

2.2 Settling velocity

In natural lakes, the settling velocity of cohesive sediment depends on the floc size, concentration of particles, organic content of the sediment, etc. The flocculation process is dynamic and complex. The median floc diameter can be estimated from the following experiment-based equation (Lick and Lick 1988; Gailani et al 1991):

$$d_m = \left(\frac{\alpha_0}{CG}\right)^{0.5} \tag{1}$$

where d_m = median floc diameter (cm); G = fluid shear stress(dyne/cm^2); C = concentration of sediment (g/cm^3); and α_0 = experimentally determined constant. For fine-grained cohesive sediments in freshwater, $\alpha_0 = 10^{-8} gm^2 / cm^3 / s^2$.

Based on laboratory experiments on flocculated, cohesive sediments in freshwater, Burban et al. (1990) proposed a formula to calculate the settling velocity:

$$w_s = a d_m^b \tag{2}$$

where $a = 9.6 \times 10^{-4} (CG)^{-0.85}$ and $b = -\left[0.8 + 0.5 \log(CG - 7.5 \times 10^{-6})\right]$

Eqs (1) and (2) address the effects of sediment concentration and flow shear stress on flocculation, and they can be used for lake simulation.

2.3 Deposition and erosion of cohesive sediment

When the bottom shear stress is less than the critical shear stress for deposition, sediment can settle down on the bed. Based on Krone (1962) and Mehta and Partheniades (1975), the deposition rate (D_b) can be calculated by:

$$D_b = \begin{cases} 0 & \tau_b > \tau_{cd} \\ w_s C \left(1 - \dfrac{\tau_b}{\tau_{cd}}\right) & \tau_b \le \tau_{cd} \end{cases} \tag{3}$$

where τ_b = bed shear stress (N/m^2); τ_{cd} = critical shear stress for deposition (N/m^2).

When the bottom shear stress is greater than the critical shear stress for erosion, cohesive sediment can be eroded. There are three erosion modes proposed by Mehta (1986). The first erosion mode is the flocs or aggregates that are eroded from bed in particles. The second mode is the mass erosion due to the failure of sediment bed along a plane below the bed surface, and the sediment above the plane is eroded in layers. The third mode is the entrainment of sediment when a fluid mud is formed at the water-sediment interface. Erosion rate is generally expressed by Partheniades (1965)

$$E_b = \begin{cases} 0 & \tau_b < \tau_{ce} \\ M\left(\dfrac{\tau_b}{\tau_{ce}} - 1\right) & \tau_b \geq \tau_{ce} \end{cases} \tag{4}$$

where τ_{ce} = critical shear stress for erosion (N/m^2); M = erodibility coefficient related to the sediment properties, the reported values are in the range of 0.00001 to 0.0004 kg/m^2/s (van Rijn 1989).

Gailani et al.(1991) and Ziegler and Nisbet (1995) studied the cohesive sediment transport in lakes and reservoirs, and found that the erosion rate is a power function of the dimensionless excessive shear stress:

$$E_b = \frac{a_0}{t_d^m}\left(\frac{\tau_b - \tau_{ce}}{\tau_{ce}}\right)^n \tag{5}$$

where a_0= site-specific coefficient; t_d = time after deposition (day); m and n = coefficients, m is about 2, and n is 2~3.

2.4 Consolidation of cohesive sediment

Consolidation is an important process in cohesive sediment transport. It is a compaction process of deposited material under the influence of gravity and water pressure with simultaneous expulsion of pore water and a gain in strength of bed material. Consolidation is generally classified in three stages. The first stage is the settlement of flocs to form a fluid mud, which occurs within several hours of deposition. The second stage is the escaping of pore water, which happens in one or two days. The third stage is the gelling of clay, which may take years to reach the final state.

The degree of consolidation is affected by the sediment size, the mineralogical composition, the thickness of deposit layer, etc. It was observed that the dry bed density varies along the depth below the bed surface, and the relationship between the bed density and consolidation time can be estimated as (Hayter, 1983):

$$\frac{\bar{\rho}_d}{\bar{\rho}_{d\infty}} = 1 - ae^{-pt_{dc}/t_{dc\infty}} \tag{6}$$

where $\bar{\rho}_d$ = mean dry bed density; $\bar{\rho}_{d\infty}$ = final dry bed density; t_{dc}= consolidation time; $t_{dc\infty}$= final consolidation time; parameters a and p are taken as 0.845 and 6.576, respectively.

Lane and Koelzer (1953) proposed a formula to estimate the dry bed density in consolidation processes:

$$\rho_d = \rho_{d0} + \beta \log t_{dc} \tag{7}$$

where ρ_d = dry bed density; ρ_{d0}= dry bed density after one year of consolidation; t_{dc}= consolidation time; and β = coefficient.
Consolidation affects the bed shear strength and also the erosion rate. The critical shear stress for erosion τ_{ce} can be estimated by considering the effect of consolidation (Nicholson and O'Connor, 1986):

$$\tau_{ce} = \tau_{ce0} + k_t(\rho_d - \rho_{d0})^{nt} \tag{8}$$

where τ_{ce0}=critical shear stress at the initial period of bed formation; ρ_d = dry bed density; ρ_{d0} = dry bed density at the initial period of bed formation; k_t and n_t are empirical parameters with the values of 0.00037 and 1.5, respectively.

3. Hydrodynamic and cohesive sediment transport models

3.1 Governing equations

A three-dimensional flow and sediment transport model is needed to study the cohesive sediment transport in surface waters. The CCHE3D hydrodynamic model (Jia et al. 2001, and 2005) and the associated transport model (Chao et al. 2006, 2009) were applied and are presented in this paper. CCHE3D is a three-dimensional model that can be used to simulate unsteady turbulent flows with irregular boundaries and free surfaces. It is a finite element model utilizing a special method based on the collocation approach called the efficient element method. This model has been successfully applied to analyze wind-driven flow, turbulent flow fields in scour holes and around a submerged training structure in a meander bend. The transport model was developed on the finite element platform of CCHE3D flow model.
The governing equations of the three-dimensional unsteady hydrodynamic model can be written as follows:

$$\frac{\partial u_i}{\partial x_i} = 0 \tag{9}$$

$$\frac{\partial u_i}{\partial t} + u_j \frac{\partial u_i}{\partial x_j} = -\frac{1}{\rho}\frac{\partial p}{\partial x_i} + \frac{\partial}{\partial x_j}\left(\nu \frac{\partial u_i}{\partial x_j} - \overline{u_i' u_j'} \right) + f_i \tag{10}$$

where u_i $(i=1,2,3)$ = Reynolds-averaged flow velocities (u, v, w) in Cartesian coordinate system (x, y, z); t = time; ρ = water density; p = pressure; ν = fluid kinematic viscosity; $-\overline{u_i' u_j'}$ =Reynolds stress; and f_i = body force terms.

The free surface elevation (η) is computed using the following equation:

$$\frac{\partial \eta}{\partial t} + u_f \frac{\partial \eta}{\partial x} + v_f \frac{\partial \eta}{\partial y} - w_f = 0 \tag{11}$$

where u_f, v_f and w_f = velocities at the free surface; η = surface water elevation.
The governing equation for cohesive sediment transport is based on the three-dimensional mass transport equation:

$$\frac{\partial C}{\partial t} + \frac{\partial (uC)}{\partial x} + \frac{\partial (vC)}{\partial y} + \frac{\partial (w - w_s)C}{\partial z} = \frac{\partial}{\partial x}(D_x \frac{\partial C}{\partial x}) + \frac{\partial}{\partial y}(D_y \frac{\partial C}{\partial y}) + \frac{\partial}{\partial z}(D_z \frac{\partial C}{\partial z}) \tag{12}$$

in which C= concentration of cohesive sediment; D_x, D_y and D_z = mixing coefficients in x, y and z directions, respectively; w_s = settling velocity.

3.2 Boundary conditions

To solve the 3D cohesive sediment transport equation (12), the boundary conditions at the free surface and bottom are needed. At the free surface, the vertical sediment flux is zero and the following condition is applied:

$$w_s C + D_z \frac{\partial C}{\partial z} = 0 \tag{13}$$

At the bottom, the following condition is applied:

$$w_s C + D_z \frac{\partial C}{\partial z} = D_b - E_b \tag{14}$$

where D_b and E_b = deposition rate and erosion (resuspension) rate at bottom, respectively (kg/m^2/s). They can be calculated using Eqs. (3) and (4).

3.3 Numerical simulation

In the CCHE3D model, the turbulence Reynolds stresses in Eq. (10) are approximated according to Bousinesq's Assumption and they are related to the rate of the strains of flow fields and a coefficient of eddy viscosity. There are several turbulence closure schemes available in the CCHE3D model, including two zero equation models (parabolic eddy viscosity model and mixing length model), a k-ε model and a nonlinear k-ε model. In this model, an upwinding scheme is adopted to eliminate oscillations due to advection, and a convective interpolation function is used for this purpose due to its simplicity for the implicit time marching scheme which was adopted in this model to solve the unsteady equations. The numerical scheme of this approach is the second order. The velocity correction method is applied to solve the dynamic pressure and enforce mass conservation. Provisional velocities are solved first without the pressure term, and the final solution of the velocity is obtained by correcting the provisional velocities with the pressure solution (Jia et al., 2001). The system of the algebraic equations is solved using the Strongly Implicit Procedure (SIP) method (Stone 1968).

The decoupled approach for flow and sediment simulation has been widely adopted in the solution of many real-life engineering problems (Wu, 2008). In general, for the lower flow regime with low sediment concentration (lower than 300 mg/l), the decoupled approach is appropriate (Willis and Krishnappan, 2004; Wu, 2008). In this study, the flow and sediment simulations were decoupled and the flow fields, including water elevation, velocity components, and eddy viscosity parameters were obtained using the CCHE3D free surface hydrodynamic model (Jia et al., 2001 and 2005). After obtaining flow fields, the settling velocity, and boundary conditions of surface and bottom, the distribution of cohesive sediment in the water column were computed by solving the 3D mass transport equation (Eq. 12) numerically. The numerical method used for solving Eq. (12) is consistent with the method employed in the CCHE3D model.

4. The effects of wind-driven flow and wind-induced wave on cohesive sediment transport

In natural lakes, wind stress and wind-induced wave are important driving forces of lake water flows. The water flows coming into and/or out of the lake are normally insignificant to the volume of the water in the lake; in comparison, the current induced by wind are much stronger and have complex distributions, because the wind forcing is acting on the entire water surface. Therefore, the flow circulations within lakes are often influenced by wind. In addition, temperature and salinity stratifications and the Coriolis force, etc., also play important roles. It is observed that in many closed inland lakes, the temperature stratification is minimal, the flow and sediment transport is dominated by wind-induced currents, and the sediment resuspension is primarily affected by wind-induced waves.

4.1 Wind-driven flow

When vertical flow circulation is formed by wind stress, the turbulence stress in the flow balances the momentum of the flow taking from the wind forcing. Because the turbulence stress is related to eddy viscosity coefficient, the distribution of vertical eddy viscosity is required for simulating the wind driven flow. A parabolic eddy viscosity distribution was proposed by Tsanis (1989) based on the assumption of a double logarithmic velocity profile. The vertical eddy viscosity was expressed as:

$$\nu_t = \frac{\lambda u_{*s}}{H}(z + z_b)(z_s + H - z) \tag{15}$$

in which λ= numerical parameter; z_b and z_s = characteristic lengths determined at bottom and surface, respectively; u_{*s} = surface shear velocity; H = water depth. To use this formula, three parameters, λ, z_b and z_s have to be determined. For some real cases with very small water depths, using this formula to calculate eddy viscosity may cause some problems.
Koutitas and O'Connor (1980) proposed two formulas to calculate the eddy viscosity based on a one-equation turbulence model. Their formulas were:

$$\nu_t = \nu_{t,\max}\eta(2-\eta) \qquad (0 \le \eta \le 0.5) \tag{16}$$

$$\nu_t = \nu_{t,\max}(1-\eta)(5\eta - 1) \qquad (0.5 < \eta \le 1) \tag{17}$$

where

$$\nu_{t,\max} = \lambda u_{*s} 0.105 H 0.3^{-0.25} = 0.142 \lambda u_{*s} H \tag{18}$$

in which, λ = numerical parameter; η = non-dimensional elevation, $\eta = z/H$.
In this paper, a new formula was proposed based on experimental measurements conducted in a laboratory flume with steady-state wind driven flow reported by Koutitas and O'Connor (1980). The form of eddy viscosity was borrowed from Koutitas and O'Connor's assumption (Eq. 16 and 17) and expressed as:

$$\nu_t = \nu_{t,\max} f(\eta) \tag{19}$$

Based on measured data, a formula was obtained to express the vertical eddy viscosity:

$$\nu_t = \nu_{t,\max}\eta(-3.24\eta^2 + 2.78\eta + 0.62) \tag{20}$$

Fig. 1 shows the vertical distributions of eddy viscosity obtained from experimental measurements and formulas provided by Tsanis (Eq.15), Koutitas (Eq. 16 and 17), and our model (Eq.20). Since there is no measured value for eddy viscosity at the water surface, the surface eddy viscosity may be estimated using Eq. 20. This formula can be used to calculate the eddy viscosity over the full range of water depth.

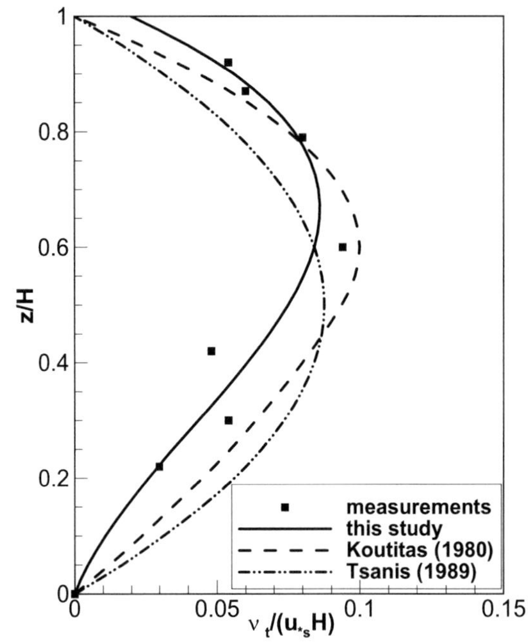

Fig. 1. Comparison of vertical eddy viscosity formulas and experimental data

The wind shear stresses (τ_{wx} and τ_{wy}) at the free surface are expressed by

$$\tau_{wx} = \rho_a C_d U_{wind} \sqrt{U_{wind}^2 + V_{wind}^2} \tag{21}$$

$$\tau_{wy} = \rho_a C_d V_{wind} \sqrt{U_{wind}^2 + V_{wind}^2} \tag{22}$$

where ρ_a = air density, U_{wind} and V_{wind} = wind velocity components at 10 m elevation in x and y directions, respectively. Although the drag coefficient C_d may vary with wind speed (Koutitas and O'Connor 1980; Jin et al. 2000), for simplicity, many researchers assumed the drag coefficient was a constant on the order of 10^{-3} (Huang and Spaulding 1995; Bailey and Hamilton 1997; Rueda and Schladow 2003; Kocyigit and Kocyigit 2004). In this study, C_d was set to 1.0×10^{-3}, and this value is applicable for simulating the wind driven flow in Deep Hollow Lake (Chao et al 2004).

In this paper, the eddy viscosity was calculated using Eq. 20. The wind shear stresses were calculated from wind speeds (Eq.21 and 22) and set as flow boundary condition at the free surface. Using the developed numerical model, the flow fields induced by wind can be simulated. Those flow fields determined the cohesive sediment transport in a water body.

4.2 Bed shear stress

In natural lakes, the interactions of high-frequency surface wind-induced waves with relatively low-frequency currents determine the structures of near-bed shear stresses in water bodies. Both mechanisms of shear stresses are considered in calculating sediment deposition rate and erosion rate. If the wave and current boundary layers are turbulent, the combined wave-current bottom shear stress is a highly nonlinear function which can be estimated based on the Grant-Madsen wave-current model (Grant and Madsen 1979; Glenn and Grant 1987).

Many investigations have shown the shear stresses exerted by circulatory currents are generally much smaller than those due to waves in shallow lakes, and can be neglected (Luettich et al. 1990; Hawley and Lesht 1992; Hamilton and Mitchell 1996). In this paper, the total bottom shear stress is computed as the sum of wave and current shear stresses (van Rijn, 1993; Hawley and Lesht, 1992; Teeter et al., 2001). In this study, the wind-induced current is not very strong, so the bed shear stresses τ_b due to waves and currents can be treated separately, and τ_b in Eq.(3) and (4) is calculated by

$$\tau_b = \tau_w + \tau_c \tag{23}$$

where τ_w and τ_c = bottom shear stresses due to waves and currents, respectively.

The current shear stress τ_c at the bottom is calculated by:

$$\tau_c = \rho u_*^2 \tag{24}$$

in which u_* = shear velocity at bottom, and can be calculated by the log law:

$$\frac{u}{u_*} = \frac{1}{\kappa} ln\left(\frac{z}{z_o}\right) \tag{25}$$

in which κ = von Karman constant; z_0= roughness length which is a function of the bed roughness height k_s and shear velocity u_* :

$$z_o = 0.11\frac{\nu}{u_*} \qquad u_* k_s/\nu \le 5 \tag{26}$$

$$z_o = 0.0333k_s \qquad u_* k_s/\nu \ge 70 \tag{27}$$

$$z_o = 0.11\frac{v}{u_*} + 0.0333k_s \qquad 5 < u_* k_s/\nu < 70 \tag{28}$$

The bottom shear stress generated by wind-induced waves can be calculated by the laminar wave theory (Luettich et al. 1990):

$$\tau_w = \frac{1}{2}\rho f_w U_b^2 \tag{29}$$

in which U_b = maximum wave orbital velocity; and f_w = friction factor. Based on the wave boundary layer (Dyer, 1986), f_w is given by

$$f_w = 2\left(\frac{U_b A_b}{\nu}\right)^{-0.5} \tag{30}$$

in which A_b = maximum wave orbital amplitude. A_b and U_b are given by CERC (1984):

$$A_b = \frac{1}{2\sinh(2\pi d / L)} \tag{31}$$

$$U_b = \frac{\pi H}{T\sinh(2\pi d / L)} \tag{32}$$

So the bottom shear stress generated by wind waves τ_w can be calculated by

$$\tau_w = H\left[\rho\frac{\left(\nu\left(\frac{2\pi}{T}\right)^3\right)^{0.5}}{2\sinh\left(\frac{2\pi d}{L}\right)}\right] \tag{33}$$

where H = wave height(m); d = water depth (m); T = wave period (s); and L = wave length (m). In shallow lakes, the wind-induced wave parameters, such as wave height H, wave period T and wave length L can be estimated using the following empirical formulas (CERC 1984):

$$\frac{gH}{U_w^2} = 0.283\tanh\left[0.53\left(\frac{gd}{U_w^2}\right)^{3/4}\right]\tanh\left[\frac{0.00565\left(\frac{gF}{U_w^2}\right)^{1/2}}{\tanh\left[0.53\left(\frac{gd}{U_w^2}\right)^{3/8}\right]}\right] \tag{34}$$

$$\frac{gT}{U_w} = 7.54\tanh\left[0.833\left(\frac{gd}{U_w^2}\right)^{3/8}\right]\tanh\left[\frac{0.0379\left(\frac{gF}{U_w^2}\right)^{1/2}}{\tanh\left[0.833\left(\frac{gd}{U_w^2}\right)^{3/8}\right]}\right] \tag{35}$$

$$L = \frac{gT^2}{2\pi}\tanh\frac{2\pi d}{L} \tag{36}$$

where g = acceleration of gravity(m/s²); U_w = wind speed at 10 m above the water surface(m/s); and F = fetch length of wind (m).

5. Model verification

5.1 Verification of the model for wind-driven flow

The actual process of wind-shear-driven flow is very complicated: waves are generated by the wind; the main flow near the free surface moves in the wind direction while the flow near the bed may move in the opposite direction to offset the surface flow. In order to check the correctness of the computational models and assure the errors between the observation and prediction are not due to mathematic derivation and numerical coding, verification of the computational models using analytical solutions is necessary. Because it is not possible to have analytical solutions under general conditions, analytical solutions obtained for a simplified problem are often used for model verification.

This test case concerns the simulation of a steady vertical two-dimensional wind-induced flow in a uniform closed rectangular basin of constant water depth with a known, constant vertical eddy viscosity coefficient. Using the no-slip condition at the channel bed, the analytical solution for the horizontal velocity component was given by Koutitas and O'Connor (1980).

In the numerical simulation, the water depth was set as 40 m, and three meshes with different vertical elements set as 6, 11 and 21 points, were used for model simulation. The other parameters were adopted as follows: water density ρ_w=1000 kg/m^3; eddy viscosity v_t =0.03 m^2/s, wind shear stress τ_{wx}=0.1 N/m^2; and gravitational acceleration g=9.817 m/s^2. Fig. 2 shows the comparison of the analytical solution and the numerical simulation results for velocity profiles along the water depth. All of the results using the three meshes are in good agreement with the analytical solution. The surface maximum velocity u_{max} obtained from numerical model using three meshes are 0.0337 m/s, 0.0334 m/s and 0.0334 m/s, respectively, and the result obtained from the analytical solution is 0.03333 m/s. At the depth of $(2/3)H$, the flow changes directions from positive to negative. The local maximum negative velocity $(u=1/3\ u_{max})$ is located at the depth of $(1/3)H$.

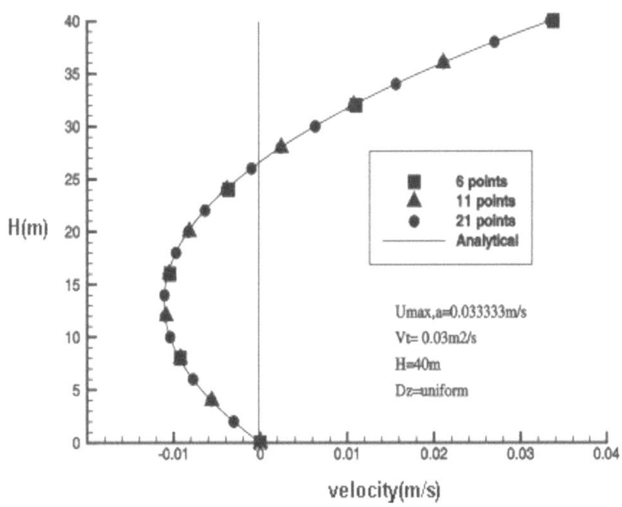

Fig. 2. Comparison of the analytical solution and the numerical simulation of wind-driven flow

5.2 Verification for the mass transport model

The proposed cohesive sediment transport model was tested against an analytical solution for predicting the concentrations of a non-conservative substance in a hypothetical one-dimensional river flow with constant depth and velocity. A continuous source of a non-conservative substance was placed at the upstream end of a straight channel for a finite period of time, τ (Fig. 3). Under the unsteady condition, the concentration of the substance throughout the river can be expressed as:

$$\frac{\partial C_s}{\partial t}+U\frac{\partial C_s}{\partial x}=D_x\frac{\partial^2 C_s}{\partial x^2}-K_d C_s \quad (37)$$

where U = velocity; C_s = concentration of substance; D_x = mixing coefficient; and K_d = decay rate. An analytical solution given by Chapra (1997) is:

$$C_s(x,t)=\frac{C_0}{2}\left[\exp\left(\frac{Ux}{2D_x}(1-\Gamma)\right)erfc\left(\frac{x-Ut\Gamma}{2\sqrt{D_x t}}\right)+\exp\left(\frac{Ux}{2D_x}(1+\Gamma)\right)erfc\left(\frac{x+Ut\Gamma}{2\sqrt{D_x t}}\right)\right] \quad (t<\tau) \quad (38)$$

$$\begin{aligned}C_s(x,t)=\frac{C_0}{2}\Bigg\{&\exp\left(\frac{Ux}{2D_x}(1-\Gamma)\right)\left[erfc\left(\frac{x-Ut\Gamma}{2\sqrt{D_x t}}\right)-erfc\left(\frac{x-U(t-\tau)\Gamma}{2\sqrt{D_x(t-\tau)}}\right)\right]\\ &+\exp\left(\frac{Ux}{2D_x}(1+\Gamma)\right)\left[erfc\left(\frac{x+Ut\Gamma}{2\sqrt{D_x t}}\right)-erfc\left(\frac{x+U(t-\tau)\Gamma}{2\sqrt{D_x(t-\tau)}}\right)\right]\Bigg\}\end{aligned} \quad (t>\tau) \quad (39)$$

where $\Gamma=\sqrt{1+4\eta}$, and $\eta=\frac{K_d D_x}{U^2}$. For the river conditions shown in Fig. 3, with a depth of 10 m, u = 0.03m/s, D_x = 30 m²/s, τ = 6 hr, and the values of K_d = 0, 1.0/day and 2.0/day, respectively. Fig. 4 shows the time series of concentration at the section x = 2000 m obtained by the numerical model and analytical solution. The maximum error is less than 2%.

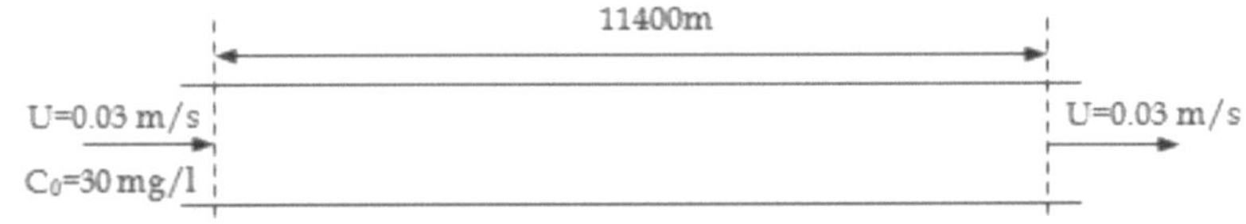

Fig. 3. Test river for verification case

6. Model application to Deep Hollow Lake

6.1 Study area

Fig.5 shows the study area of Deep Hollow Lake. It has a typical morphology of an oxbow lake, with a length of about 1 km and a width of about 100 m. Lake water depth ranges from 0.5 m to 2.6 m, with the greatest depth in the middle. The lake receives runoff from a two square kilometer watershed that is heavily cultivated. This lake is located in Leflore County, Mississippi, and it was one of three natural lakes in the Mississippi River Alluvial Plain monitored under the interagency Mississippi Delta Management System Evaluation Area Project (MDMSEA). This project was part of a national program designed to evaluate the

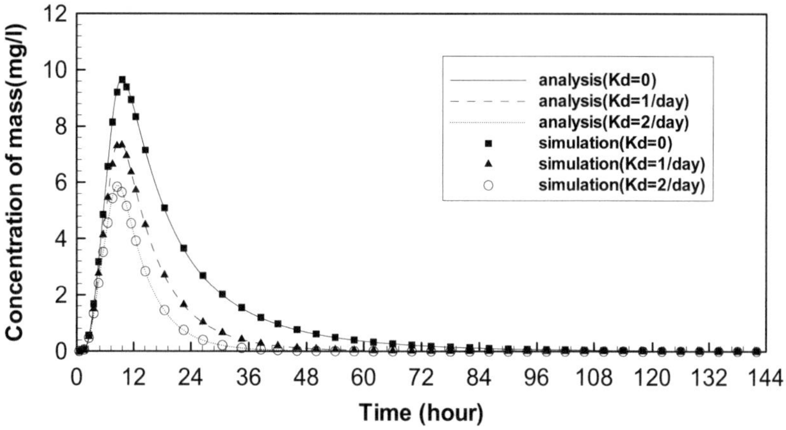

Fig. 4. The time series of concentration at the section x = 2000m obtained from the numerical model and analytical solution

impact of agricultural practices on water quality and to develop best management practices (BMPs) to minimize adverse effects of agricultural activities on water quality of the lakes (Rebich and Knight, 2001; Locke 2004).

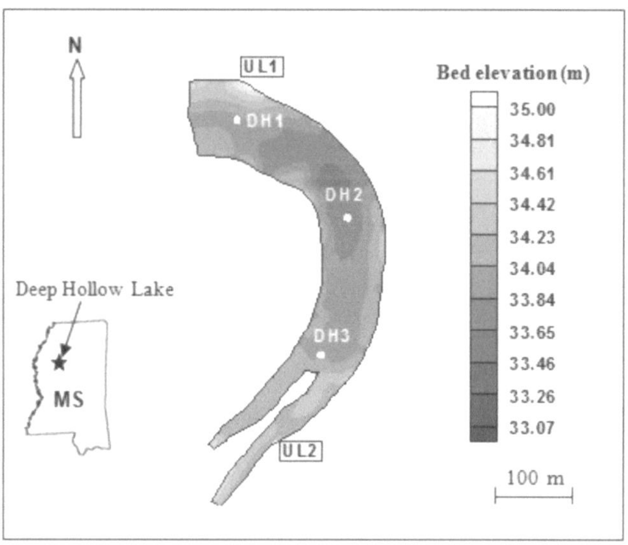

Fig. 5. Study Site - Deep Hollow Lake

Weekly or biweekly samples of suspended sediment, nutrients, chlorophyll, bacteria, and other selected water quality variables were collected at Stations DH1, DH2 and DH3. Two of the major inflows, located at the Stations UL1 and UL2, were monitored for water quality and quantity by the U.S. Geological Survey. The nutrient concentrations in Deep Hollow Lake are mainly dependent on the fertilizer loadings in the surrounding farmland and the quantity of runoff. Field measurements show that the concentrations of nitrate and ammonia in the lake are very low, while the concentration of phosphorus is relatively high in comparison with other areas nationwide. Suspended sediment concentrations are relatively high, exceeding published levels known to adversely impact fish growth and health (Rebich and Knight, 2001).

Based on bathymetric data, the computational domain was discretized into a structured finite element mesh using the CCHE Mesh Generator (Zhang 2002). In the horizontal plane, the irregular computational domain was represented by a 95 × 20 mesh. In the vertical direction, the domain was divided into 8 layers with finer spacing near the bed. This grid system has been successfully applied to simulate the flow and water quality in Deep Hollow Lake (Chao et al 2004 and 2006).

6.2 Model calibration

The cohesive sediment transport model was calibrated using weekly field data and analysis of lake water samples obtained between October and December, 1999. Although there was no water flow discharged into the lake during this period, the measured suspended sediment concentrations varied from 20 to 90 mg/l (Rebich and Knight, 2001). Evidently these concentrations were reflecting the influence of wind-induced currents and waves.

Based on field measurements conducted by the USDA, National Sedimentation Laboratory, the median diameter of sediment particles d_{50} in Deep Hollow Lake is ~ $2.5\times10^{-6} - 3\times10^{-6}$ m, well within the clay size range. The settling velocity of cohesive sediment was estimated using Eqs.(1) and (2) to be on the order of 10^{-4} m/s.

Fig. 6 shows observed wind speeds and directions at the 3-meter level during the calibration period. The flow currents induced by wind during this period were obtained by the CCHE3D hydrodynamic model. The flow patterns were mainly determined by wind stresses. Wind shear on the surface of the lake forced upper layers of the water column to move in the direction of the wind, and produced opposite movement in deeper layers. The flow model was first calibrated using field measurements obtained from Deep Hollow Lake using an Acoustic Doppler Current Profiler (Shields et al. 2003), and then it was applied to simulate the flow fields during the simulation period. The surface velocities were measured using floating tracer particles. Fig. 7 shows the comparison of simulated flow currents with field measurements. It can be seen that the numerical results are generally in good agreement with field measurements. Figs. 8a and 8b show the simulated flow currents during the calibration period.

In order to test the model simulation capability of mass transport under the wind driven condition, the field observation data of a slug-injection dye tracer study at Deep Hollow Lake conducted by National Sedimentation Laboratory was compared with model results (Chao et al. 2007). Fig. 9 shows the dye concentration distribution in the lake at a depth of 1.25m and 24hr after injection. During this period, the prevailing wind direction was from southwest to northeast. The computed dye concentrations are generally in good agreement with the observations.

The bed shear stress generated by wind-induced currents can be obtained using CCHE3D, and the values of the bed shear stress were in the order of 10^{-3} N/m^2. Computed results show that the bed shear stresses generated by wind waves were generally in the ranges of 10^{-2} to 10^{-3} N/m^2, about one order of magnitude greater than those produced by wind-induced currents. Evidently, sediment in Deep Hollow Lake is primarily re-suspended by wind wave action and transported primarily by currents induced by wind.

For the sediment resuspension or deposition to occur, bed shear stress must be greater than the critical erosion shear stress (τ_{ce}), or less than the critical deposition shear stress (τ_{cd}),

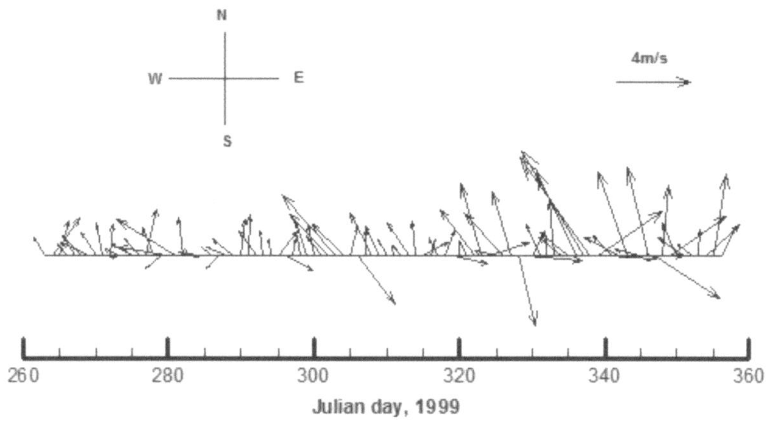

Fig. 6. Observed wind speeds and directions at 3m level

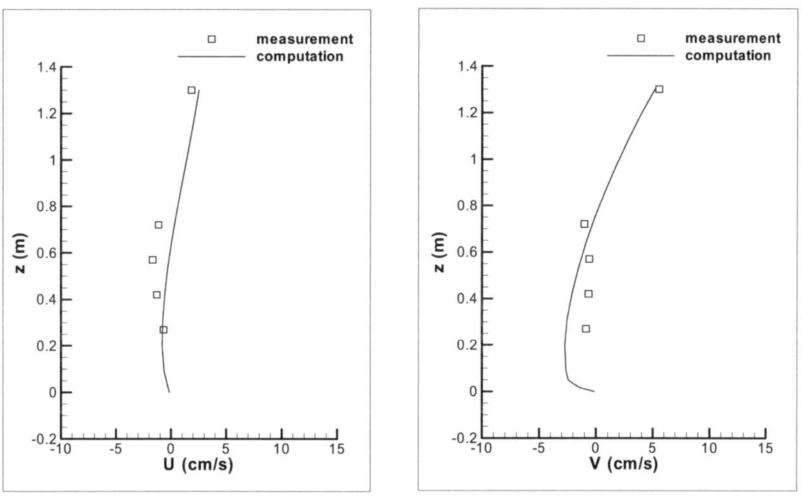

Fig. 7. Observed and simulated velocity at Station DH1 (11:05 am, 11/12/03)

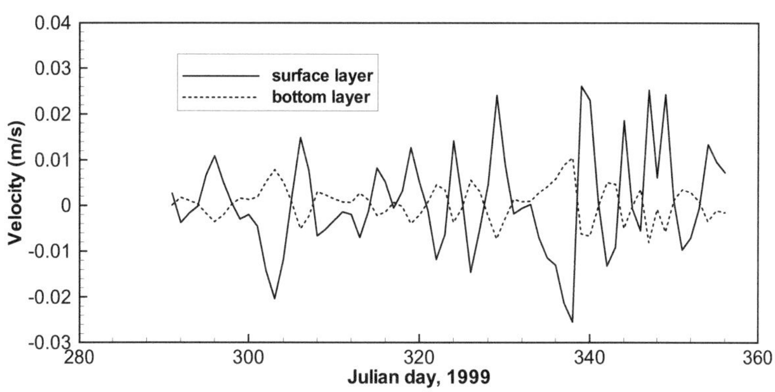

Fig. 8a. Simulated east-west velocity components at Station DH1

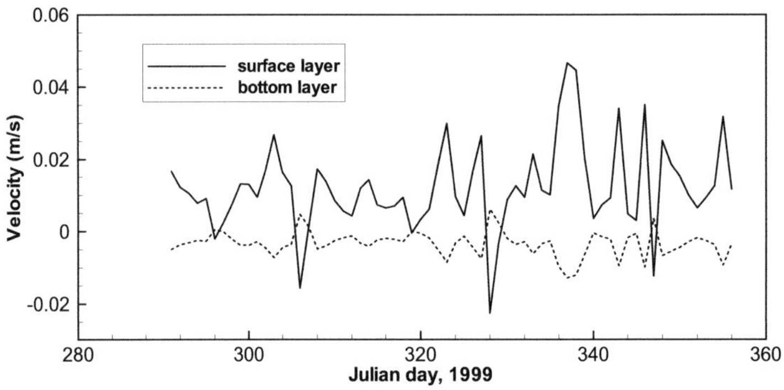

Fig. 8b. Simulated north-south velocity components at Station DH1

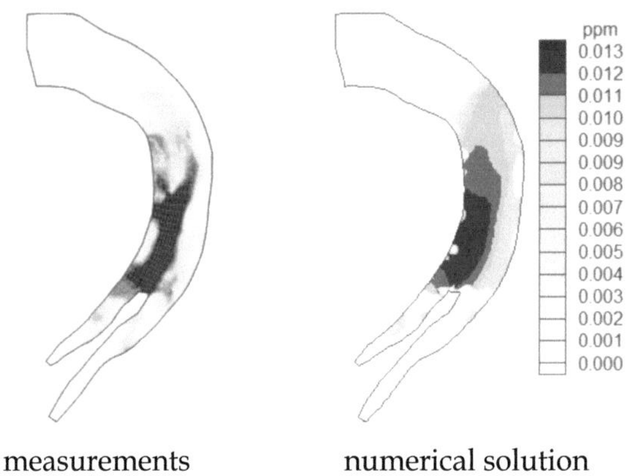

Fig. 9. Dye concentration distribution at Deep Hollow Lake (T=1500 minutes, H=1.25m)

respectively. Previous studies on cohesive sediment transport in lakes, have reported that values for τ_{ce} and τ_{cd} in the range of 0.009 N/m^2 to 0.25 N/m^2 and 0 to 0.18 N/m^2, respectively (Lou et al. 2000; Hamilton and Mitchell 1996; Ziegler and Nisbet 1995; Mehta and Partheniades 1975).

Since these two critical stresses were not measured in Deep Hollow Lake, sensitivity analyses were conducted to determine the values of critical shear stresses for erosion and deposition (τ_{ce}, τ_{cd}). Fig. 10 and Fig.11 show the cohesive sediment concentration at DH2 Station under different values of τ_{ce} and τ_{cd}, respectively. In general, the suspended sediment concentrations increase with the decrease of critical shear stresses for erosion and deposition (τ_{ce}, τ_{cd}). By comparing the numerical results of sediment concentration with field measurements shown on Fig. 10 and Fig. 11, the values of τ_{ce} and τ_{cd} for the model calibration run were set to 0.02 N/m^2 and 0.01 N/m^2, respectively.

Fig. 12 and Fig.13 show the simulated and observed concentrations of cohesive sediment at Stations DH1 and DH2. Some differences between measurement and prediction may arise from the fact that measurements occurred weekly while the time step for the simulation was 300 seconds. However, trends obtained from the numerical model were generally in agreement with the observations. Fig. 12 and Fig.13 also show the concentration of cohesive sediment at Stations DH1 and DH2 without considering the effects of wind-induced waves. Big errors between simulations and measurements were observed without considering the wave effects.

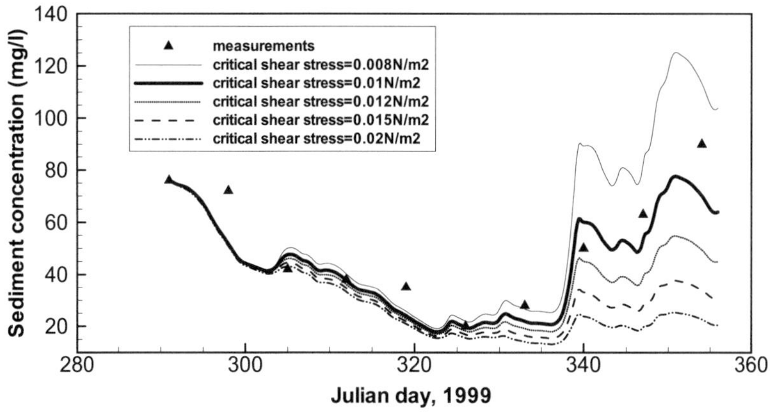

Fig. 10. The time series of sediment concentration with different critical shear stress for deposition at Station DH2

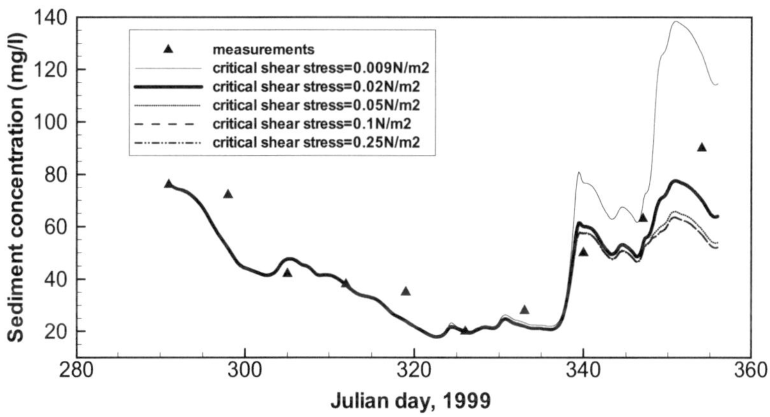

Fig. 11. The time series of sediment concentration with different critical shear stress for erosion at Station DH2

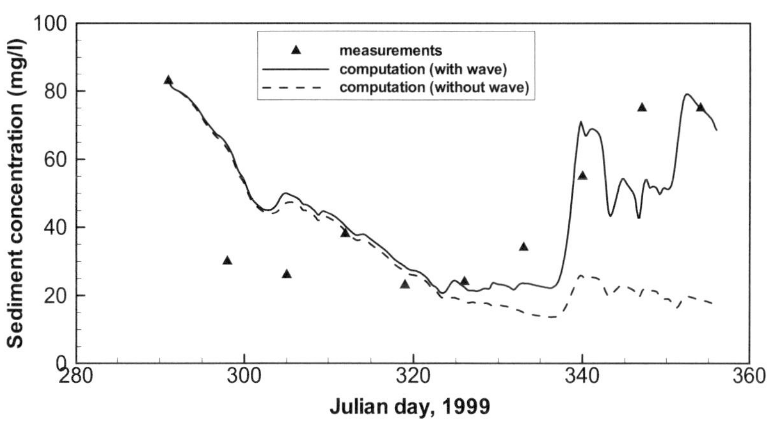

Fig. 12. Suspended sediment concentration at Station DH1

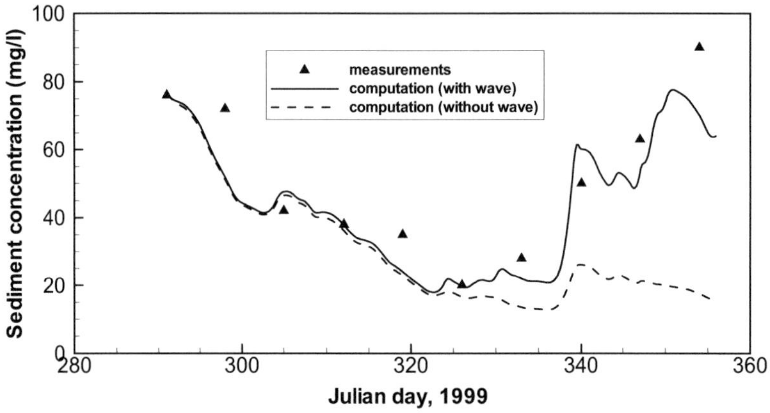

Fig. 13. Suspended sediment concentration at Station DH2

A set of statistics was used to assess the performance of the model (Stow et al 2003). The mean error and root mean square error (RMSE) of the model predictions and observations are summarized in Table 1. Without considering the effects of wind-induced waves, the percentages of mean errors of cohesive sediment concentration at Stations DH1 and DH2 were increased from 8% to 24%, and from 11% to 38%, respectively; Root Mean Square Errors (RMSE) of cohesive sediment concentration for DH1 and DH2 Stations were increased from 15.2 to 29.5 mg/l, and from 11.7 to 29.4 mg/l, respectively.

Fig. 6 shows that from the beginning to Julian day 322, winds were not strong and wind wave effects were negligible. After the Julian day 322, wind speeds (3m above surface) were generally greater than 4 m/s, and the major direction was southeast. Due to the relative long wind fetch and shallow water depth at the upstream end of the lake, the wave - induced bottom shear stress was greater there and caused sediment resuspension in that area. As shown in Figs. 12 and 13, wind-driven waves played a significant role in sediment resuspension and the maximum additional sediment concentration due to wave effects could reach 60 mg/l.

Stations	Mean observations(mg/l)	Wave Effects	Mean prediction (mg/l)	Mean error(mg/l)	Mean error(%)	Root mean square error
DH1	46.30	Yes	50.02	3.72	8	15.2
		No	35.27	-11.03	-24	29.5
DH2	51.4	Yes	45.69	-5.71	-11	11.7
		No	32.09	-19.31	-38	29.4

Table 1. Calibration statistics of cohesive sediment concentration in Deep Hollow Lake

6.3 Model validation

The period from August to October 2000 was chosen for model validation. Just as for the calibration period, there was no water flow discharged into the lake during this period, and wind and wind-induced waves were the major factors for cohesive sediment movement.

Parameter values in the simulation were based on those calibrated values during the period of October to December, 1999. Fig. 14 shows the observed wind speeds and directions at the 3-meter level during the validation period. Fig. 15 shows the simulated and observed concentrations of cohesive sediment at Station DH2. Although there were some differences between measurement and prediction, trends and quantities of concentration of cohesive sediment from the numerical model were generally in agreement with the observations.

During the validation period, the effects of wind-induced waves were not as significant as in the calibration period. It can be seen from Fig.15, due to the effects of wind-induced waves, the sediment concentrations have some differences after the Julian day 270. Fig. 14 shows strong northwest (NW) winds (>5m/s) occurred for a few days during a period from Julian day 270 to 290. However, due to the relative short wind fetch, the wave-induced bottom shear stresses were not large, and sediment resuspension due to the wave effects may not be so significant. In another short period from Julian day 310 to 318, wind speeds were generally greater than 4 m/s, and the major direction was southeast (SE). Due to the relative long wind fetch and shallow water depth at the upstream of the lake, the wave-induced bottom shear stresses were great enough to cause sediment resuspension in that area. Both field measurements and numerical results show sediment concentration in this period increased gradually.

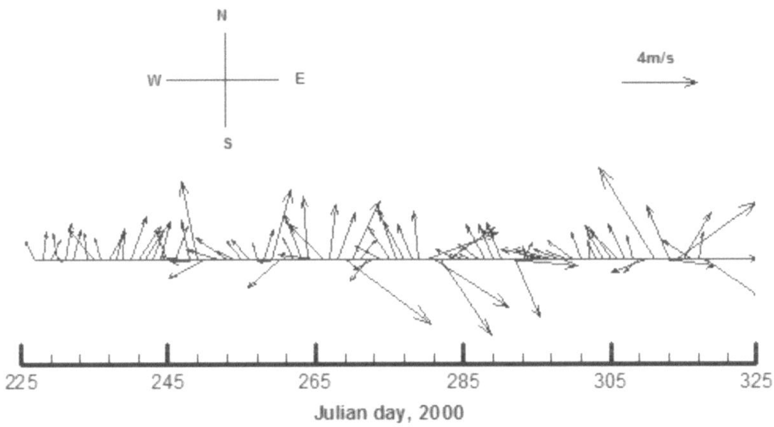

Fig. 14. Observed wind speeds and directions at 3-meter level

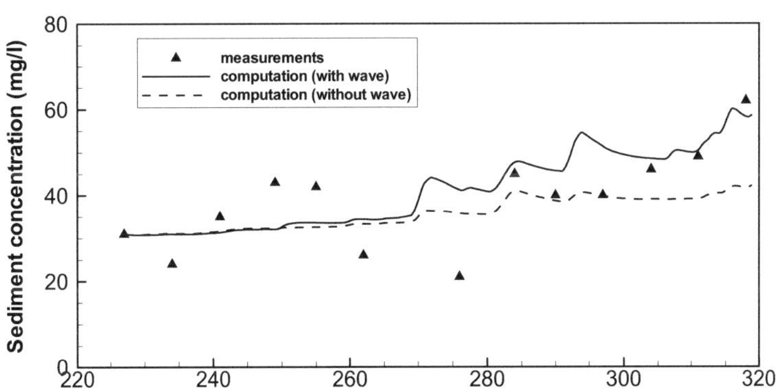

Fig. 15. Suspended sediment concentration at Station DH2

7. Discussion

7.1 Effects of wind induced current and wave on sediment resuspension in Deep Hollow Lake

Field observations of Deep Hollow Lake indicated that the suspended sediment concentration is affected strongly by wind-induced currents and waves. To better understand this dynamic process in Deep Hollow Lake, wind-induced currents and waves were simulated for some hypothetical cases. It was assumed the wind speed at 10-meter level was 10m/s, and the directions were S, SE, SW, N, NE and NW. Table 2 shows the comparisons of simulation results of wind-induced currents and waves for all the cases. On the water surface, the mean velocities of wind-induced currents were from 0.078 to 0.083 m/s, about 0.8% of wind speed. The maximum bottom shear stresses due to current were from 0.02 to 0.024 N/m^2, slightly greater than the critical shear stress (0.02 N/m^2). In most areas of the lake, the wind current-induced bottom shear stresses were less than critical shear stress for erosion. For the S, SE and SW wind, computed results show that the maximum bed shear stresses generated by wind waves were generally about one order of magnitude greater than bed shear stresses due to currents. For the N, NE and NW wind, the maximum bed shear stresses due to waves were about 2 to 4 times greater than those due to currents. So in Deep Hollow Lake, if there is no runoff discharged into the lake, suspended sediment is transported primarily by wind- induced currents and resuspended primarily by wind wave action.

Figure 16 shows regions where the wind wave-induced bottom shear stress exceeds critical shear stress for erosion under the actions of wind from S, SE, SW, N, NE, and NW directions in Deep Hollow Lake. Those regions might be the potential sediment resuspension area in the lake due to the effects of wind- induced waves.

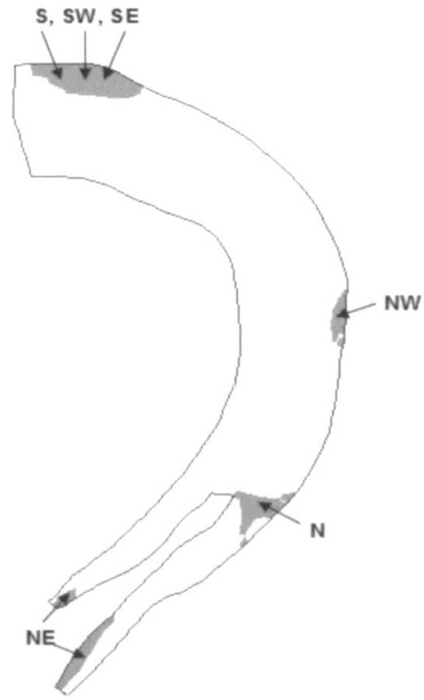

Fig. 16. Regions where the wind wave-induced bottom shear stresses exceed critical shear stress under the actions of wind from S, SE, SW, N, NE, and NW directions

Wind direction	Wind speed *(m/s)*	$\tau_{w,max}$ *(N/m²)*	$\tau_{c,max}$*(N/m²)*	$\overline{u}_s$ *(m/s)*
S	10	0.2	0.024	0.078
SE	10	0.423	0.023	0.083
SW	10	0.213	0.02	0.079
N	10	0.065	0.024	0.080
NE	10	0.095	0.024	0.081
NW	10	0.033	0.021	0.081

Table 2. Simulation results for wind-induced currents and waves for all the cases.

In Table 2, $\tau_{w,max}$ *= maximum wind wave induced bottom shear stress;* $\tau_{c,max}$ *= maximum current induced bottom shear stress; and* $\overline{u}_s$ *= averaged surface velocity.*

8. Conclusions

A three-dimensional numerical model for simulating the concentration of cohesive sediment influenced by currents and wind waves in natural lakes has been presented. The bottom shear stresses induced by wind driven flow and waves were calculated, and the processes of resuspension, deposition, settling, etc., were considered. This model was first verified using analytical solutions of flow and mass transport, and then it was applied to simulate the concentrations of suspended sediment in a closed inland lake, Deep Hollow. Trends and magnitudes of cohesive sediment concentration obtained from the numerical model were generally in good agreement with field observations. Field measurements and model results show that the sediment is resuspended primarily by the actions of wind waves and transported by wind driven flow in the lake. This model provides a useful tool for predicting the cohesive sediment resuspension and transportation in a natural lake, which is an important component for studying the lake water quality and ecology system.

9. Acknowledgments

This work is a result of research sponsored by the USDA-ARS National Sedimentation Laboratory and the University of Mississippi. The suggestions and advice provided by Dr. F. Douglas Shields, Jr and Dr. Charles M. Cooper of the USDA-ARS National Sedimentation Laboratory, and Dr. Sam, S.Y. Wang, Dr. Weiming Wu, Dr. Yan Ding and Ms. Kathy McCombs of NCCHE, The University of Mississippi are highly appreciated.

10. References

Bailey, M.C. and Hamilton D.P. (1997). Wind induced sediment resuspension: a lake-wide model. *Ecological Modeling*, 99, 217-228.

Burban, P.Y., Xu, Y.J., Mcneil, J. and Lick, W.(1990). Settling speeds of flocs in fresh water and seawater. *J.Geophys. Res.*,95(C10), 18213-18220.

CERC (1984). *Shore Protection Manual,* Vol.1, US Army Corp of Engineering, Vicksburg, MS, USA.

Chao, X., Jia, Y. and Shields, D.(2004). Three dimensional numerical simulation of flow and mass transport in a shallow oxbow lake. *Proc., World Water & Environmental Resources Congress 2004,* ASCE, Resyon, Va. (CD-Rom).

Chao, X., Jia, Y., Cooper, C.M., Shields Jr., F.D., and Wang, S.S.Y. (2006). Development and application of a phosphorus model for a shallow oxbow lake. *J. Environmental Engineering*, Vol. 132, No.11, 1498–1507.

Chao. X.B., Jia,Y. , Shields, D. and Wang, S.S.Y.(2007), Numerical modeling of water quality and sediment related processes, *Ecological Modelling*, 201, 385–397.

Chao, X., Jia, Y., and Wang, S.S.Y.(2009), 3D numerical simulation of turbulent buoyant flow and heat transport in a curved open channel, *Journal of Hydraulic Engineering*, Vol. 135, No.7, 554-563.

Cozar, A., Galvez, J.A., Hull, V., Garcia, C.M., and Loiselle, S.A. (2005). Sediment resuspension by wind in a shallow lake of Esteros del Ibera (Argentina): a model based on turbidimetry. *Ecological Modelling*, 186, 63-76.

Chapra, S.C., 1997. *Surface Water-Quality Modeling*, The Mcgraw-Hill Companies, Inc, New York.

Dyer, K.R.(1986). *Coastal and Estuarine sediment dynamics*, John Wiley & Sons Inc.

Gailani, J., Ziegler, C.K. and Lick, W. (1991). Transport of suspended solids in the Lower Fox River. *J. Great Lakes Res.*, 17(4), 479-494.

Glenn, S.M. and Grant, W.D. (1987). A suspended sediment stratification correction for combined waves and current flows. *J.Geophys. Res.*,92(C8), 8244-8264.

Grant, W.D. and Madsen, O.S.(1979). Combined wave and current interaction with a rough bottom. *J.Geophys. Res.*,84(C4), 1797-1808.

Hamilton, D.P. and Mitchell, S.F. (1996). An empirical model for sediment resuspension in shallow lakes. *Hydrobiologia*, 317, 209-220.

Hamm, L. and Migniot, C. (1994). Elements of cohesive sediment deposition, consolidation, and erosion, *Coastal, Esturial, and Harbor Engineer's Reference Book*, M.B.Abbott and W.A.Price (Ed.), Chapman and Hall, London, pp 93-106.

Hawley, N. and Lesht, B.M.(1992). Sediment resuspension in Lake St.Clair. *Limnol. Oceanogr.*, 37(8), 1720-1737.

Huang, W. and Spaulding, M. (1995). 3D model of estuarine circulation and water quality induced by surface discharges. *Journal of Hydraulic Engineering*, 121(4), 300-311, 1995.

Jia, Y., Kitamura, T., and Wang, S.S.Y.(2001). Simulation scour process in a plunge pool with loose material. *J. Hydraul. Eng.*, 127, 219-229.

Jia, Y., Scott, S., Xu, Y., Huang, S. and Wang, S.S.Y.(2005). Three-Dimensional numerical simulation and analysis of flows around a submerged weir in a channel bendway. *J. Hydraulic Engineering*, 131(8), 682–693.

Jin, K. R.,Hamrick, J. H., and Tisdale, T. (2000). Application of three dimensional hydrodynamic model for Lake Okeechobee. *J. Hydraulic Engineering*, 126(10), 758–771.

Jin, K.R., and Ji, Z.G. (2004). "Case study: modeling of sediment transport and wind-wave impact in Lake Okeechobee." J. Hydraulic Engineering, 130(11), 1055-1067.

Jin, K.R., and Sun D. (2007). Sediment resuspension and hydrodynamics in Lake Okeechobee during the late summer. *J. Engineering Mechanics*, 133(8), 899-910.

Hayter, E.J. (1983), *Prediction of Cohesive Sediment Movement in Estuarial Waters*, Ph.D. Dissertation, University of Florida, Gainesville, Florida, USA.

Kocyigit, M.B., and Kocyigit, O. (2004). Numerical study of wind-induced currents in enclosed homogeneous water bodies. *Turkish J. Engineering & Environmental Science*, 28, 207- 221.

Koutitas, C., and O'Connor, B., (1980). Modeling three-dimensional wind-induced flows. ASCE, *Journal of Hydraulic Division*, 106 (11), 1843-1865.

Krone, R.B. 1962. *Flume Studies on the Transport of Sediment in Estuarine Shoaling Processes*, Hydraulic Engineering Laboratory, University of California, Berkeley.

Lane, E.W. and Koelzer, V.A. (1953). Density of sediments deposited in reservoirs, *Report No.9, A study of Methods Used in Measurement and Analysis of Sediment Loads in Streams*, Engineering District, St.Paul, MN,USA.

Li, Y. and Mehta, A.J.(1998). Assessment of hindered settling of fluid mudlike suspensions. *J. Hydraulic Engineering*, 124(2), 176-178.

Lick, W. and Lick, J. (1988). Aggregation and disaggregation of fine-grained lake sediments. *J. Great Laks Res.*, 14(4), 514-523.

Liu, W.C. (2007). Modelling the effects of reservoir construction on tidal hydrodynamics and suspended sediment distribution in Danshuei River estuary, *Environmental Modelling & Software*, 22(11), 1588-1600.

Locke, M. A. (2004). Mississippi Delta Management Systems Evaluation Area: Overview of water quality issues on a watershed scale. M.T. Nett, M.A.Locke, and D.A.Peninngton (Ed.), *Water Quality Assessment in the Mississippi Delta: Regional Solutions, National Scope.* ACS Symposium Series 877, American Chemical Society, Washington, D. C.

Lou, J., Schwab, D.J., Beletsky, D. and Hawley, N. (2000). A model of sediment resuspension and transport dynamics in southern Lake Michigan. *Journal of Geophysical Research*, 105, (C3), 6591-6610.

Luettich, R.A., Harleman, D.R.F. and Somlyody, L. (1990). Dynamic behavior of suspended sediment concentrations in a shallow lake perturbed by episodic wind events. *Limnol. Oceanogr.*, 35(5), 1050-1067.

McConnachie, G. L. (1991). Turbulence intensity of mixing in relation to flocculation, *J. Environ. Eng.*, 117(6), 731-750.

Mehta, A.J. and Partheniades, E.(1975). An investigation of the depositional properties of flocculated fine sediment. *J. Hydraulic Research*, 13(4), 361-381.

Mehta, A.J. (1986). Characterization of cohesive sediment properties and transport processes in estuaries, *Estuarine Cohesive Sediment Dynamics*, A.J.Mehta (Ed.), Springer-Verlag, 290-325.

Nicholson, J. and O'connor B.A. (1986). Cohesive sediment transport model. *J. Hydraulic Engineering*, 112(7), 621-640.

Normant, C.L. (2000). Three-dimensional modeling of cohesive sediment transport in the Loire estuary. *Hydrol. Process.* 14, 2231-2243.

Partheniades, E.(1965). Erosion and deposition of cohesive soils. *J. Hydraulic Division*, ASCE, 91(HY1).

Rebich, R.A and Knight, S. S. (2001). The Mississippi Delta Management Systems Evaluation Area Project, 1995-99. *Mississippi Agriculture and Forestry Experiment Station, Information Bulletin 377*, Division of Agriculture, Forestry and Veterinary Medicine, Mississippi State University.

Rueda, F. J. and Schladow, S. G. (2003). Dynamics of large polymictic lake. II: Numerical Simulations. *J. Hydraulic Engineering*, 129(2), 92-101.

Shields, F. D. Jr., S. S. Knight, C. M. Cooper, and S. Testa. (2003). Use of acoustic Doppler current profilers to describe velocity distributions at the reach scale. *Journal of the American Water Resources Association*, 39, 1397–1408.

Stone, H. L. (1968). Iterative solution of implicit approximation of multidimensional partial differential equations. *SIAM (Soc. Ind. Appl.Math.) J. Numer. Anal.*, 5, 530–558.

Stow, C. A., Roessler, C., Borsuk, M. E., Bowen, J. D., and Reckhow, K. H. (2003). Comparison of Estuarine Water Quality Models for Total Maximum Daily Load Development in Neuse River Estuary. *J. Water Resour. Plan. Manage.*, ASCE, 129(4), 307-314.

Teeter A.M. et al. (2001). Hydrodynamic and sediment transport modeling with emphasis on shallow-water, vegetated area (lakes, reservoirs, estuaries and lagoons). *Hydrobiologia*, 444, 1-23.

Thomann, R.V. and Mueller J.A. (1988). *Principles of Surface Water Quality Modeling and Control*, Harper&Row Publication, New York.

Thorn,M.F.C.(1981). Physical processes of siltation in tidal channels. *Proc. Hydraulic Modelling Applied to Maritime Engineering Problems*, ICE, London, 47-55.

Tsanis, I.K. (1989). Simulation of wind-induced water currents. *J. Hydraulic Engineering*, 115(8), 1113-1134.

Van Rijn, L. C. (1989). *Handbook Sediment Transport by Currents and Waves*, Report H461, Delft Hydraulics.

Van Rijn, L.C. (1993). *Principles of Sediment Transport in River, Estuaries and Coastal Seas*, AQUA Publication, Amsterdam, Netherlands.

Willis, D. and Krishnappan, B.J.(2004). Numerical modelling of cohesive sediment transport in rivers. *Can. J. Civil Engineering*, 31, 749-758.

Wu, W. (2008). *Computational River Dynamics*, Taylor & Francis Group, London, UK.

Wu, W. and and Wang, S.S.Y.(2004). Depth-averaged 2-D calculation of tidal flow, salinity and cohesive sediment transport in estuaries. *Int. J. Sed. Res.*, 19(3), 172-190.

Ziegler, C.K. and Nisbet, B.S.(1995). Long-term simulation of fine-grained sediment transport in large reservoir. *J. Hydraulic Engineering*, 121(11), 773-781.

Zhang, Y.X.(2002). *CCHE Mesh Generator User Manual*, Technical Report, The University of Mississippi.

8

Sediment Transport Dynamics in River Networks Induced by Water Diversion

Xu-Ming Wang, Jie Huo, Rui Hao and Shi-Feng Cao
NingXia University
P.R.China

1. Introduction

A river network is a typical completely open system formed by interconnected river channels. The interactions between the imports of a river (water and sediment) and the channel will cause the change of the channel patterns. This is actually a feedback between the erosion and the sedimentation states of the channel via an adjustment of the sediment-carrying capability of the stream. This feedback mechanism indicates that river network is a self-organized system (Rodriguez-Iturbe, 1997), and some dynamical laws lead its evolution and some statistic laws dominate its steady state (Leopold, 1953; Leopold & Maddock, 1953; Dodds & Rothman, 2000). So, river networks have attracted, in decades past, a good much attention of physicists and geophysicists (Banavar, et al., 1997; Manna & Subramanian, 1996; Manna, 1998; Sinclair & Ball, 1996; Kramer & Marder, 1992; Takayasu & Inaoka, 1996; Maritan, et al., 1996; Caldarelli, et al., 1997; Giacometti, 2000; Somfai & Sander, 1997; Rinaldo, et al., 1996, and so on). They focused mainly on the distributions of river parameters or the scaling relations between them, as well as the evolutionary mechanism, that is, what creates the distributions and scaling relations?

The pioneering field investigation executed by Leopold revealed that the slope, width and depth of a channel respectively depend on the discharge in power functions (Leopold & Maddock, 1953). From then, some other scaling relations, in power functions, between river parameters were found (Hack, 1957; Flint, 1974). The theoretical studies for the purpose of getting deeper understandings of the reasons why the nature has selected these laws were conducted by the dynamic modeling based on the local erosion rules (for instance, Banavar, et al. ,1997) or the other considerations, such as erosion process based on the minimum energy dissipation (Sun, 1994; Dhar, 2006), evolution of a quasi-random spanning tree (Manna, 1996), statistical physics method based on the self-similarity theory (Banavar, et al. ,1997), and so on.

Our previous modeling studies were focused on another process, that is, sediment transport in river networks (Wang, et al. 2008; Hao, et al., 2008; Huo, et al., 2009). The core spirit of the models embodies the feedback mechanism between erosion and sedimentation via the adjustment of sediment-carrying capability (SCC) of runoff. A steady state shows scaling law that the quantity of erosion or sedimentation (QES) distributes exponentially along the channel in the downriver direction. The response of a river to the abrupt change of the input shows self-organized and self-adaptive behaviors. The former is represented by opposite variation of the SCC to the QES as the discharge changes, which shows that the response of the river trends to depress the increase of erosion as water flow increases and that of sedimentation as water flow decreases, the latter is denoted by the deformation of the channel forced by sediment erosion or sedimentation. Sediment concentration is a key factor that influences the sediment transport dynamics in higher-water seasons for its higher value. A higher sediment concentration implies that the stream will inject more sediment into its downriver neighbouring segment of the channel. It follows that the SCC of the stream flowing in this neighbouring segment will be compelled to improve. In other words, the SCC of a channel will be driven by higher sediment concentration of injecting stream (Huo, et al., 2009). The basic natural process associated with sediment transport in a river network is streams confluence occurs from the top rank segments to the bottom rank ones. The change of water flux caused by the confluence will lead to the deformation of the channel due to erosion or sedimentation. In contrast to the confluence, water diversion is often related to the hydraulic engineering, and therefore of practical significance. It may be similar to confluence that diversion will cause the change of the discharge, and then result in the variation of the QES. In this article, a sediment transport dynamical model, for confluence and/or diversion, is proposed to investigate the influence of water diversion on the mainstream. The model is based on the mechanisms that the SCC of a stream is modified by the QES and driven by the sediment concentration of the injecting streams. As far as these two adjustment mechanisms of SCC are concerned, the former is positive and the latter is passive. The numerical results of this model can simulate the dynamic behavior in real river networks: water diversion can cause sedimentation to increase (for instance, Zhang & Liang, 1995). The transient dynamics shows some interesting characteristics. (Huo, et al., 2009).

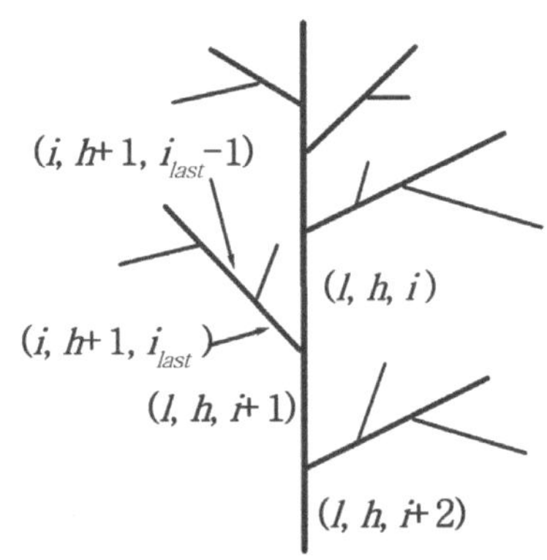

Fig. 1. Schematic drawing of ranks of a river and the segments

In our previous works, we introduced a new ordering scheme of river network (Wang, et al. 2008; Hao, et al., 2008; Huo, et al., 2009). As shown by Fig.1, the rank of a stream can be defined as the times of the confluence while it drains to the sea. If one goes against a stream, say rank h, at each of nodes where the confluences occur, the laterally injecting stream belongs to the branches of rank h+1, but the non-laterally injecting stream is actually one segment of this stream. So the rank of a stream can be uniquely determined by the times of its laterally injecting confluences as it drains to the sea. And the segments of a stream can be numbered with the times of its non-laterally injecting confluences as it drains to the sea. The first segment is at the upper end, which is actually taken as the source. And then the segment number increases one by one as branches of rank i+1 inject into the stream in turn. For convenience, we denote the stream flowing on a channel segment by the segment itself.

So one segment can be denoted by three characters, for instance, (l,h,i), they represent the 'i th' segment of the 'lth' channel of rank 'h', respectively.

2. The model

As discussed above, a confluence of streams is a fundamental process in natural river networks, while a diversion is often related to the hydraulic engineering. Both of the confluence and diversion will cause change of the discharge, and then result in the variation of the QES. Let's firstly consider the confluence of streams on the segments and diversion on some one segment in a time interval, $t \rightarrow t+dt$, which can be simply denoted by t (one of the discrete time variables, i.e., t=1,2,...). So the confluence or diversion, in the t th time step, can be directly expressed as

$$Q_h^l(i+1,t+1) = Q_h^l(i,t) \pm Q_{h+1}^i(i_{last},t) \quad h = 0,1,2\cdots \tag{1}$$

Where "+" and "-" is applicable to confluence and diversion, respectively. i,l=1,2,$\cdots$, and i=1 denotes the source of the stream. The denotation, i_{last}, represents the last segment for a confluence, or the diversion channel for a diversion. This equation indicates that the flux of the stream that is running on segment i+1 in time step t+1, $Q_h^l(i+1,t+1)$, is equal to the flux of the outflows of segment i and the last segment of the ith branch of rank h+1 in time step t, $Q_h^l(i,t)$ and $Q_h^i(i_{last},t)$. It's obvious that $Q_h^l(i+1,t+1)$ actually denotes the outflow of segment i+1 in time step t+1, and the outflow of segment i is also the inflow of the segment i+1. The transported sediment, S, carried by water in a higher-water season (usually has higher sediment concentration, q, in compare with that in a lower-water season) not only depends on the flux of the stream but is also driven by the sediment concentration of the injecting stream from its upriver neighbour. The higher the sediment concentration of the stream flowing in segment i is, the more sediment is injecting into segment i+1, and the more sediment has to be transported in this segment. Thereupon the following relation can be taken to express the aforementioned passive adjustment mechanism of SCC (Huo, et al. 2009):

$$S_h^l(i+1,t) = A_h^l(i+1,t) Q_h^l(i+1,t)^a q_h^l(i,t)^b \tag{2}$$

It means that the transported sediment is nonlinearly determined by both stream flow and the sediment concentration of the injecting stream. Where a and b are exponents, which are related to the landform and scale of basin, etc.. A is the so-called sediment-carrying coefficient, which denotes the positive part of the SCC and is adjusted by the undergone erosion or sedimentation state.

Generally, the sediment carried by outflow of a segment may not be equal to that carried by the inflows. So erosion or sedimentation will occur in the segment. Based on the confluence and/or diversion expressed by equation (1) and the definition presented above, the QES in this time step can be determined by equation

$$\Delta S_h^l(i+1,t) = S_h^l(i+1,t) - (S(_h^l(i,t) \pm S_{h+1}^i(i_{last},t)) \tag{3}$$

Where "+" and "-" is for confluence and diversion, respectively. The relation expresses QES on the i+1th segment in the tth time step. $\Delta S_h^l(i,t) > 0$ indicates that the segment is eroded, while $\Delta S_h^l(i,t) < 0$ means that it is deposited.

The positive adjustment of SCC, A, is described as follows:

1. Confluence: As presented in our previous studies (Wang, et al., 2008; Hao, et al., 2008 and Huo, et al., 2009), sediment-carrying coefficient A of segment i+1 will increase in time step t+1 if it is deposited but decrease if it is eroded in time step t due to a basic fact that the sedimentation implies both of the deposited and suspended sediment particles are relatively smaller, which leads to enhancement of the viscosity intensity of fluid and increase in the friction of river-bed against the flow, so the stream can carry more sediment. On the contrary, the erosion implies the deposited material over the river-bed and the suspended particles in the water become coarser. It causes the friction to reduce, thus the stream can transport lesser sediment. So this adjustment of SCC via the variation of A is conducted by the QES on the segment itself.

In fact, A of the stream is also adjusted by QES on its upriver neighbouring segment in the reversed way of that just mentioned above. If a segment is in scouring state, the increasing sediment may cause A of its downriver neighbouring segment to increase so that the stream can discharge more sediment, while if the segment is in deposited state, the decreasing sediment may let A of the downriver neighbouring segment decrease. Naturally, the QES on the last segment of a higher rank branch may work upon A in the same way.

2. Diversion: diversion differs from confluence mainly in the direction of water flowing, and the influence of it on the sediment transport dynamics is principally determined by the so-called water diversion ratio (WDR) that is defined as

$$\kappa = Q_{h+1}^{i}(i_{last},t) / Q_{h}^{l}(i,t) \tag{4}$$

and the corresponding sediment diversion ratio(SDR) that is expressed as

$$\lambda = S_{h+1}^{i}(i_{last},t) / S_{h}^{l}(i,t)\,. \tag{5}$$

An experiential equation obtained by a fitting of the field measurement in the Yellow River may generally describe the relation between κ and λ (Zhao, et al., 1997), that is, $\lambda = B\kappa^{\alpha}Q_{h}^{l}(i,t)^{\beta}$, which means that the diverted sediment nonlinearly depends on WDR and discharge, so we have

$$S_{h+1}^{i}(i_{last},t) = B\kappa^{\alpha}Q_{h}^{l}(i,t)^{\beta}S_{h}^{l}(i,t)\,. \tag{6}$$

As $\kappa > \lambda$ there should be an inequality, $q_{h+1}^{i}(i_{last},t) < q_{h}^{l}(i,t)$, which indicates that the stream, injecting the main channel after diversion, is of lesser water but more sediment. It may compel the stream to improve the SCC, while $\kappa < \lambda$, $q_{h+1}^{i}(i_{last},t) > q_{h}^{l}(i,t)$ means that the injecting stream after diversion becomes lesser water and lesser sediment. This may abate the burden of transporting sediment for the stream, so the SCC may decrease. Such adjustment of SCC via sediment concentration is the same of the aforementioned passive one in the case of confluence. Meanwhile, the abrupt change of the sediment concentration induced by water diversion will lead to the rapid response of the channel via variation of QES at once. Then the positive adjustment of SCC conducted by QES works here in the same way of confluence. So the positive adjustment of the SCC, A, being fit for the description of confluence and/or diversion, can be expressed as

$$\begin{aligned} A_{h}^{l}(i+1,t+1) = A_{h}^{l}(i+1,t) - k_1\Delta S_{h}^{l}(i+1,t) / (Q_{h}^{l}(i+1,t+1)^{a}q'(t+1)^{b}) \\ -k_2\Delta S'(t) / (Q_{h}^{l}(i+1,t+1)^{a}q'(t+1)^{b}) \end{aligned} \tag{7}$$

Where k_1,k_2 denotes the strength of the positive adjustment of SCC respectively from the QES on segment i+1 and that of both the QES on segment i and on the last segment of rank h+1. $\Delta S^{'}(t)$ and $q^{'}(t+1)$ in Equ.(7) have different meaning for confluence and diversion. According to the discussions above, equations $q^{'}(t+1)=q_h^l(i,t+1)+q_{h+1}^i(i_{last},t+1)$ and $\Delta S^{'}(t)=\Delta S_h^l(i,t)+\Delta S_h^i(i_{last},t)$ satisfy for confluence. As far as the diversion, if the influence of the state of the diversion channel on the main channel can be neglected, we can draw the conclusion that as soon as the diversion is executed the diverted water drops out of the river network. Therefore, there are relations $q^{'}(t+1)=q_h^l(i+1,t+1)$ and $\Delta S^{'}(t)=\Delta S_h^l(i,t)$. The former implies that the SCC of the stream flowing in the segment below and near to the diversion mouth is actually driven by the sediment concentration of the injecting stream.

In our model, any one upriver segment can be chosen as the source of the stream. The regular or irregular change of the discharge $\xi(t)$ represents rainfall. Rainfall usually results in the fluctuation of sediment concentration $\zeta(t)$, and then we have

$$Q_h^l(1,t)=Q_{h0}+\xi(t), \tag{8}$$

$$q_h^l(1,t)=q_{h0}+\zeta(t), \tag{9}$$

$$\Delta S_h^l(1,t)=\Delta S_{h0} \tag{10}$$

for all possible h, l and t. Here we assign a fixed value to ΔS_{h0} for simplicity.

The core spirit of our model consists in the description of self-adjustability of a river in the sediment transport process, which is manifested by the positive adjustment of SCC that is expressed by Equ. (7). Similar self-adjustability is often related to the dynamics of a completely open system. The simulation work may not only help us to understand the nature of sediment transport in river networks but also provide us with some illumination for general dynamics of self-adaptive systems else.

When a diversion is executed, there are many factors that influence the dynamics of sediment transport such as WDR, SDR, the discharge and the sediment concentration of the injecting fluid. The interaction of these factors will induce much more complex behaviour of the dynamics, which usually differs from that without diversion. Here we present the discussions of the simplest case, $h=0,l=1$, so the superscript l can be deleted in the aforementioned equations. Please note that in our simulation, the diversion mouth is located at the node that connects the 20th segment and the 21st segment. In our study, the values of the parameters and variables are set as $\Delta S_0(i,1)=15.0$, $S_0(i,1)=10.0$, $k_1=6.0\times10^4$, $k_2=9.0\times10^5$, $A(1,t)=5.7\times10^2$, $Q_1(i,t)=80.0$, $q_1(i,t)=10.0$, $a=1.1$, $b=0.7$, $\alpha=1.2$, $\beta=-0.2$, B=8.0 and $Q_0(i,1)=rand(i=1\sim30,100.0\sim150.0)$ (random function, $rand(\bullet)$, denotes the operation that assigning 30 random numbers, produced between 100.0 and 150.0, to $Q_0(i,1)$).

3. Influence of WDR on sediment transport

As discussed above, the relative distance of the value of κ and that of λ can influence the state of the main channel via changing the SCC of the stream. Fig. 2 shows this relation as the diversion is executed for 6000 time steps, Fig. 3 (a) presents QES on the segments below the diversion mouth and Fig.3 (b) denotes the corresponding SCC. Fig. 2 indicates that there is a critical value of WDR, $\kappa_c=0.051$. As $\kappa<\kappa_c$, $\kappa>\lambda$, the injecting stream of the 21st segment will become lesser water and more sediment, which implies that the sedimentation will occur due to the forced increase of SCC. This is the passive adjustment of SCC driven by sediment concentration. In addition, the positive adjustment will lead the dynamics to be more complex, and induce the so-called diversion-sedimentation effect, which means that water diversion can cause sedimentation or cause sedimentation quantity to get more (see Fig. 3 (a)). So one can say it is that the positive and passive adjustments of SCC together determine the dynamic behaviour of the 21st segment and those below it even if for the case of $\kappa > \kappa_c$, $\kappa < \lambda$. In this case, the injecting stream of the 21st segment also has a higher SCC. It is partly shown by Fig. 3 (b), and differs sharply from the other segments in the values of A_0. It follows that the 21st segment will respond intensely to the change caused by diversion.

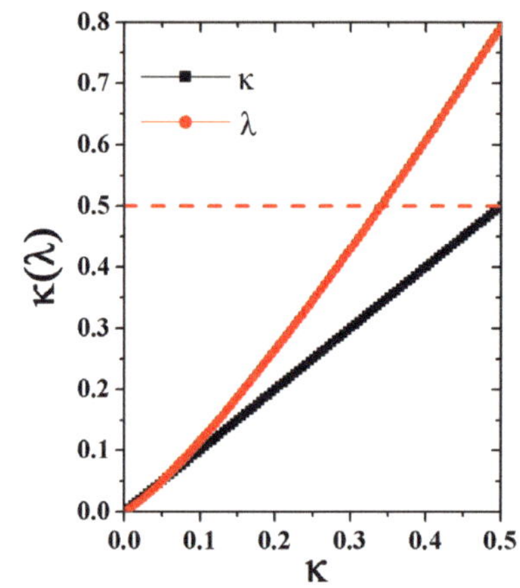

Fig. 2. The relation between WDR and SDR

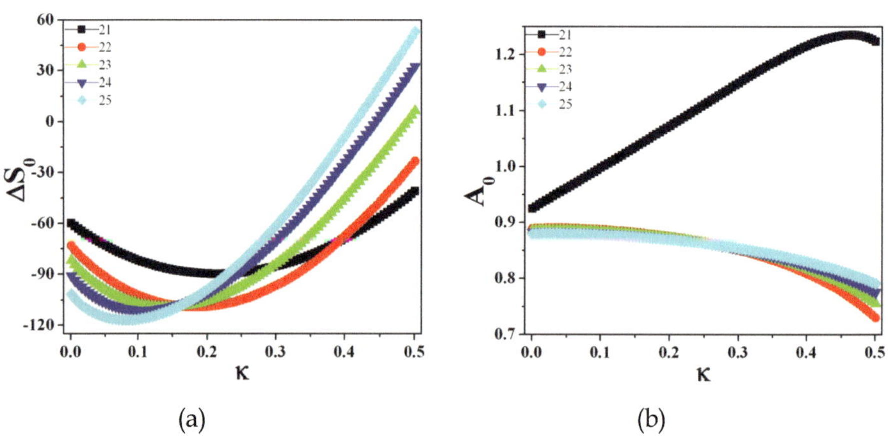

Fig. 3. Influence of the WDR, κ, on the dynamics of sediment transport: (a) the QES, ΔS; (b) the positive part of SCC, A_0

As shown in Fig. 3 (a), there is an another critical value of κ, $\kappa_c^{'}$, for each of the segments below the diversion mouth (the values of $\kappa_c^{'}$ of the 21st segment, the 22nd segment, ..., is

0.216, 0.186, …, respectively) at which the sedimentation quantity on the segments reach the maxima. The maximum indicates that sedimentation will get less with the increase of κ as it is beyond $\kappa_c^{'}$, so one can infer that there is the third critical value of κ at which the state of the segments will transform from sedimentation to erosion. It is very interesting that this transform may take place in the segments in turn in the upstream direction. It is noticeable that the maximum of the sedimentation quantity on the segments decreases in downstream direction, and proves that the sedimentation mainly occurs on the segments below and near to the diversion mouth. This behaviour can be attributed to the strong response of the SCC of the 21st segment to the diversion.

To test the aforementioned calculation and get more understandings to the dynamics, we give the QES on the 21st segment obtained analytically. As a matter of convenience, on the 21st segment the increment of the QES, from time step t to time step $t+1$, is taken into account, that is,

$$\delta\Delta S_0(21,t+1) = \Delta S_0(21,t+1) - \Delta S_0(21,t)\,. \tag{11}$$

This equation can be easily expressed as

$$\delta\Delta S_0(21,t+1) = (u\kappa^{0.2} - v(1-\kappa)^{0.1})\Delta\kappa \; - k_1\Delta S_0(21,t) + k_2\Delta S_0(20,t)\,. \tag{12}$$

Where $u = 9.6Q_0^{-0.2}(20,t{+}1)q_0(20,t{+}1)$, $v = 1.1q_0^{0.7}(20,t{+}1)Q_0^{0.4}(20,t{+}1)A_0(21,t)$ and $\Delta\kappa$ is the increment of κ in any two neighboring time steps and is assigned a value, 0.0005. As the parameters and variables take the above values and $Q_1(i,t) = 295.0$, the comparison between the result obtained analytically and that obtained numerically is shown in Fig.4, and indicates that the result calculated analytically is, on the whole, in agreement with that one obtained numerically. One may note that the value of $\delta\Delta S_0(21,t+1)$ changes from the negative to the positive with the increase of κ. This signifies that the sedimentation quantity on the 21st segment firstly gets more and then becomes less. The variation trends of $\delta\Delta S_0(21,t+1)$ is mainly determined by two terms $\kappa^{0.2}$ and $(1-\kappa)^{0.1}$ in Equ. (12), which shows that when κ takes a small value $(1-\kappa)^{0.1}$ takes a relative greater value. However, with the increase of κ, $\kappa^{0.2}$ increases but $(1-\kappa)^{0.1}$ decreases. It is the reason why the value of $\delta\Delta S_0(21,t+1)$ is firstly negative and then become positive, and also the reason why there is the critical WDR, $\kappa_c^{'}$.

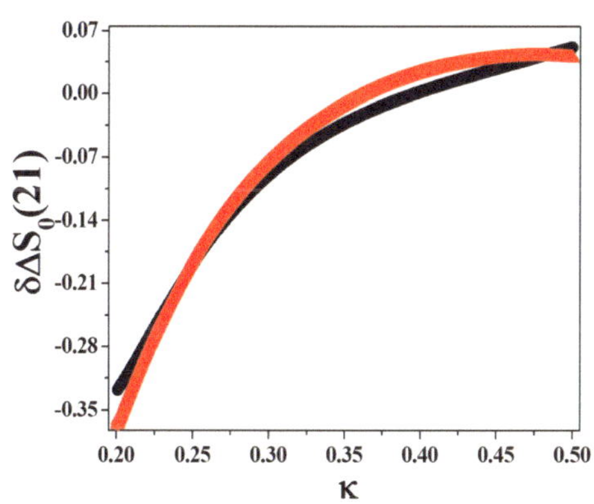

Fig. 4. Comparison of $\delta\Delta S_0(21,6000)$ obtained analytically with that obtained numerically

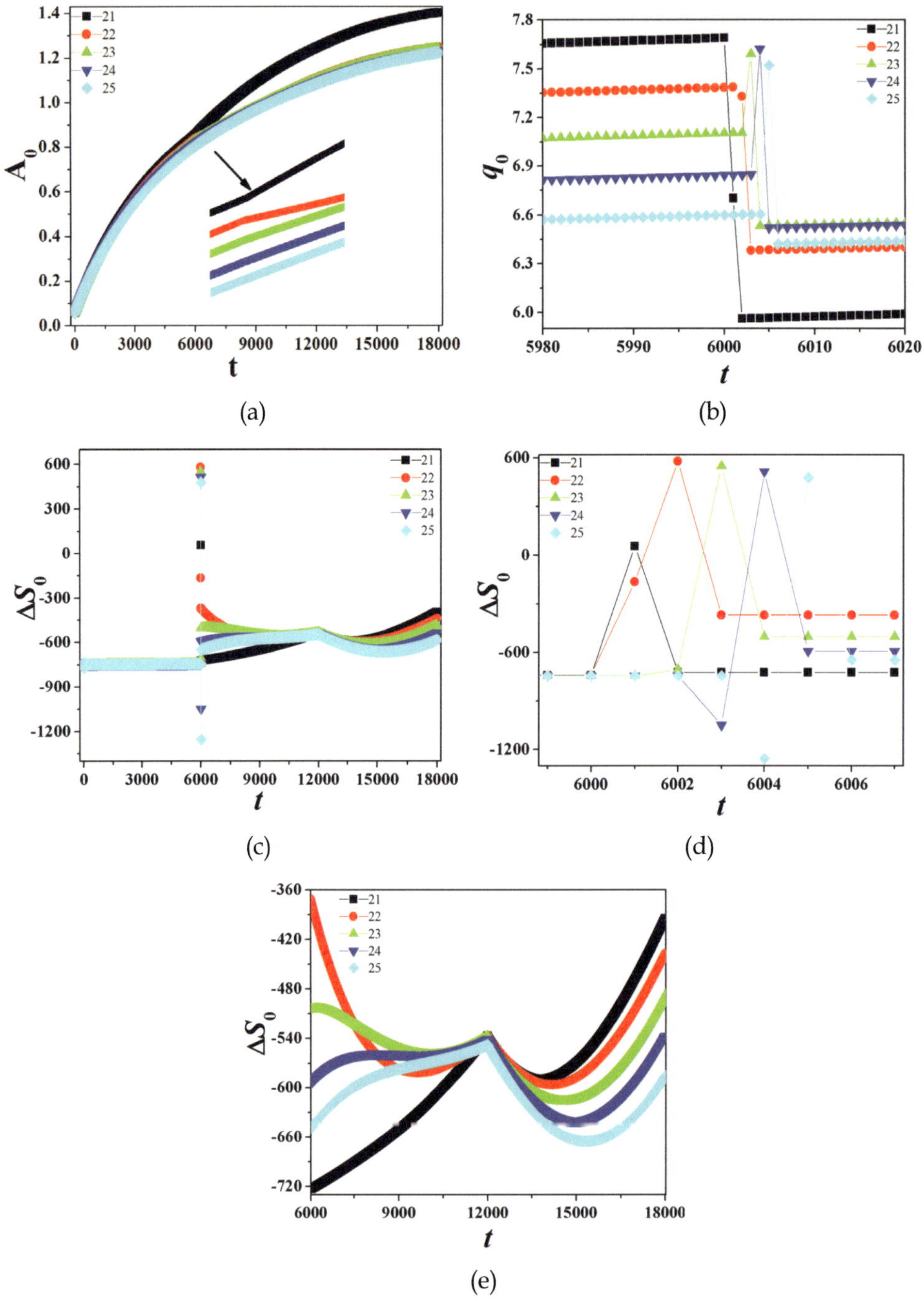

Fig. 5. Variations of the QES and SCC on the mainstream segments that below the diversion mouth. The details are presented in text. (a) the positive part of SCC, A_0 ; (b) sediment concentration, q_0 ; (c) QES, ΔS_0 ; (d) a partial magnification of (c) shows initial influence of the steady diversion; (e) another partial magnification of (c) shows the instantaneous process while the discharge of the first rank rivers evenly increase

4. Complexity of the sediment transport dynamics induced by diversion

Since diversion will cause a drastic change of the state, which is described by SCC and QES, of the segment below and near to the diversion mouth, the dynamics of the sediment transport will become more complex. In order to facilitate discussion, we only consider the confluence of the mainstream with the first rank branch rivers. These branch rivers are characterized by $Q_1(i,t)$=25.0 and $q_1(i,t)$=30.0 also for simplicity. The other parameters and variables are set as those in Fig. 3. The simulation is fallen into three phases. The first phase (1-6000 time step) is confluence with fixed imports so as to avoid the unreasonable transience and obtain reasonable SCC and QES for each segment. The second one (6001-12000 time step) is diversion. It is conducted with the fixed WDR, κ = 0.1, so that we can track the dynamics of the erosion-sedimentation state under steady diversion conditions. After this, the third one runs with the increase of the discharge of each branch of rank 1, that is, $Q_1(i,t) = 25.0 + \xi(t)$, $\xi(t) = (t - 12000)\delta Q_1$. Here $\delta Q_1 = 0.0042$, it is the increment of the discharge of the first rank rivers in a time step. In this phase, we mainly investigate the variations of SCC and QES on the segments those below and near to the diversion mouth. Fig. 5 shows these simulation phases, where (a) , (c) denotes the variation of SCC and QES, respectively, in the process; (b) presents the sediment concentration before and after the diversion; (d) is the local magnification of (c), and displays the great changes of the QES caused by the diversion; (e) is also the partial magnification of (c), and demonstrates, in detail, the variation of the QES when it goes to the third phase(the discharge of the main stream and the diverted water increases due to the even increase of the first rank streams) from the second one.

4.1 Sediment transport dynamics when a steady diversion is conducted

As shown by Fig.5 (a), the diversion (starts at the fore-end of time step 6001) strengthens the SCC of the stream flowing on the 21st segment immediately. However, the SCC of the stream running on the 22nd segment decreases with a time lag of one step in comparison with the time at which the diversion starts. These changes are denoted by the inflection points on the corresponding curves. Meanwhile, the inflow of the 21st segment decreases suddenly due to the diversion, and as $\lambda > \kappa$, the sediment concentration of the inflow of the 21st segment also decreases suddenly (Fig.5 (b)). The decrease results in the decrease of the sedimentation quantity on this segment (please see Fig.5 (c) and (d)). The decrease of the sediment concentration of the stream flowing on the 21st segment lowers the driving action on the SCC of the stream flowing on the 22nd segments. Therefore, sediment carried by the inflow now gets lesser than that before the diversion is conducted, and then the sedimentation quantity decreases. In time step 6002, the inflow of the 22nd segments decreases since in this time step the outflow of the 21st segment becomes the inflow of the 22nd segment. The sediment carried by these inflow decreases greatly, so a great deal of erosion occurs on this segment. In time step 6003, the stream flowing on this segment becomes steady, it returns to a sedimentation state. The similar changes of the state take place in the segments, 23rd, 24th, ... in turn in the downriver direction. This likes a propagation process in which the amplitude of the change gradually reduces in the downriver direction, and indicates that the influence of the diversion on the segments below the diversion mouth weakens in this direction. It is also demonstrated by Fig.5 (a) that the SCC of the stream, from the 22nd segment to its downriver neighbors, goes down. Fig.5 (e) also proves that our simulation is in qualitative accord with the field observation that sedimentation is enhanced, especially on the segments near to the diversion mouth.

One may note if a steady diversion lasts for some time steps, an interesting distribution of the QES will be established. By a careful investigation, we find, via magnifying the part of Fig.5 (a) to which the arrowhead points in the inset, that the SCC of the stream flowing on the 22nd segment decreases suddenly in time step 6003 and then increases gradually. The SCC of the stream flowing on the 23rd segment decreases slightly in time step 6004. The SCC of the stream of the rest nearly remains changeless. Correspondingly, the variation of erosion-sedimentation state is demonstrated by Fig.5 (e). The monotonously ascending curve means that sedimentation on the 21st segment gets lesser gradually. The sedimentation on the 22nd segment firstly goes down and then goes up. This implies that there is a maximum of the sedimentation quantity on this segment. The curves of QES for the rest of the segments show different changes at first, whereas they tend to be the same of that on the 21st and 22nd segments. So one can infer that on all of the segments, the dynamics of sediment transport is inclined to remove the difference of the QES on the segments, and reach a common state, a so-called nearly erosion-sedimentation equilibrium (NESE) state, at which the state is closed to that one of QES being equal to aero, and the channel keeps unchanged. This is sharply different from that dynamics of sediment transport in river network without diversion, and embodies the complexity of the dynamics caused by diversion.

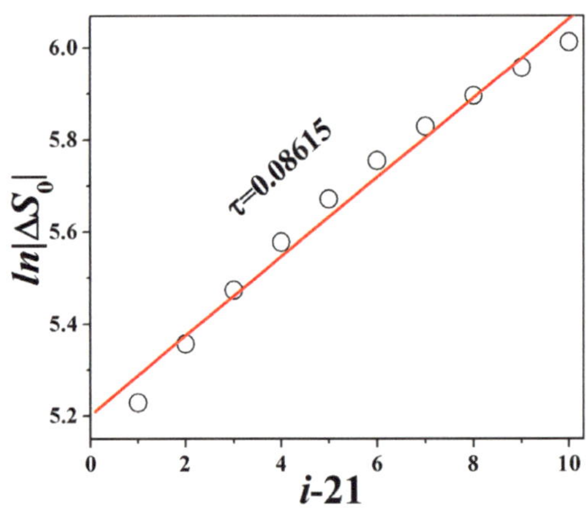

Fig. 6. Scaling law of instantaneous sedimentation distribution on the segments that are below the diversion mouth when they tend to the NESE state

While the dynamics approaches to the NESE, a scaling law that dominates the distribution of the QES can be gradually established, which is described by an exponential function

$$|\Delta S_0(i-21,t)| \propto e^{\tau(i-21)} \quad i>21 \tag{13}$$

The scaling function can be obtained by linear fitting in space $\ln\Delta S(i-21,t)$ versus $i-21$, and shown by Fig.6 (presents the calculation at $t=20000$). In fact, this scaling function is suitable to the instantaneous distribution, and that the longer the diversion lasts for, the better the distribution accords with an exponential function. The scaling behavior actually embodies the characteristics of the self-adaption and the self-organization. The former refers to the adjustment of the channel via changing the QES to adapt to the import, the latter drives at the feedback mechanism between the SCC of the stream and the state of the channel. The former is usually an effect, while the latter is the cause why there exists the result. It is just the feedback mechanism that the behavior shown in Fig.5 from time step 6001 to 6005 when both of the SCC and the QES respond intensely to the sudden change of the imported water and sediment. The intense oscillation between erosion and sedimentation (shown in Fig.5 (d)) denotes this response, the so-called rapid response (Huo, et al., 2009).

4.2 Influence of the changing discharge on sediment transport dynamics

As shown in Fig.5 (with fixed WDR), the sudden change of the streams flowing on the first rank branches interrupts the way to the NESE state, and therefore the scaling law expressed by Equ.(13) is destroyed. Since time step 12001, the discharge of the inflow of each mainstream segment increases due to the even increase of the discharge of the first rank rivers. However, the discharge increase of the outflow of these mainstream segments is postponed for one time step in comparison with that of the inflow. These cause the sedimentation quantity on these segments to get more (Fig. 5 (e)). The detail of the response of the QES on these mainstream segments to the discharge increase of the first rank branches can help us to obtain more understandings to the dynamics. For each segment, there is a special section of time series on the curve of ΔS_0 near t = 12000. For a certain segment, it corresponds to the changes resulted from the confluence of the first rank streams, one by one, of which discharge have increased. When the first (top) branch of the first rank that has its discharge increased injects this segment for the first time, the section ends up. The section length of the 21st segment, the 22nd segment … prolongs for one time step in turn from above to bellow. In the section, the discharge increment of the stream on a segment, for instance, the 21st segment, in each one time step comes from the change of the discharge on the 20th segment after diversion. This increment can be expressed as

$$\delta Q(21,t+1)=(1-\kappa)\delta Q(20,t)=(1-\kappa)(t-12001)\delta Q_1 \qquad t=12002,12003,\cdots,12020\,. \quad (14)$$

Since time step 12021, the increment remains changeless. It follows that there is the following relation,

$$\delta Q(20,t+1)=19\delta Q_1 \qquad t>12020\,. \quad (15)$$

Obviously, the discharge increment of the stream on the 21st segment reaches the maximum in time step 12020, and then the discharge increases evenly. This change may alter the variation trend of the QES: the increase of sedimentation quantity becomes slower relatively.

One can see from the discussions above, water diversion will give rise to complexity of the sediment transport dynamics, and can further infer a conclusion from this fact that the dynamics will become more complex if a random fluctuation is introduced to mimic a natural rainfall.

Interestedly, if the even increase of the discharge goes on long enough, the transition from sedimentation increasing to sedimentation decreasing will occur in the segments below the diversion mouth in turn from above to below, which indicates that there is an instantaneous maximum of the sedimentation quantity, denoted by $\Delta S_{0\min}$, for each one of the segments. Fig. 7 suggests the following scaling function:

$$\left|\Delta S_{0\min}(i-21)\right| \propto e^{\gamma(i-21)} \qquad i>21\,. \quad (16)$$

This scaling relation reflects the fact that a river network tends to constitute an order to dominate the dynamics in the self-adaption process when the import varies regularly.

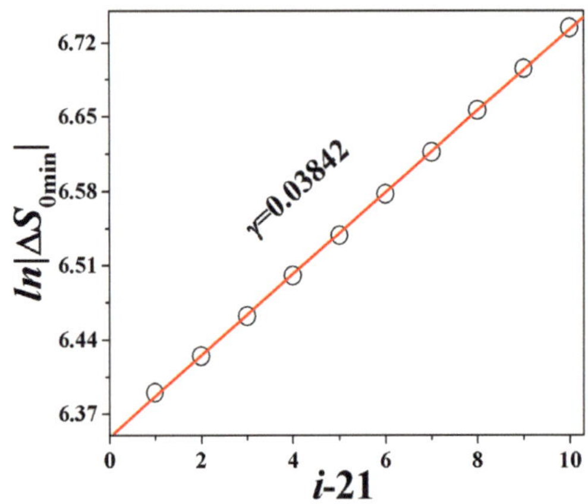

Fig. 7. Scaling law of the maximum sedimentation quantity on the segments that are below the diversion mouth as the flux of water increases linearly

5. Conclusions and discussions

In this article, we suggest a dynamical model for sediment transport in river network to study the characteristic of the dynamics induced by water diversion. The model is based on the mechanisms that the SCC of a stream is adjusted by the QES and driven by the sediment concentration of the injecting streams. The numerical results of this model can simulate the dynamical behaviour in real networks: water diversion can cause sedimentation quantity to increase. Some details of the dynamics may help us to understand the nature of river networks and the sediment transport in them.

The transient dynamics shows some interesting characteristics. 1) There is a critical value of WDR at which the state of the segments those are below the diversion mouth will transform from sedimentation getting more to sedimentation getting lesser. 2) Water diversion will cause the dynamics of sediment transport to be more complex. It will lead to the abrupt change of the erosion-sedimentation state. On the segments that are below and near to the diversion mouth sedimentation quantity will oscillate between increase and decrease. This is the rapid response of the channel to the diversion. 3) After the rapid response, if the WDR is fixed the oscillation extent will decay in the downriver direction. The sedimentation quantity on each of the segments will exponentially decreases with time if the steady diversion goes on. This may lead the dynamics to the NESE state near which a scaling law that governs distribution of the QES is established. 4) There is a maximum quantity of sedimentation on each of segments, which is dominated by a scaling law, that is, the maximum quantity of sedimentation depends exponentially on the number of the segments. The aforementioned results may give some enlightenment, especially from the scaling laws. In a river network, the inputs may be random and complex, the outputs (after the transience or in the transient process) may be, however, regular and simple. This dynamic behavior can be attributed to the feedback between the input (stream) and internal factor (the state or the pattern of channel) of the system, and may be common for a kind of completely open system that is endued with self-organizing capacity by the feedback mechanism.

Of cause, these scaling laws can, generally, not observed in natural river networks since they are destroyed by the large-amplitude random fluctuations in the stream. It follows that the random fluctuation of the discharge caused by rainfalls may play an important role in keeping a relative steadier conformation of the channel.

6. Acknowledgements

This study is supported by the National Natural Science Foundation of China under Grant Nos. 10565002 and 10965004, by the Program for New Century Excellent Talents in University of China under Grant No. NCET-06-0914, and by the Ningxia NSF under Grant No. NZ0944.

7. References

Banavar, J. R. ; Colaiori, F.; Flammini, A.; Giacometti, A.; Maritan, A. & Rinaldo, A. (1997). Sculpting of a Fractal River Basin, *Phys. Rev. Lett.*, 78, 23, 4522-4525.

Caldarelli, G.; Giacometti, A.; Maritan, A.; Rodriguez-Iturbe, I. & Rinaldo, A. (1997). Randomly pinned landscape evolution, *Phys. Rev. E*, 55, 5, R4865-R4868.

Dhar, D. (2006). Theoretical studies of self-organized criticality, *Physica A*. 369, 1, 29-70.

Dodds, P. S.& Rothman, D. H. (2000). Geometry of river networks. I. Scaling, fluctuations, and deviations, *Phys Rev. E* 63, 016115. and the following papers

Dodds, P. S.& Rothman, D. H. (2000). Geometry of river networks. II. Distributions of component size and number, *Phys Rev. E* 63, 016116.

Dodds, P. S.& Rothman, D. H. (2000). Geometry of River Networks III: Characterization of Component Connectivity, *Phys. Rev. E* 63, 016117.

Flint, J. J. (1974). Stream gradient as a function of order, magnitude, and discharge, *Wat. Resour. Res.*, 10, 5, 969-973.

Giacometti, A. (2000). Local minimal energy landscapes in river networks, *Phys. Rev. E*, 62, 5, 6042-6051.

Hack, J. T. (1957). Studies of longitudinal stream profiles in Virginia and Maryland, *U. S. Geol. Surv. Prof. Pap*. 294B, 45-97.

Hao , R.; Wang , X.-M.; Huo, J. & Zhang, J.- F. (2008). Simulating sediment transport on river networks, *Mod. Phys. Lett. B*, 22, 2, 127-137.

Huo, J.; Wang, X.-M.; Hao, R. & Zhang, J.-F. (2009). Sediment transport dynamics in river networks: A model for higher-water seasons, *Lecture Notes of the Institute for Computer Sciences, Social Informatics and Telecommunications Engineering*, 4, Part I, 832-840.

Kramer, S. & Marder, M. (1992). Evolution of river networks, *Phys. Rev. Lett.*, 68, 2, 205-208.

Leopold, L. B. (1953). Downstream change of velocity in rivers, *Am. J. Sci.*, 251, 606.

Leopold, L. B. & Maddock, T. (1953). The hydraulic geometry of stream channels and some physiographic implications.*US Geol. Surv. Prof. pap.*, 252, 57.

Manna, S. S. (1998) Branched tree structures: From polymers to river networks, *Physica A.*, 254, 1-2, 190-197.

Manna, S. S. & Subramanian, B. (1996). Quasirandom spanning tree model for the early river network, *Phys. Rev. Lett.*, 76, 18, 3460-3463.

Maritan, A.; Colaiori, F.; Flammini, A.; Cieplak, M. & Banavar, J. R. (1996). Universality classes of optimal channel networks, *Science*, 272, 5264, 984-986.

Maritan, A.; Rinaldo, A.; Rigon, R.; Giacometti, A. & Rodriguez-Iturbe, I. (1996). Scaling law for river networks, *Phys. Rev. E.* 53, 2, 1510-1512.

Rinaldo, A.; Maritan, A.; Colaiori, F.; Flammini, A.; Rigon, R.; Rodriguez-Iturbe, I. & Banavar, J. R. (1996). Thermodynamics of fractal networks, *Phys. Rev. Lett.* 76, 18, 3364-3367.

Rodriguez-Iturbe, I.; & Rinaldo, A. (1997). *Fractal River Basins:Chance and Self-Organization*, Cambridge University Press, Cambridge.

Sinclair, K. & Ball, R. C. (1996). Mechanism for global optimization of river networks from local erosion rules, *Phys. Rev. Lett.* 76, 18, 3360-3363.

Somfai, E. & Sander, L. M. (1997). Scaling and river networks: A Landau theory for erosion, *Phys. Rev. E* , 56, 1, R5-R8.

Sun, T.; Meakin, P. & Jøssang, T. (1994). A minimum energy dissipation model for drainage basins that explicitly differentiates between channel networks and hillslopes, *Physica A* , 210, 1-2, 24-47.

Sun, T.; Meakin, P. & Jøssang, T. (1994). Minimum energy dissipation model for river basin geometry, *Phys. Rev. E*, 49, 6, 4865-4872.

Takayasu, H. & Inaoka, H. (1992). New type of self-organized criticality in a model of erosion, *Phys. Rev. Lett.*, 68, 7, 966-969.

Wang, X.-M.; Hao, R.; Huo, J. & Zhang, J.-F. (2008). Modeling sediment transport in river networks, *Physica A*, 387, 25, 6421-6430.

Zhang, D. & Liang ,Z., (1995). Overview on the relation of water and sediment diversion with erosion/sedimentation of channel of the Lower Yellow River, *J. Sediment Res.*, 2, 32-42. (in Chinese).

Zhao, W.-L. (1997). *Sediment in the Yellow River*, The Yellow River Water-Conservancy Publishing Company, 7-80621-106-3/TV·82, Zhengzhou, (in Chinese).

9

Sediment Transport in Kulim River, Kedah, Malaysia

Chun Kiat, Chang and Aminuddin, Ab. Ghani
Universiti Sains Malaysia
Malaysia

1. Introduction

Rivers are dynamic by nature; they adjust their characteristics in response to any change in the environment. These environmental changes may occur naturally, as in the case of climatic variation or changes in vegetative cover, or may be a result of human activities. Human factors influence channel changes, both directly by engineering projects including channelization, dredging, snag removals, dam construction and bridge construction, and indirectly through altering floodplain landuse such that erosion is more likely to occur during flood events more likely to occur during flood events (Ab. Ghani et al., 2010). These changes to river hydrology and sedimentation will in turn modify the channel morphology, which include changes to channel cross section, stability and capacity. Otherwise, hazard flood increases with the sedimentation and damages exceeded with muddy water. Consequently, it is necessary to study river channel behaviour and evaluate the river channel stability for its natural state and response to human modification due to the existing and future developments.

2. Study site

Kulim River catchment (Figure 1) is located in the southern part of the state of Kedah and in the northwestern corner of Peninsular Malaysia. Kulim River is a natural stream in Kedah state, Malaysia. Kulim River drains 130 km^2 of the surface area of southern part of the state of Kedah, is in the northwestern corner of Peninsular Malaysia. Kulim River emanates from the western slopes on Gunung Bangsu Range and flows in a north-westerly direction. The river slopes are steep and channel elevations drop from 500 meter to 20 meter above mean sea level (AMSL) over a distance of 9 kilometer. The central area of the catchment is undulating with elevation ranging from 100 meter down to 18 meter above mean sea level.
The study area has a tropical climate influenced by the movement of Inter-Tropical Convergence Zone. Its passage over the area results in two wet periods during the year which occur from April to May and from September to November. There is a transitional period of moderate rainfall during June to August and dry during December to March. Rainfall is generally convective and increase from around 2000 mm a year at the downstream to over 3200 mm a year on the mountainous area.
The Kulim River has experienced severe environmental damages, mostly related to significant erosion and sedimentation. Anthropogenic activities and natural events cause

changes in river morphology and stability of Kulim River. The human activity include the development to the year 2010 of Kulim district based on the Kulim Structure Plan, 1990-2010 (MDK, 1993), rapid urbanization at Kulim River catchment especially construction for housing state, the on-going 145 km^2 Malaysia's first and fully integrated Kulim Hi-Tech Industrial Park and river sand mining activities which may maximize the disturbance to river equilibrium and environment. Frequent flood occurrences in Kulim River catchment have significantly affected the community because of extensive damage in built up and agriculture areas especially the flood event in October 2003, which is an event slightly lower than the 100-year ARI based on the frequency analysis. Finally, these changes to the river hydrology and sedimentation will in turn alter the channel morphology, which can include changes to channel cross section, stability and capacity (Chang et al., 2005). The study reach covers about 14.4 km of Kulim River, from the upstream (CH 14390) to the state boundary between Kedah and Penang (CH 1900) and further downstream at the Ara Kuda gauging station (CH 0).

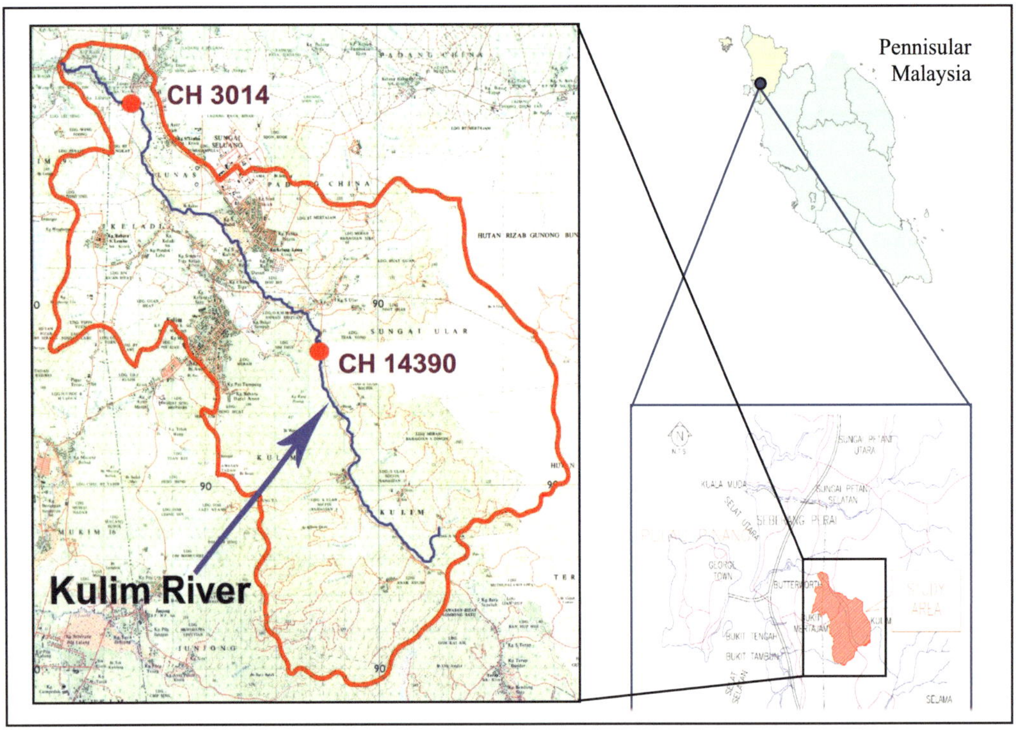

Fig. 1. Delineated Kulim River Catchment and Study Reach

3. Flood frequency analysis

The Kulim River benefited from 46-year period (1960–2005) of daily discharge measurements at Ara Kuda streamflow station, which include two major floods have occurred in 2001 and 2003 within the period. The annual peak discharges ranked in Table 1 indicate that the ten largest floods have been measured since 1961. This can be considered that the discharge of 92.90 m^3/s measured in 5 October 2003 is the highest during that period of record.

Rank	Discharge, Q (m^3/s)	Year	Date
1	92.90	2003	05-Oct
2	89.90	2001	22-Jan
3	67.90	1998	16-Nov
4	65.00	2000	22-Sep
5	62.30	1963	13-Nov
6	61.20	1999	05-Sep
7	58.50	2004	23-Sep
8	57.90	1962	21-Oct
9	56.90	1964	26-Sep
10	55.10	1987	09-Nov

Table 1. Flood Ranking for Kulim River at Ara Kuda

A flood frequency analysis was carried out for the 42-year period of streamflow data using Gumbel Extremal Type I. It was found that the result shows the better agreement to the measured streamflow data (Figure 2). The flood frequency analysis provided by the present study is also given in Table 2. It is therefore concluded that the 2003 flood discharge of 92.90 m^3/s is slightly lower than the 100-year peak discharge.

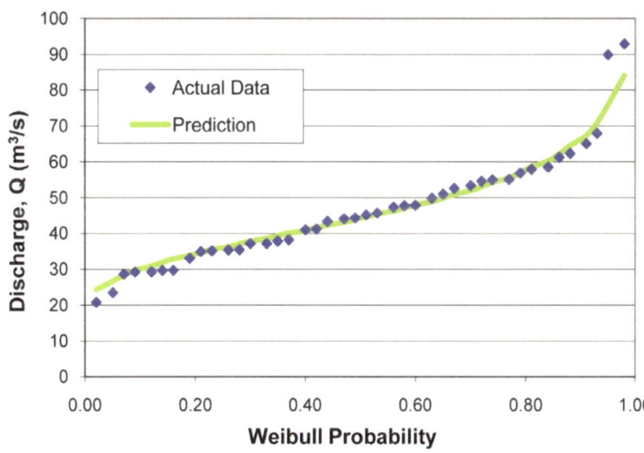

Fig. 2. Flood Frequency Analyses Using Gumbel Extremal Type I Distribution

Return Period	Discharge (m^3/s)
200	102.27
100	94.08
50	85.86
25	77.58
10	66.42
5	57.59
3	50.58
2	44.25

Table 2. Summary of Flood Frequency Analyses using Gumbel Extremal Type I

4. Sediment data collection and analysis

4.1 Field measurement

River surveys, flow measurement and field data collection provide the basic physical information such as sediment characteristics, discharge, water surface slope; which is needed for the planning and design of river engineering. In addition to the data needed for sediment transport studies, use of a sediment transport model also requires field data such as channel configuration before and after the changes, a flow record and sediment characteristics, which are generally used for test and calibration of a model. Field measurements were obtained at the selected cross sections (CH 14390 and CH 3014) from October 2004 to November 2006 along Kulim River by using Hydrological Procedure (DID, 1976; DID, 1977) and recent manuals (Yuqian, 1989; USACE, 1995; Edwards & Glysson, 1999; Lagasse et al., 2001; Richardson et al., 2001). The data collection includes flow discharge (Q), suspended load (T_s), bed load (T_b) and water surface slope (S_o). The water-surface slopes of the study reaches were determined by taking measurements of water levels over a distance of 200 m where the cross section was located (FISRWG, 2001). Besides that, bed elevation, water surface and thalweg (the minimum bed elevation for a cross section) measurement were also carried out at the selected cross sections. Details and examples of the measurement methodology for rivers in Malaysia can be found in Ab. Ghani et al. (2003) and DID (2009).

In this study, the water-surface slopes were found to be mild, where the average slope, S_o for CH 14390 and CH 3014 is 0.001 m/m. Bed material samples were also collected at the selected cross sections including bank samples. This data were analyzed to determine the distributions of the mean sediment size or d_{50} and used to characterize the physical characteristics of the sediment responsible for sediment transport, which determines the river response in terms of erosion and deposition. Low sediment transport rate for Kulim River occurred during the field measurements. The mean sediment sizes show that Kulim River is sand-bed streams where d_{50} ranges from 1.00 to 2.40 mm.

A summary with ranges for hydraulics and sediment data collection is shown in Table 3 (Chang et al., 2008). The surveyed cross sections for the Kulim River show that it is a single

Study Site	CH 14390	CH 3014
No. of Sample	10	12
Discharge, Q (m^3/s)	0.73 – 3.14	3.73 - 9.98
Bankfull width, TW (m)	25.0	50.0
Water surface width, B (m)	9.0 - 13.0	13.0 - 19.0
Flow depth, y_o (m)	0.20 - 0.54	0.36 - 0.58
Hydraulic radius, R (m)	0.23 - 0.57	0.40 - 0.63
Water surface slope, S_o	0.001	0.001
Mean sediment size, d_{50} (mm)	1.00 – 2.40	1.10 - 2.00
Manning *n*	0.029 - 0.072	0.024 - 0.037
B/y_o	23.4 - 44.8	26.0 - 52.5
y_o/d_{50}	126.9 - 369.01	240.0 - 550.9
R/d_{50}	141.4 - 406.6	266.5 - 570.9
Bed load, T_b (kg/s)	0.06 - 0.33	0.11 - 0.36
Suspended load, T_s (kg/s)	0.02 - 0.27	0.03 - 1.21
Total load, T_j (kg/s)	0.09 - 0.56	0.27 - 1.35

Table 3. Range of Field Data for Kulim River Catchment (Chang et al., 2008)

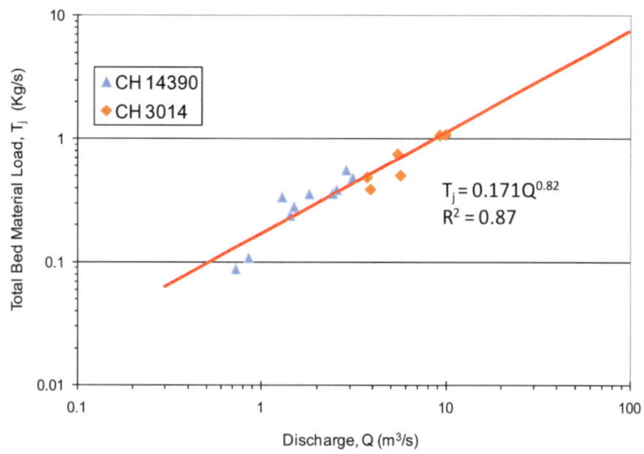

Fig. 3. Sediment Rating Curve along Kulim River

thread channel with the bankfull width ranging between 25 and 50 m, and aspect ratio (B/y_o) was between 23 and 53 indicating that it is a moderate-size channel. The total bed material load (T_j) is composed of the suspended load and bed load, representing the ability of the river to replenish the sediment and it must be specified for sediment transport, scour and deposition analysis. The measured total bed material load rating curves for these two sites at the Kulim River are illustrated in Figure 3. A mild curve is obtained for Kulim River indicating that a longer time is required for the replenishment before it is viable for sand extraction purpose. Therefore it is recommended that deposition should be allowed to occur first after a major flood before any river sand mining activity is allowed between these two sites. Based on the Table 1 and Figure 3, it is estimated that the 2003 flood with a discharge of 93 m^3/s will transport 7 kg/s of sand during the flood.

4.2 Channel morphology

Water and sediment transport through the Kulim River increase with time due to the reduction of river capacity that resulted from reclamation and sedimentation along the river. River bank erosion, river bed degradation, river buffer zone encroachment and deterioration of river water quality cause a serious and regular hazard in urban settlements at Kulim town. The powerful water currents wear away at the edges of these settlements during the wet periods and sometimes entire settlements established near the bank are washed away.

Figure 4 shows the channel planform modification along Kulim River due to an event slightly lower than the 100-year ARI flood during October 2003 at two urbanized areas of Kulim town and its surrounding areas. The main and most urbanized area of Kulim River pass the Kulim town and its surrounding. The channel widening can be observed in many locations along the Kulim River such as CH 10000, CH 12000 and CH 13500. The depositional areas are mainly located at the outer banks of meanders, while the erosion areas are at the opposite banks. These changes in channel river planform may cause extensive damages and inconvenience to the community (Sirdari, 2009).

Field measurements results including bed elevation, thalweg and water surface were carried out at several cross sections were compared to the river survey geometry data in September 1991. However, the comparison between cross sections provided by Department of Irrigation and Drainage (DID) and field measurement after October 2003 flood shows that there has been a change in cross section. The channel bed profile has gradually reduced

(a) CH 9000 - CH 11000 (b) CH 13000 - CH 14000

Fig. 4. Channel Planform of Kulim River (Sirdari, 2009)

within 13 years period, which proves that channel degradation occurred at most cross sections at Kulim River after the flood event (Chang et al., 2005). Thalweg at the CH 14390 has changed from 24.58 m to 23.50 ± 0.5 m and thalweg at the CH 3014 has changed from 8.38 m to 6.45 ± 0.5 m (Figure 5). From these results of cross section changes, it's shown that steep slope in Kulim River has induced higher discharge, and it was associated with the spatial variation in sediment transport and sediment size. The changes in river bed profile may be attributed to the erosion or deposition along the banks or the channel width. As a result, the study of changes in channel-bed profile, width variation and changes in bed topography, bank erosion, changes in channel degradation and aggradation, changes in channel curvature and river meandering were also carried out using sediment transport model.

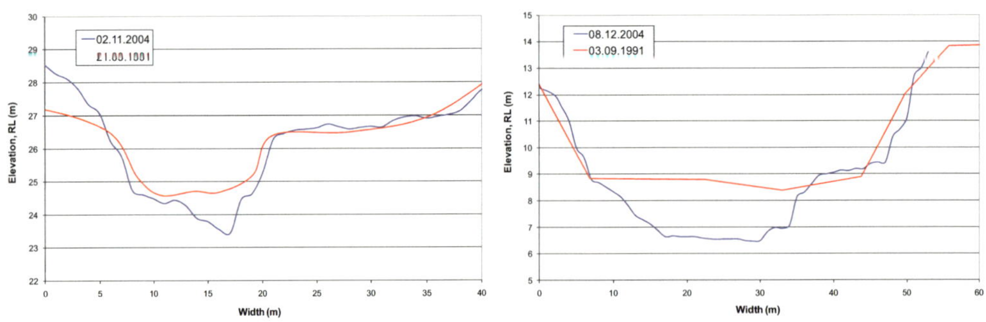

(a) CH 14390 (b) CH 3014

Fig. 5. Cross Section Changes

4.3 Sediment transport equation assessment

The analysis for a total of 22 sets of data based on averaged size of sediment (d_{50}) have been obtained for nine sediment transport equations including five bed load equations namely Einstein bed load function (Einstein, 1942, 1950), Einstein-Brown's equation (Brown, 1950), Meyer-Peter-Muller's equation (1948), Shields' equation (1936), Duboys' equation (1879) and four total load equations namely Yang's equation (1972), Engelund-Hansen's equation (1967), Ackers-White's equation (1973) and Graf's equation (1971). The performances of the equations were measured using the discrepancy ratio (DR), which is the ratio of the predicted load to measured load (DR=predicted/measured). In this study, a discrepancy ratio of 0.5 to 2.0 (DR = 0.5-2.0) was used as a criterion in the evaluation of the selected equations. However, the evaluation of these equations shows that all the existing equations, in most cases, over-predicted the measured values, as shown in Table 4. The result shows that Engelund & Hansen equations gives better prediction of measured data and yielded the highest percentage of data sets within discrepancy ratio of 0.5 to 2.0 at CH 14390 (50 %) and CH 3014 (41.67 %). The analysis also shows that all of the bed load equations gave unsatisfactory performance to predict the sediment load compared to total load equations.

Sediment Transport Equations	CH 14390		CH 3014	
	Total of Data	Total of Data Falls within 0.5-2.0	Total of Data	Total of Data Falls within 0.5-2.0
Einstein Bed Load Function (1942, 1950)	12	0	10	0
Einstein-Brown Equation (1950)		0		0
Meyer-Peter-Müller Equation (1948)		0		0
Shields Equation (1936)		0		0
Duboys' equation (1879)		0		0
Yang's equation (1972)		3		5
Engelund-Hansen equation (1967)		6		5
Ackers-White's equation (1973)		0		1
Graf equation (1971)		4		2

Table 4. Summary of Sediment Transport Assessment (Chang, 2006b; Ab. Ghani et al., 2007)

5. Sediment transport modeling

5.1 Software used

Studies of sediment transport, scour and fill, aggradation and deposition analyses can be performed by computer model simulation. The rapid pace of computer technology has been a milestone for mathematical models in sediment transport. As a result, the high demand on the models resulted in development of many models and the selection of the right model under certain constraints requires a comprehensive knowledge of the capabilities and features of available models. Recently, wide acceptance of a community sediment transport model would make the model a more effective tool for research, planning and design of river engineering, therefore numerous sediment models are available in the study of hydraulic and sediment transport modeling.

The review of capabilities and performance of sediment transport models has been discussed by the National Research Council (1983), Fan (1988), American Society of Civil Engineers Task Committee on Hydraulics, Bank Mechanics, and Modeling of River Width Adjustment (ASCE, 1998), Federal Interagency Stream Restoration Working Group (FISRWG, 2001) and Department of Water Resources, Resource Agency State of California (DWR, 2004). In addition, applications of the several commonly used sediment transport models have been described by Ab. Ghani et al. (2003) and Chang (2006b). These applications illustrate various capabilities of different models and each sediment transport model has its own limitations. The selection of the right model under certain constraints requires a comprehensive knowledge of the capabilities and features of available models.

The sediment transport model, FLUVIAL-12 (Chang 1982, 1984, 1988), which was first developed in 1972, has been selected for the Kulim River study. FLUVIAL-12 is developed for water and sediment routing in natural and man-made channels. The combined effects of flow hydraulics, sediment transport and river geomorphic changes are simulated for a given flow period. FLUVIAL-12 model is an erodible-boundary model that includes the width adjustment component, which simulates inter-related changes in channel-bed profile, width variation and changes in bed topography induced by the channel curvature effect. Besides that, bank erosion, changes in channel curvature and river meandering can also be modeled (Chang, 2006a).

The applicability of the FLUVIAL-12 model for the river channel responses under its existing conditions and proposed conditions in response to human intervention and the environmental impacts has confirmed by Chang et al. (2002), where FLUVIAL-12 simulations were made based on a 100-year flood as well as a long-term flood series. Besides that, several case studies of FLUVIAL-12 model applications as discussed by Chang (2006b) and Chang et al. (2008) also showed that FLUVIAL-12 was capable to predict river changes caused by nature and human activities, including general scour at bridge crossings, sediment delivery, channel responses to sand and gravel mining and channelization. Sediment delivery is defined as the cumulative amount of sediment that has been delivered passing a certain channel section for a specified period of time (Chang, 2006a).

5.2 Model input and output

The study reach covers approximately 14.5 km of Kulim River, from the upstream (CH 14390) to the Ara Kuda streamflow station (CH 0). The inputs to the FLUVIAL-12 model are described in detail in FLUVIAL-12 Users Manual (Chang, 2006a).

The geometry data consists of existing survey cross-sections in September 1991 between CH 1900 to CH 14390 at the upstream of Kulim River. However, the survey of CH 0 cross section in December 1995 was provided by DID Hydrology Division for the FLUVIAL-12 modeling requirement. In this study, a total of 120 existing survey cross sections were selected along the study reach to define the channel geometry as the input for FLUVIAL-12 model. FLUVIAL-12 has been used to simulate the channel geometry, lateral and vertical elevation changes for the flood events from 1991 to 2006.

The input hydrograph at Ara Kuda for year 1991 (Figure 6) was used for model sensitivity analysis whilst model calibration and validation was using the hydrograph from year 1991 to June 2006 (Figure 7). The rating curve which is used to define discharge variation of stage (water surface elevation) for the downstream boundary condition is shown in Figure 8; the

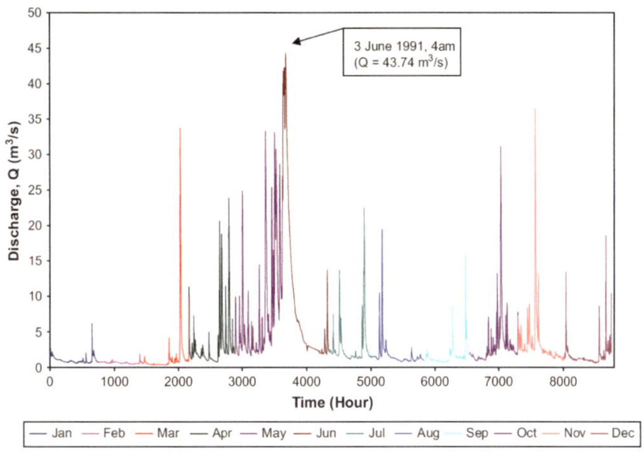

Fig. 6. Input Hydrograph for Year 1991

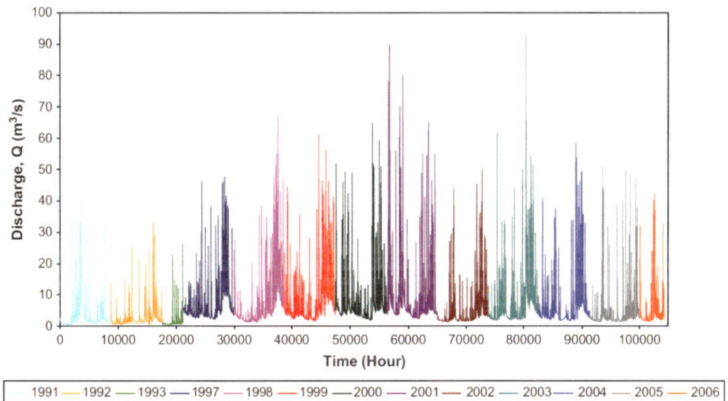

Fig. 7. Input Hydrograph for Year 1991 to June 1993, 1997 to June 2006

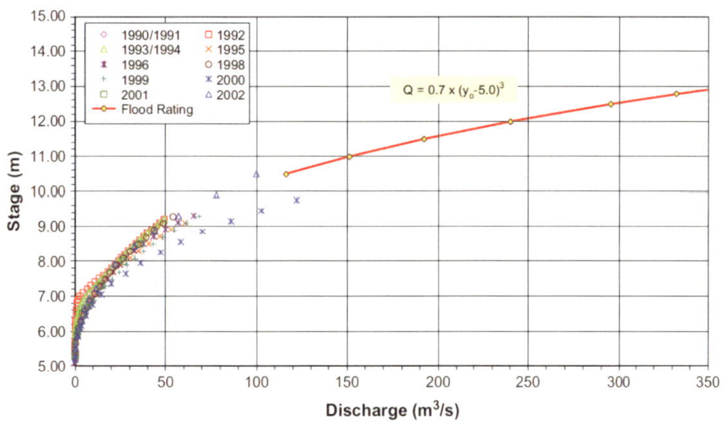

Fig. 8. Flood Rating Curve at Ara Kuda (CH 0)

shifts in stagedischarge relationships reflect the variability at Ara Kuda streamflow station derived from the past 12-year rating curve for Kulim River. The geometric mean of the bed material size fractions is adequately described from the sediment size distribution. Two sediment size distributions of such samples based on sieve analysis are required at the upstream (d_{50} = 1.50 mm) and downstream (d_{50} = 0.75 mm) cross sections to specify initial bed material compositions in the river bed (Figure 9). These input data, can be grouped into the categories of geometry, sediment and hydrology. A summary of the input and output parameter for each category is shown in Table 5.

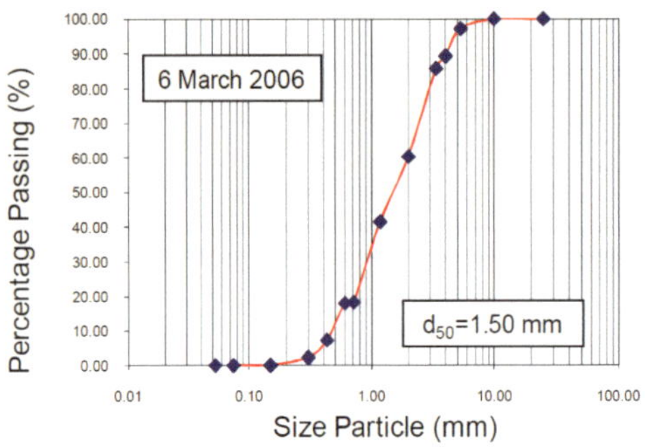

(a) Upstream (CH 14390)

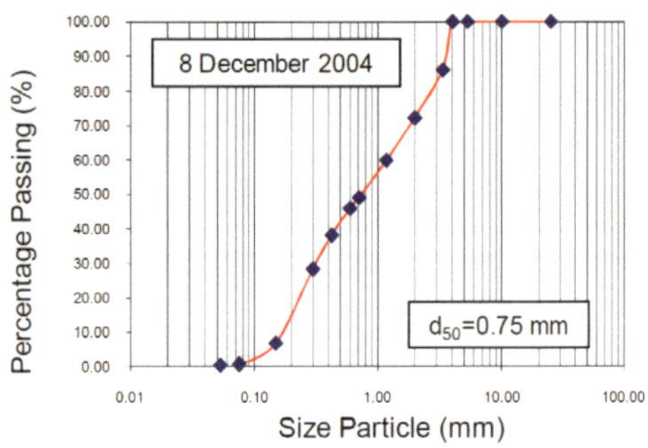

(b) Downstream (CH 3014)

Fig. 9. Initial Bed Material Size Distributions

Category	Category	Parameter	Value	Source
Input Parameter	Geometry	Cross section	per section (Total of cross sections = 120)	CH 1900 – CH 14390 (DID 1991 Survey) CH 0 (DID 1995 Survey)
		Reach lengths	per section (Total length = 14.4km)	1991 (DID 1991 Survey)
		Roughness coefficient	Same by cross section (n = 0.020, 0.025, 0.030, 0.035, 0.040 were evaluated during the sensitivity analysis)	Values static at all levels of flow.
		Radius of curvature	per section	1991 (DID 1991 Survey)
	Sediment	Sediment samples	2 sediment size distributions of such samples are required (upstream and downstream section)	Data sampling at CH 14390 (Year 2006) and CH 0 (Year 2004)
		Regular non-erodible bank	Generally fix at left and right bank, varies by cross section	DID 1991 Survey
		Sediment transport formula	Seven sediment transport formulas were evaluated during the sensitivity analysis	Graf's sediment formula (1971) Yang's unit stream power formula (1972) Engelund-Hansen sediment

				formula (1967) Parker gravel formula (1982) Ackers-White sediment formula (1973) Meyer-Peter Muller formula (1948) Singer-Dunne formula (2004)
		Specific Gravity	2.65	Default (Soulsby, 1997)
	Hydrology	Discharge hydrograph	Varies by hydrograph	Historical hydrograph for Kulim River at Ara Kuda streamflow station Design Hydrograph from past study (DID, 1996)
		Rating Curve	Year 1991 to Year 2002	Developed by DID Hydrology Division
Output parameter	Geometry	Width Depth Cross-sectional area Slope	Changes over time in water surface, bed elevation and thalweg profiles. Simulation of curvature induced aggradation and deposition.	
	Sediment	Mean sediment size (d_{50}) Bed material size fractions	Changes over time in sediment transport, channel scour and fill, aggradation and degradation	
		Sediment concentration Sediment yield	Sediment delivery or the total bed material yield during the study period	
	Hydraulic	Water surface	Simulated water surface based on input hydrograph	
		Mean velocity Froude number	Flow data sets for representative cross sections in the study reach	

Table 5. Summary of Input and Output Parameter for FLUVIAL-12 in Present Study (Chang et al., 2008)

5.3 Sensitivity analysis

An analysis was conducted to evaluate the sensitivity of the modelling results to changes in input parameters. To determine the sensitivity of FLUVIAL-12, which including flow, sediment transport and the channel geomorphic changes caused by the variation of each parameter, different values of the parameter were used in simulation runs and the results obtained are compared. Sensitivity analysis is an important step to be taken for more effective use of a model. Major items that required sensitivity test include roughness coefficient, sediment transport equations, channel curvature and number of cross section (reach length between two sections). This sensitivity analysis was carried out using the existing survey cross section and hydrograph for the year of 1991. However, the accuracy of the model is limited to the quality and quantity of the input data. Therefore, using available hydraulic and hydrology data including cross section spacing will affect the quality of the output data. Besides that, selection of the sediment transport formula and model calibration for roughness coefficient are also essential. Table 6 shows the summary of the sensitivity analysis for Sungai Kulim using FLUVIAL-12.

Parameter	Values Tested	Comments
Roughness Coefficient	Range: 0.020 – 0.050	Started out with values recommended in the 0.020. Some values were then changed. Water Surface is increasing when roughness coefficient increasing.
Sediment Transport Equation	7 sediment transport equations	All equations were tested due to field observation. Selection of the proper and applicable sediment transport formula is essential.
Channel Curvature	Zero curvature and curvature	Simulation of curvature induced aggradation and deposition in the model based on the flow curvature.
Number of Cross Sections	120, 62 and 32	Shortened distance between cross section or closely spaced along a reach produce more accuracy result in channel geometry changes.

Table 6. Summary of Sensitivity Analysis

5.4 Model calibration and validation

The simulation of the FLUVIAL-12 was obtained using 1991 cross section survey and hydrograph. Based on measured water levels, predictions using both roughness coefficients are close to the observed data during low flow. However, as the field data was not available from year 1991 to 2003, a long-term simulation has been carried out to calibrate and validate the model based on the recent measured water level and bed level data that were obtained from 2004 to 2006. Therefore, the calibration of the roughness coefficient using measured water level and bed level in November 2004 is done. As a part of the calibration procedure, the model was run for 12-year period between 1991 to 1992 and 1997 to 2006.

The results of the model simulation during the calibration period agree very well (Table 7 and Figure 10), and it can be concluded that prediction using roughness coefficient n = 0.030 and Engelund-Hansen formula were in good agreement with measured water levels and bed profiles and used for model validation. As a part of validation, measured water levels and bed profiles, during September 1991, January 2005 and March 2006 was compared to the predicted water levels and bed profiles by FLUVIAL-12 (Table 8). Longterm simulations including of the historical flood events showed very good results for both calibration and validation. Good agreements were obtained for both water level and bed levels between the measured and predicted by FLUVIAL-12 model.

Roughness coefficient *n*	Location	Water Level (m)					Thalweg Level (m)				
		Measured	Yang fomula		Engelund-Hansen fomula		Measured	Yang fomula		Engelund-Hansen fomula	
			Predicted	Difference	Predicted	Difference		Predicted	Difference	Predicted	Difference
0.025	CH 0	7.45	7.80	+0.35	7.80	+0.35	5.05	5.34	+0.29	5.29	+0.24
	CH 3014	8.61	8.14	-0.47	8.13	-0.48	6.66	6.79	+0.13	6.96	+0.30
	CH 8185	13.55	12.47	-1.08	13.67	+0.12	12.27	11.68	-0.59	12.85	+0.58
	CH 14390	25.61	26.00	+0.39	25.99	+0.38	23.45	24.58	+1.13	24.58	+1.13
0.030	CH 0	7.45	7.80	+0.35	7.82	+0.37	5.05	5.40	+0.35	5.27	+0.22
	CH 3014	8.61	8.36	-0.25	8.29	-0.32	6.66	6.57	-0.09	6.69	+0.03
	CH 8185	13.55	13.00	-0.55	13.36	-0.19	12.27	11.87	-0.40	12.16	-0.11
	CH 14390	25.61	26.17	+0.56	26.14	-0.53	23.45	24.58	+1.13	24.58	0.00
0.035	CH 0	7.45	7.80	+0.35	7.93	+0.48	5.05	5.33	+0.28	5.26	+0.21
	CH 3014	8.61	8.59	+0.02	8.52	-0.09	6.66	7.30	+0.64	7.53	+0.87
	CH 8185	13.55	13.55	0.00	13.45	-0.10	12.27	12.26	-0.01	12.31	+0.04
	CH 14390	25.61	26.20	+0.59	26.29	+0.68	22.85	24.58	+1.73	24.58	+1.73

Table 7. Comparison of Simulated Water Level and Bed Profile with Measured Data during 2 Nov 2004 for Roughness Coefficient n = 0.025, 0.030 and 0.035 (Chang et al., 2008)

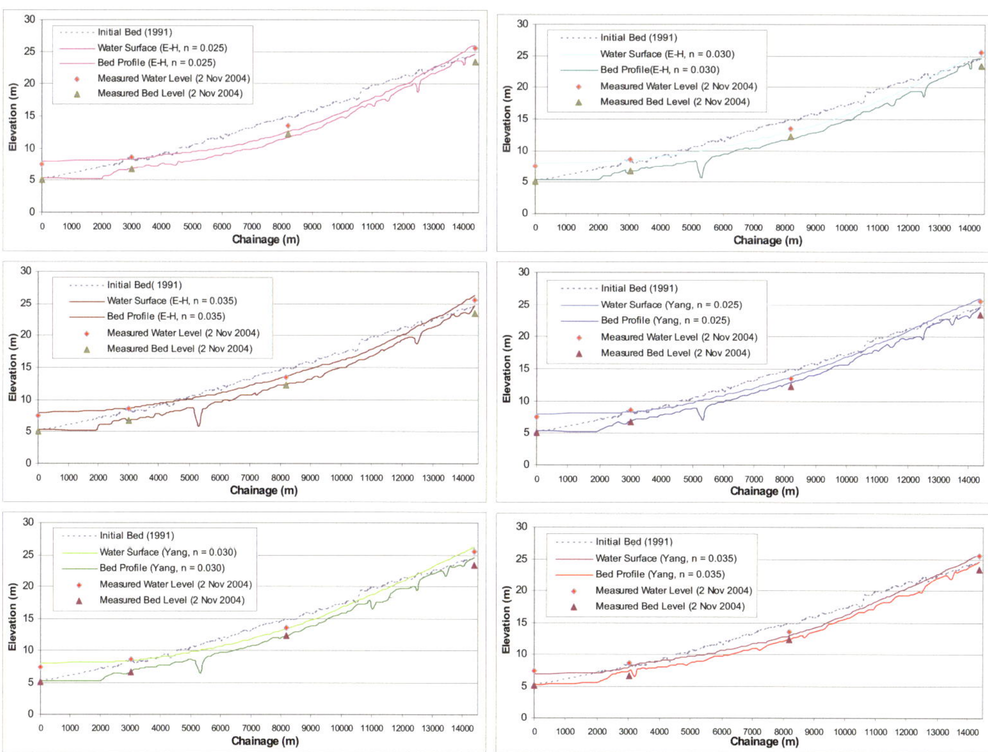

Fig. 10. Comparison of Water Level and Bed Profile for Roughness Coefficient n = 0.025, 0.030 and 0.035 (2 Nov 2004)

Date	Location	Water Level (m)			Thalweg Level (m)		
		Measured	Predicted	Difference	Measured	Predicted	Difference
20 September 1991	CH 0	6.40	6.00	-0.40	5.05	5.34	+0.29
	CH 10195	18.25	18.07	-0.18	6.66	6.79	+0.13
	CH 10438	18.53	18.39	-0.14	12.27	11.68	-0.59
	CH 14091	24.84	24.67	-0.17	23.45	24.58	+1.13
	CH 14206	24.86	24.80	-0.06	N/A	N/A	N/A
11 January 2005	CH 0	6.10	6.09	-0.01	5.20	5.26	+0.06
	CH 10195	7.03	7.19	+0.16	6.50	6.59	+0.09
	CH 10438	17.32	17.14	-0.18	16.81	16.91	+0.10
	CH 14091	24.52	25.09	+0.57	23.42	24.58	+1.16
8 March 2006	CH 0	5.90	5.96	+0.06	5.20	5.27	+0.07
	CH 10195	7.38	7.05	-0.33	6.48	6.46	-0.02
	CH 10438	24.36	24.99	+0.63	24.04	24.58	+0.54
	CH 14390	25.61	26.20	+0.59	22.85	24.58	+1.73

Table 8. Comparison of Simulated Water Level and Bed Profile with Measured Data

5.5 Model simulation

Engelund-Hansen formula and roughness coefficient n = 0.030 were found to be the best combination to represent the sediment transport activity in the study reach throughout the model calibration and validation. The sediment transport modeling was conducted based on three scenarios. These include the existing condition modeling using October 2003 flood hydrograph, future condition modeling by using the design flood hydrograph for the Kulim River based on 2010 landuse (DID, 1996) and long-term modeling by relicensing the time frame using hydrograph for year 1991-1992 and 1997-2006 to predict future ongoing changes for the next 10 years.

The peak discharge of 92.90 m^3/s measured on 5 October 2003, which was the highest discharge measured in a 42-year period since 1960 is adopted as the design peak discharge for existing condition. Consequently, sediment transport modeling was carried out for this flood event (3 to 19 October 2003) as shown in Figure 11. Spatial variations of the sediment delivery during the October 2003 flood are shown in Figure 12. Sediment delivery generally decreased towards downstream especially near to the river sand mining site at CH 5064. This pattern indicated that erosion occurred at upstream and more sediment deposited at downstream of Kulim River.

Peak water surface and changes of the channel geometry due to scour and fill were depicted by the simulated changes in channel bed profile as illustrated in Figure 13. From the simulation results, flood level was higher at the downstream compare to the upstream of Kulim River. Whilst, the results also show that scour of the bed occurred at upstream and the cross sections near to the sand mining area (CH 5064) were subjected to greater changes than other cross sections. Commonly, channel degradation was predicted at most cross sections at Kulim River after the flood event.

Figure 14 shows the sediment transport rates at peak discharge during 2003 flood along the river. Figure 15 shows the example of cross section changes for several locations along Kulim River. In general, the river is stable at most locations after October 2003 flood with the exception of CH 5306 and CH 12490 where lateral migration is predicted at these two locations.

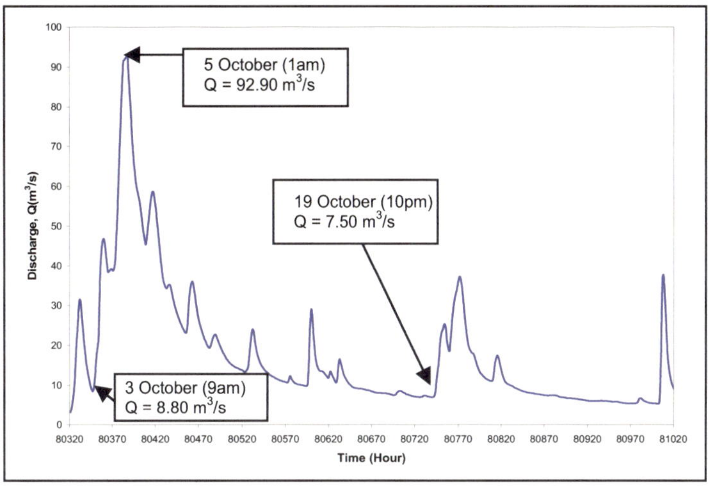

Fig. 11. Hydrograph of the October 2003 Flood at Ara Kuda (CH 0)

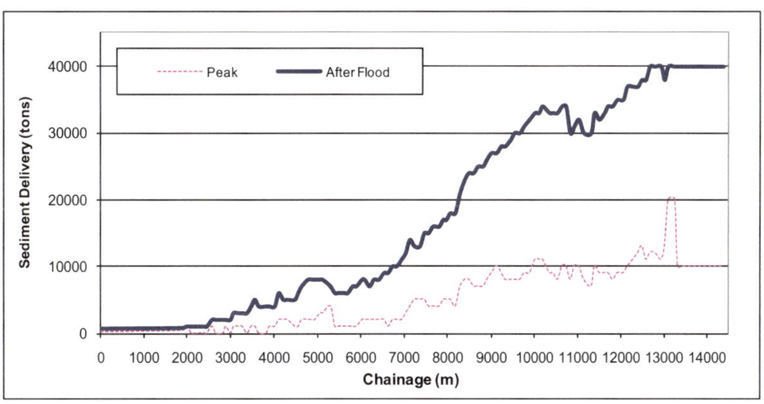

Fig. 12. Spatial Variations of the Sediment Delivery during the October 2003 Flood

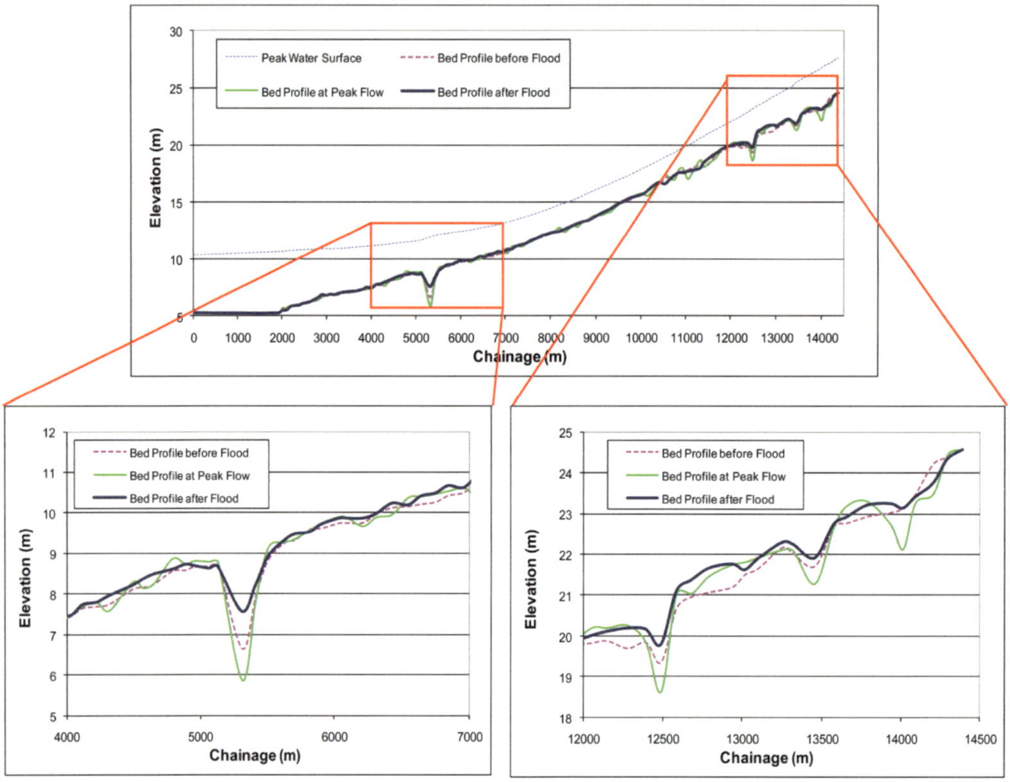

Fig. 13. Prediction of Water surface and Bed Profile Changes during October 2003 Flood

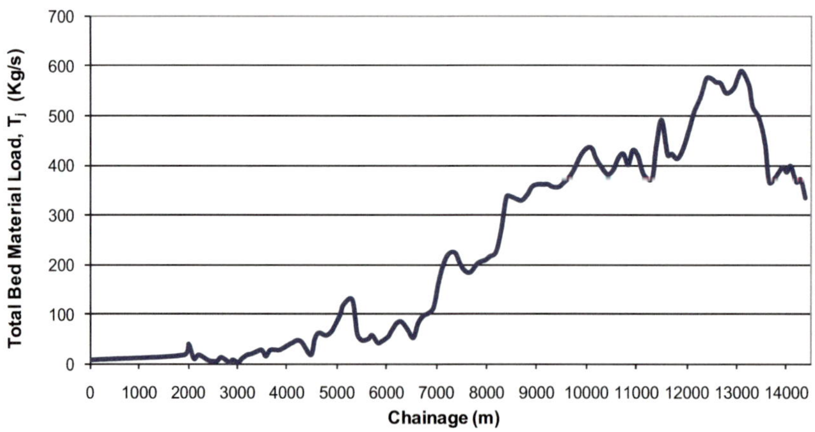

Fig. 14. Sediment Transport Rate at Peak during October 2003 Flood

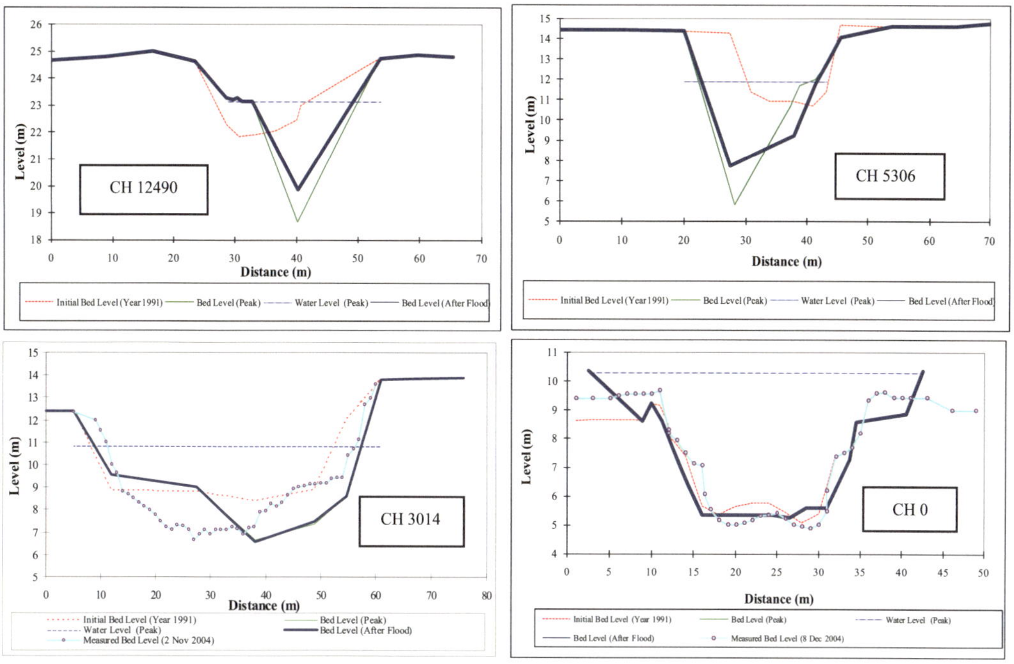

Fig. 15. Modeled Cross Section Changes before and after October 2003 Flood

The design flood hydrograph for the Kulim River based on 2010 landuse (DID, 1996) is shown in Figure 16. The critical peak flow of the event is 306.6 m^3/s (18-hour rainfall duration). Simulated peak water surface and channel bed changes for Kulim River based on design hydrograph are shown in Figure 17. The cross sections especially near to the sand mining area and few cross sections especially CH 10000 to CH 14390 were subjected to greater changes than other cross sections. In spite of this, channel degradation was predicted at most cross sections after the peak. Figure 18 shows the cross section changes for two selected locations along Kulim River.

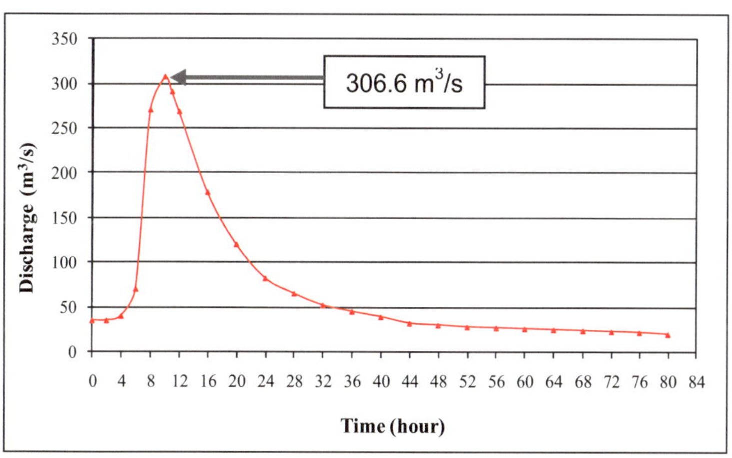

Fig. 16. Design Hydrograph for 2010 Landuse (DID, 1996)

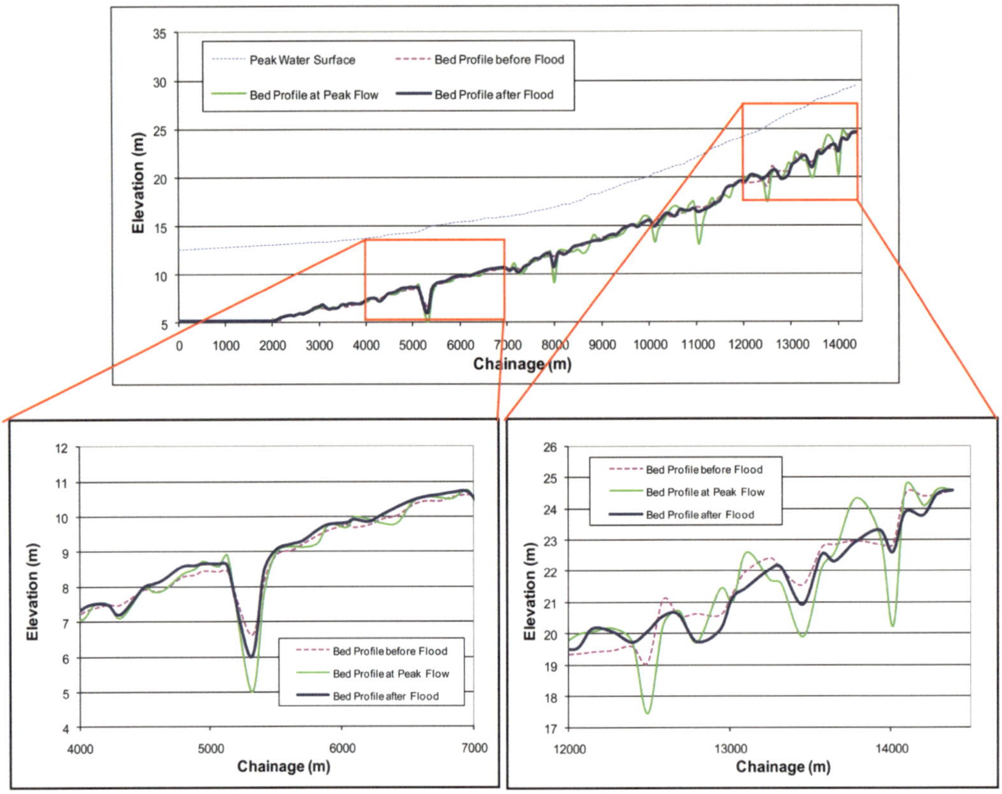

Fig. 17. Water Surface and Bed Profile Changes based on Design Hydrograph

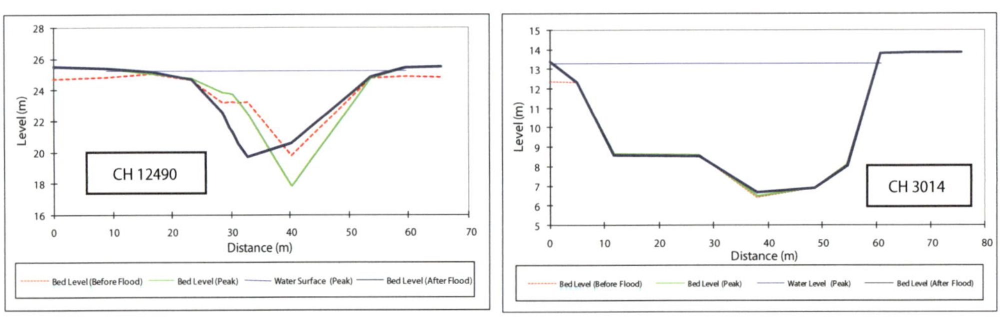

Fig. 18. Modeled Cross Section Changes before and after Design Flood

FLUVIAL-12 model was run to predict the channel geometry changes and sediment delivery for the next 10 years. Future changes for the next 10 years were simulated by using hydrograph as shown in Figure 7, which consists of 50-year ARI and 100-year ARI flood events. Sediment delivery or the amounts of sediment moving past each cross section predicted for the next 10 years (Year 2016) is shown in Figure 19. The simulation results show that the amount of sediment delivery was twice for year 2016 compared to the year 2006, but lesser sediment delivery at the downstream of Kulim River. The decreasing trend of sediment delivery indicates long-term sediment deposition at the downstream of Kulim River.

Simulation for Kulim River based on the time series illustrated the changes of the channel geometry as shown in Figure 20. The cross sections especially CH 10000 to CH 14000 are subjected to change with sediment aggradation, whilst sediment deposition occur at CH 6000 to CH 10000. Figure 21 shows the spatial variations of the predicted median grain size in year 2006 and 2016. The model run shows a large decrease in the sediment size at middle reach of Kulim River between years 2006 to 2016; where the reach-mean sediment size

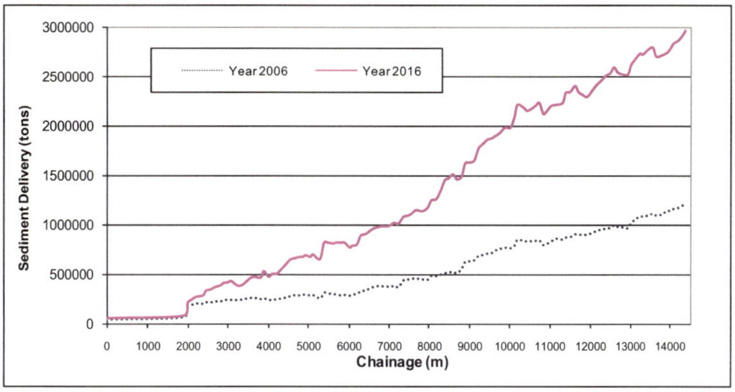

Fig. 19. Spatial Variations of the Predicted Sediment Delivery

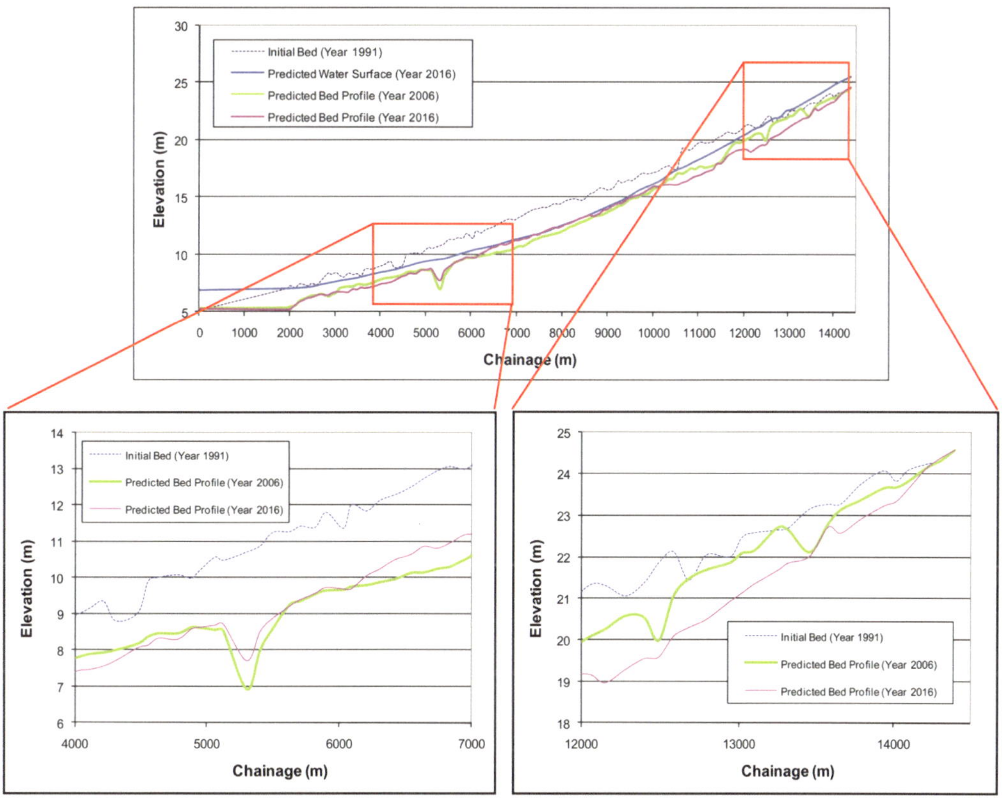

Fig. 20. Water Surface and Bed Profile Changes based on Design Hydrograph

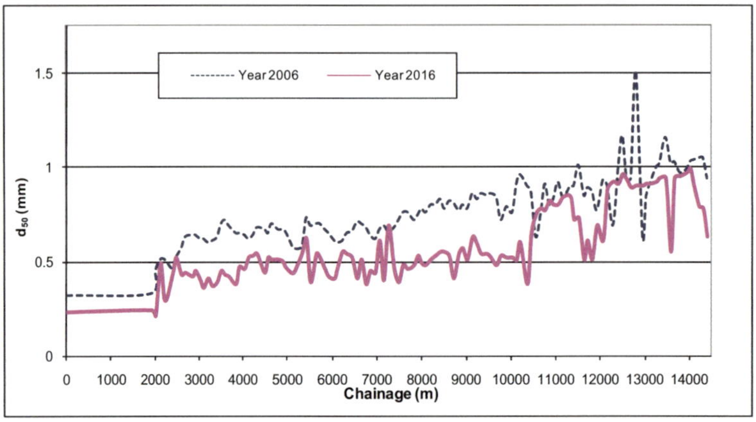

Fig. 21. Spatial Variations of the Predicted Median Grain Size for Year 2006 and 2016

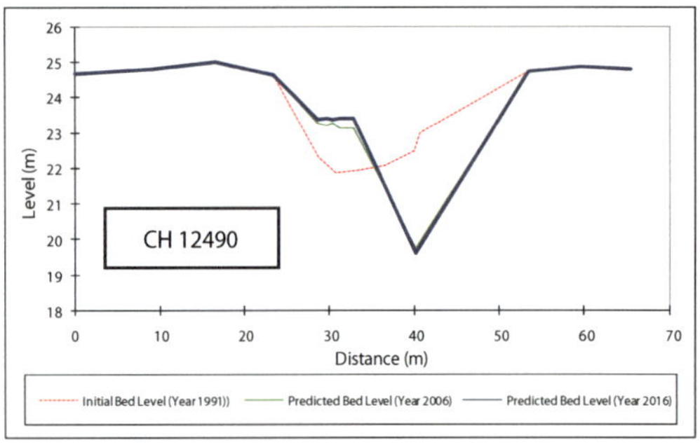

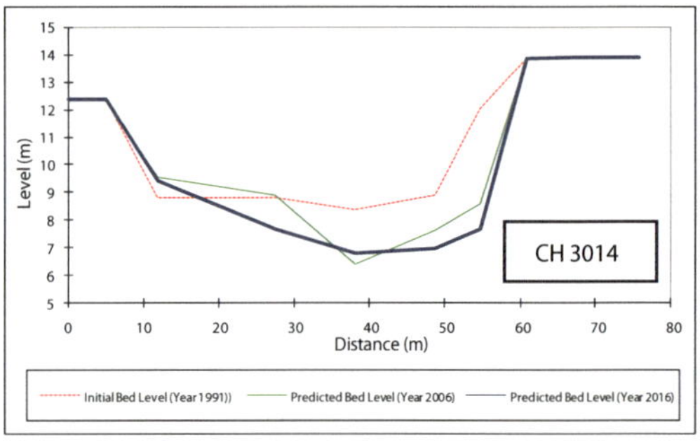

Fig. 22. Predicted Cross Section Changes for Year 2006 and 2016

decrease from 0.77 mm to 0.58 mm. As the channel bed became finer, more sediment was removed by erosion. Figure 22 shows the example of cross section changes for three locations along Kulim River.

In general, it is found that Kulim River will be in equilibrium conditions with slight degradation or erosion which deepen the river. The modeled results show that future changes in cross sectional geometry will generally be limited and erosion along the reach will be slowed down in the simulation period from 2006 to 2016. Thus, Kulim River was predicted to be stable at most locations.

6. Conclusion

Flooding in Kulim River is found to affect channel geometry, cross sectional geometry, sediment size and sediment delivery, which consists of scour and fill. Three scenarios was evaluated for Kulim River; the model simulation results for existing conditions, future conditions and long-term modeling show that the sediment size and channel geometry in Kulim River changed significantly and the amount of sediment delivery trend decrease with time indicates that long term sediment aggradation occurred at upstream and deposition occurred at downstream of Kulim River. However, modeled results show that future changes in cross sectional geometry will be limited and erosion along the reach will slow down from 2006 to 2016. The results based on the water surface profile simulated from the model should also be considered that the proposed bund level and bank protection should stay above the predicted water surface to avoid overtopping and reduce the flooding impact. The present study provides an estimate of sediment transport in moderate sandy stream and serves as a reference for sediment transport modeling of sandy streams in Malaysia and overseas.

7. References

Ab. Ghani, A., Ali, R., Zakaria, N.A., Abu Hasan, Z., Chang, C.K. & Ahamad, M.S.S. (2010). A Temporal Change Study of the Muda River System over 22 Years. *International Journal of River Basin Management*, IAHR & INBO. Vol. 8, No. 1, 25-37. ISSN: 1571-5124.

Ab. Ghani, A., Zakaria, N.A., Abdullah, R., Chang, C.K., Sinnakaudan, S.K., Mohd Sidek, L. (2003). Guidelines for Field Data Collection and Analysis of River Sediment. Department of Irrigation and Drainage Malaysia, Kuala Lumpur. 35 pp.

Ab. Ghani, A., Zakaria, N.A., Chang, C.K., Ariffin, J., Abu Hasan, Z., Abdul Ghaffar, A.B. (2007). Revised Equations for Manning's Coefficient for Sandbed Rivers. *International Journal of River Basin Management,* Vol. 5, No. 4, 329-346. ISSN: 1571-5124.

American Society of Civil Engineers or ASCE (1998). River width Adjustment II: Modeling, by the ASCE Task Committee on Hydraulics, Bank Mechanics, and Modeling of River Width Adjustment. *Journal of Hydraulic Engineering,* 124 (9), 903-917. ISSN: 0733-9429.

Chang, H.H. (1982). Mathematical Model for Erodible Channels. *Journal of the Hydraulics Division,* 108 (HY5), 678-689. ISSN: 0733-9429.

Chang, H.H. (1984). Modeling of River Channel Changes. *Journal of Hydraulic Engineering* 110 (2), 265-267. ASCE157-172. Closure in 113(2), 1987. ISSN: 0733-9429.

Chang, H.H. (1988). *Fluvial Processes in River Engineering*. John Wiley and Sons, ISBN: 0-471-63139-6, New York. 432pp.

Chang H. H., Pearson, D. & Tanious, S. (2002). Lagoon Restoration near Ephemeral River Mouth. *Journal of Waterway, Port, Coastal, and Ocean Engineering,* Vol. 128, No.2, 79-87. ISSN: 0733-950X.

Chang, H.H. (2006a). *Generalized computer program: FLUVIAL-12 Mathematical Model for Erodible Channel Users Manual*. San Deigo State University, San Deigo, California.

Chang, C.K., Ab. Ghani, A., Zakaria, N.A., & Abdullah, R. (2005). *Sediment Transport in Kulim River, Malaysia*. In: Proceedings of XXXI IAHR Congress: Water Engineering for the Future - Choice and Challenges, September, 2005. Seoul, Korea, pp. 1154-1162, ISBN: 89-87898-24-5.

Chang, C.K. (2006b). *Sediment Transport in Kulim River, Kedah*. M.Sc. Thesis. Universiti Sains Malaysia. Penang.

Chang, C.K., Ab. Ghani, A., Abdullah, R., Zakaria, N.A. (2008). Sediment transport modeling for Kulim River : a case study. *Journal of Hydroenvironment Research,* Vol. 2, No. 1, 47-59. IAHR & KWRA. ISSN: 1570-6443.

Department of Irrigation and Drainage Malaysia or DID. (1976). *River Discharge Measurement by Current Meter - Hydrological Procedure No. 15*. DID, Kuala Lumpur.

Department of Irrigation and Drainage Malaysia or DID. (1977). *The Determination of Suspended Sediment Discharge - Hydrological Procedure No. 19*. DID, Kuala Lumpur.

Department of Irrigation and Drainage Malaysia. (1996). *Study on Flood Mitigation and Drainage Master Plan for Kulim and its Surroundings*. Final Report. DID, Kedah Darul Aman.

Department of Irrigation and Drainage Malaysia or DID. (2009). *Study on River Sand Mining Capacity in Malaysia*. DID, Kuala Lumpur.

Department of Water Resources or DWR (2004). *Effects of Project Operations on Geomorphic Processes Downstream of Oroville Dam: Task 7 - Hydraulic and Sediment Transport Modeling With FLUVIAL-12*. Draft Final Report. Oroville Facilities Relicensing FERC Project No. 2100, The Resources Agency, State of California.

Edwards, T.K., Glysson, G.D. (1999). *Field Methods for Measurement of Fluvial Sediment*. U.S. Geological Survey Techniques of Water-Resources Investigations. Book, Chapter C2.

Fan, S.S. (1988). *Twelve Selected Computer Stream Sedimentation Models Developed in the United State*. In: Proceeding of the Interagency Symposium on Computer Stream Sedimentation Model, Denver, Colorado. Published by the Federal Energy Regulatory Commission, Washington, DC.

Federal Interagency Stream Restoration Working Group or FISRWG. (2001). *Stream Corridor Restoration: Principles, Processes, and Practices*. Federal Interagency Stream Restoration Working Group (15 Federal agencies of the US Government), Portland. ISBN: 0-934213-59-3.

Lagasse, P.F., Schall, J.D., Richardson, E.V. (2001). *Stream Stability at Highway Structures*, US Department of Transportation, Federal Highway Administration. Publication No. FHWA NHI 01-002 (Hydraulic Engineering Circular No. 20), 3rd Edition, Washington, D.C.

Majlis Daerah Kulim or MDK (1993). *Kulim Structure Plan*, 1990-2010. Kuala Lumpur.

National Research Council. (1983). An Evaluation of Flood-level Prediction using Alluvial River Model. National Academy Press, Washington, D.C

Richardson, E.V., Simons, D.B., Lagasse, P.F., (2001). *River Engineering for Highway Encroachments - Highways in The River Enviroment*, US Department of Transportation, Federal Highway Administration. Publication No. FHWA NHI 01-004 (Hydraulic Design Series Number 6), Washington, D.C.

Sirdari, Z.Z. (2009). *Effects of Channelization on the Stability of Kulim River*. M.Sc. Thesis. Universiti Sains Malaysia. Penang.

United States Army Corps of Engineers or USACE (1995). *Sedimentation Investigations of Rivers and Reservoirs. USACE Engineering and Design Manual*. Publication No. EM 1110-2-4000, Washington, D.C.

Yuqian, L. (1989). *Manual on Operational Methods for the Measurement of Sediment Transport*. Operational Hydrology Report No. 29, World Meteorological Organisation, Geneva, Switzerland, 169 p. ISBN: 92-63-10686-X.

10

Changes in Sediment Transport of the Yellow River in the Loess Plateau

Faye Hirshfield and Jueyi Sui
University of Northern British Columbia
Canada

1. Introduction

Sediment erosion is a pressing problem throughout the world as it leads to loss of resources such as agricultural land. Soil erosion most commonly occurs as a result of the forces exerted by wind and water. Human induced landscape change can expedite soil erosion due to removal of vegetation, urbanization and rangeland grazing (to name a few). Sediment erosion can lead to increased sediment input to nearby rivers which can alter river channel morphology through increased sediment deposition. Sediment transport in rivers is also important on a global scale as sediments carry organic carbon from the land to oceans via river channels (Ludwig et al., 1996).

River sediment levels depend largely on the surrounding landscape. Areas where the soil is being impacted directly through activities such as cultivation and urbanization will generally contribute large amounts of sediment to a nearby channel. As seen over the last century, high river sedimentation can lead to issues with drinking water quality and engineering structures such as reservoirs. In order to manage the landscape effectively, soil conservationists in the United States developed a universal soil loss equation during the 1950's (Wischmeier, 1976). The soil loss of an area is calculated as follows,

$$A = RKLSCP \tag{1}$$

where A is the computed soil loss per unit area in tons per acre, R is the rainfall factor, as it defines the erosive force of a specific rainfall, K is the soil erodability factor, L is the slope length factor, S is the slope gradient factor, C is the cropping management factor as it defines the rate of soil loss from a field with specific cropping practice and P is the erosion control factor as it defines whether any erosion management factors have been implemented The universal soil loss equation has since been modified a number of times and still acts as a powerful tool to calculate soil movement across a landscape.

Alternative techniques that are applied in the study of sediment transport are the interpretation of rainfall data and watershed area to calculate runoff depth and the use of river discharge data along with sediment level measurement to calculate river sediment yield. Annual runoff is given by the following equation,

$$Runoff\ Depth = \frac{Q(365)(24)(3600)}{A(1000)} \tag{2}$$

where A is the drainage area in kilometres squared and Q is river discharge in cubic metres. River sediment yield is given by the following equation,

$$\frac{\textit{suspended sediment flux}}{\textit{drainage area}} \tag{3}$$

The Yellow River in China has long been known for its sediment problems. During the past 50 years, industrial development (hydropower) and urbanization in China have led to drastic changes in topography. Decreases in vegetation cover have led to increased soil erosion through wind and runoff processes. Vast amounts of research have examined the sediment levels of the Yellow River. The purpose of the chapter is to review characteristics of the Loess Plateau region of the Yellow River watershed, examine how sediment transport has changed over the past 50 years and assess whether sediment conservation initiatives have been successful. The chapter is outlined as follows: (a) an introduction to the Yellow River watershed and Loess Plateau, (b) review government policy that has been developed in order to manage soil erosion of the region, (c) an in depth look at research surrounding sedimentation in the Yellow River's Middle Reach, and (d) success in sediment management of the Loess Plateau.

2. The Yellow River watershed

The Yellow River, the second largest river in China, and 6th longest in the world is an area of historical significance as it was once the most prosperous region in China. The area is both a place of prosperity and tragedy as over the centuries flooding has caused devastation in the area. The Yellow River headwaters are located in the Bayan Har Mountain region of western China. The river flows east, crossing 9 provinces, and empties into the Bohai Sea. Since the Yellow River watershed is so large, approximately 742,440 square kilometres in area and a channel length of 5464 kilometres, the basin is typically divided into 3 geographic areas: the Upper, Middle and Lower reaches (Figure 1). The Upper Reach originates in China's Qinghai province and travels approximately 3,500 kilometres before the channel turns south in Inner Mongolia. The Upper Reach constitutes just over 50 percent of the total basin area and varies in topography from mountain regions to swamps to grasslands. Much of the Upper Reach channel morphology is dominated by deep canyons and steep channel gradients which make the area suitable for hydro electric power stations

The Middle Reach of the Yellow River begins in Inner Mongolia and stretches approximately 1,200 kilometres to the Province of Henan. The Middle Reach acts as a natural border for the provinces of Shaanxi and Shanxi. The Middle Reach cuts through the Loess Plateau, an area dominated by highly erosive loess sediments. The Loess Plateau covers an area of approximately 640,000 square kilometres. The Yellow River's name refers to the often yellowish color the water has from the high levels of loess. Loess is windblown sediment that contains various levels of sand, silt and clay. Loess grains can be angular in shape which often leads to the formation of loess ridges and banks which dominate the Middle Reach. The Loess Plateau is the main source of sediment for the Yellow River. Due to the large amount of sediment transported from the Loess Plateau into the Yellow River, the Yellow River is considered to be one of the most sediment laden rivers in the world.

The Lower Reach is approximately 790 kilometres long and flow northeast across the North China Plain and into the Bonai Sea. The Lower Reach has little elevation change, the channel widens considerably and river flows are calm and slow. The Lower Reach is an area of sediment deposition as a result of the sediment loading that takes place in the Loess Plateau region. Sediment deposition in the Lower Reach has caused the river level to rise several metres above the surrounding land elevation. As reported by the Chinese Association of Hydraulic Engineering, during the past 50 years the river bed in the Yellow River Lower Reaches has been raised from 1.9 metres to 3 metres. Dike construction in the Lower Reach is important for flood prevention and allows for human settlement. Dike construction, while it helps to prevent flooding it does not eliminate the risk. Historical records indicate that in 1,500 of the past 2,540 years dike breaches have occurred (Shi & Ye, 1997). Dike failure has caused property devastation and loss of human life.

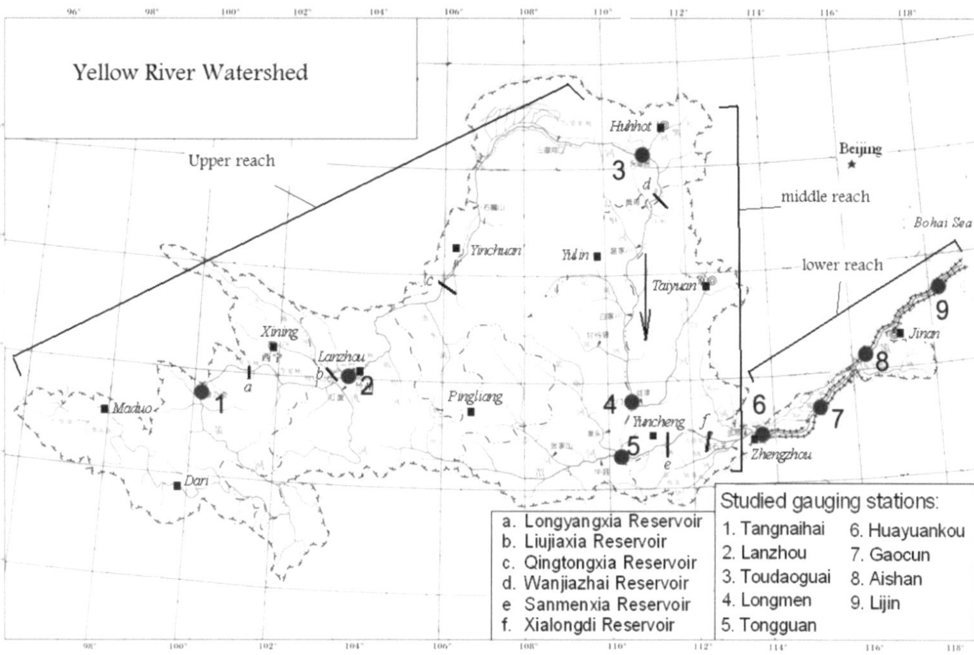

Fig. 1. Yellow River watershed (modified from Lui et al., 2008).

Prior to 1960, no hydropower development existed on the Yellow River. During the 1960's and 1970's, the Yellow River watershed experienced economic prosperity which led to the construction of hydropower stations. As reported by Xu (2003), the Yellow River watershed currently has over 3,300 reservoirs and 17,000 hydraulic structures for water pumping and transferring. There are 9 major hydroelectric stations on the main stem of the Yellow River. The hydropower stations serve not only to provide electricity but are also designed to reduce sedimentation, provide flood control and provide irrigation water supply. Following the extensive hydro development on the Yellow River, soil conservation programs have grown from 8,000 square kilometres in the 1950's to 171,300 square kilometres in the 1990's (Gu, 1994, 2002).

2.1 The Loess Plateau

The Loess Plateau covers an area of approximately 640,000 square kilometres, similar to the size of France (Figure 2). It spans the Middle Reach of the Yellow River and is the world's largest deposit of loess. Historically, the Loess Plateau was highly fertile area which contributed to the establishment of ancient Chinese civilizations originating in the region. Due to the angular characteristics of loess sediments, inhabitants of the region were able to construct dwellings in the form of caves. The Loess soil is rich in organic content and ranges from 100 to 300 metres in depth (Cai, 2001). The north and east regions of the Loess Plateau are home to mixed deciduous forests. In the eastern province of Shanxi is the Li Shan Nature Reserve. This nature reserve covers an area of approximately 250 square kilometres, supports a temperate forest and is home to rare species of salamander, deer and pheasant. High sediment levels in the Yellow River are a direct result of erosion from the Loess Plateau. The Loess Plateau is thought to have one of the highest erosion rates in the world (Fu et al., 2000). The annual soil loss from the Loess Plateau is estimated to range from 200 to 30,000 tonnes/square kilometres (Liu, 1985; Zhu et al., 2004). The Yellow River transports approximately 30 times the sediment of the Nile and 98 times the sediment of the

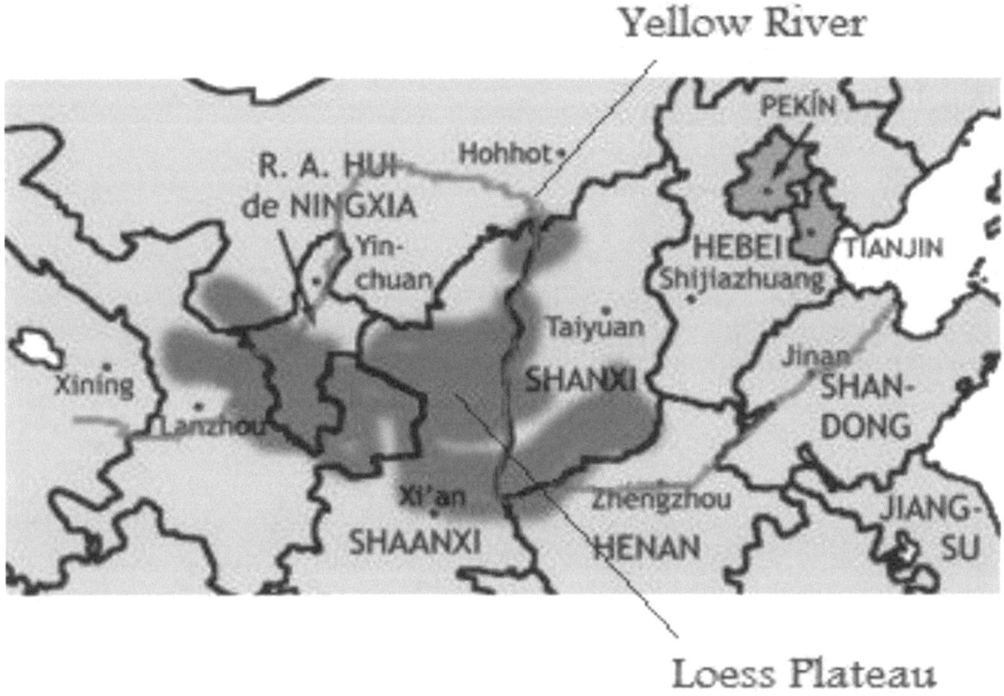

Fig. 2. China's Loess Plateau as indicated by dark grey shaded area.

Mississippi River (Kleine, 1997). The fine sand and silt particles that comprise loess are highly susceptible to wind and water erosion. Loess soils can be eroded at a faster rate than any other soil type (Pye, 1987). Human activity and expansion of farmland have removed a lot of the areas forests and led to increased soil erosion. As China's population grows, the government is planning to intensify agricultural production of the Loess Plateau region.

2.2 Loess Plateau Watershed Rehabilitation Project

In 1994, in response to growing concern over desertification of the plateau, the Loess Plateau Watershed Rehabilitation Project was launched and co-ordinated by the Chinese

Fig. 3. Loess Plateau and the Yellow River.

Government and World Bank. The goal of the project was to increase agricultural productivity and income of the area while also decreasing the sediment transport to the Yellow River. After a three year planning phase the following policy guidelines were introduced:

1. Tree cutting banned-people in China were not allowed to undertake indiscriminate tree cutting.
2. Tree planting banned on steep slopes-historical planting of crops on hillsides and gullies was no longer permitted.
3. Land Tenure guidelines-specific rules were set out for farmers designating their rights and responsibilities for each developed terrace and field.
4. Banning of free range grazing-sheep and goats were no longer allowed to wander freely and graze, but rather had to be contained in pens or fenced areas.

In addition to policy implementation, a number of engineering structures were created in order to reduce sedimentation. The most important engineering component was the development of terraces on the loess slopes. This was achieved mainly by bulldozers, where terraces were developed so farmers could plant crops in level fields that would retain a greater amount of water and be less subject to erosion. A second key engineering component of the project was the creation of sediment dams. Three types of sediment dams, key dams, warping dams and check dams were all designed to either intercept sediment or to decrease water flow there by reducing the erosive channel characteristics.

Outcomes of the project according to the World Bank (World Bank, 2003) are as follows:

a. approximately 1 million farmers have benefited from increased agricultural sustainability. Average agricultural income has increased by up to 4 times.

b. grain output from the region has increased 1.5 times and annual fruit output has increased four fold.
c. annual sediment transport to the Yellow River has decreased by approximately 57 million tonnes.

Overall, the Loess Plateau watershed rehabilitation project was the largest of its kind to date. Both the Chinese government and World Bank rated the project as successful as all project targets were met.

3. Sediment transport and the Loess Plateau

There are a number of tributaries in the Middle Reach of the Yellow River all with varying topography and drainage areas. When examining sediment transport in any river, it is important to consider not only the river main stem but also the tributaries in order to specify sediment source locations. The following sections identify research done by the authors and others which examine main stem gauging stations, tributaries and reservoirs in the Loess Plateau. Variations in sediment transport under winter ice conditions are also examined.

3.1 General trends in sediment transport, Yellow River's Middle Reach, Loess Plateau

This section looks at general sediment characteristics of the middle reach (Figure 4) and how it compares to the Upper and Lower reaches; this section is based largely upon Liu et al. (2008). There are two gauging stations on the Yellow River's Middle Reach. The most upstream station is the Longmen station, and downstream is the Tongguan gauging station (Figure 4). Both gauging stations in the Middle Reach revealed decreasing sediment yields over the past 50 years (Liu et al., 2008). Long term and short term runoff depths are Lower in the Middle Reach in comparison to the Upper Reach. The long term average runoff depths as reported by Liu et al. (2008) have decreased by approximately 40 percent during the past ten years (Table 1). Similarly, annual sediment yield in the Middle Reach has also decreased between 50 to 60 percent in the last ten years (Table 1).

Station	Drainage area (km^2)	Data period	Annual runoff depth Q (mm)		Annual sediment yield S (t/km^2)	
			LTA Q	RA Q_{10}	LTA S	RA S_{10}
Longmen	497,552	1950-2005	53.68	32.47	1543.65	612.48
Tongguan	682,141	1952-2005	51.29	29.79	1631.39	766.13

Table 1. Long term average (LTA) and recent 10 year average (RA10) runoff depths and annual sediment yield for the middle reach of the Yellow River, as reported by Liu et al., (2008).

Even though the sediment yield is decreasing, the sediment yield in the Middle Reach during the past 10 years is still 4 to 7 times the Upper Reach and 2 to 3 times the Lower Reach (Liu et al., 2008). The Middle Reach, while it accounts for only 15 percent of the drainage area, approximately 90 percent of the sediment comes from this region (Liu et al., 2008). The influence of the Loess landscape definitely has a direct impact on sediment levels in the middle reach. Poor grazing and cultivation practices in the Loess Plateau can lead to soil disturbance. The majority of the precipitation for the region falls between the months of June to September (Cai, 2001); during such time disturbed soils can be transported to the river channel by overland flow. The decrease in sediment yield experienced in the past 10 years may be due to the implementation of terraces and sediment dams throughout the region as part of the Loess Plateau watershed rehabilitation project.

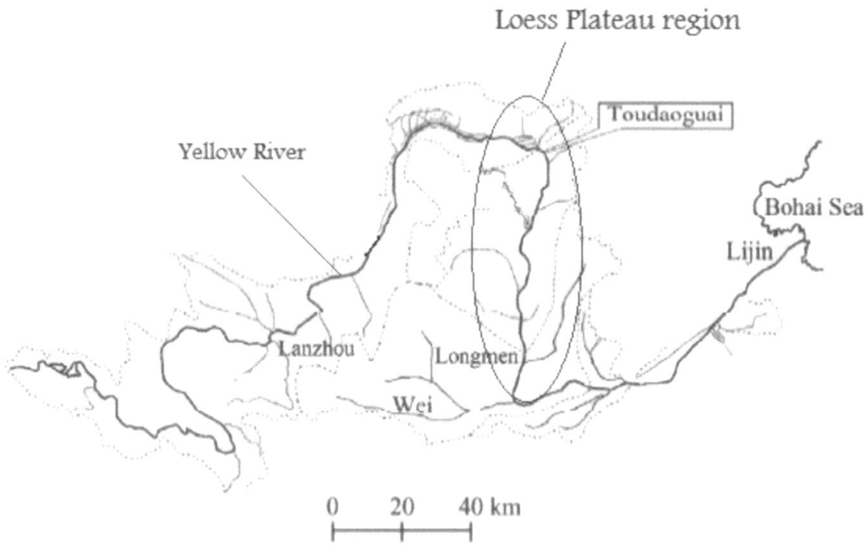

Fig. 4. Loess Plateau region along the middle reach of the Yellow River (modified from Sui et al., 2008).

3.2 Sediment transport in the Huangfuchuan watershed along the Loess Plateau

The Huangfuchuan watershed, which is located in the centre of the middle reach in the Inner Mongolia region of China, is an important contributor of sediment to the Yellow River's main stem (Figure 5). The following examination of sediment contributions from the Huangfuchuan watershed is based upon Sui et al. (2008).

The Huangfuchuan watershed covers an area of 3199 square kilometres. The main stem is 137 kilometres long, and the average channel slope is 2.7 percent. The watershed landscape consists of many hills and gullies with the majority of the soil type being loess. The thickness of the loess layers ranges from 20 to 80 centimetres. Three gauging stations are used to examine sediment transport in the Huangfuchuan watershed: (1) the Huangfuchuan station, located approximately 14 kilometres upstream from the Huangfuchuan's confluence with the Yellow River, (2) the Toudaoguai gauging station, located on the Yellow River mainstream, 178 kilometres upstream of the Huangfuchuan River and (3) the Fugu gauging station, located on the Yellow River main stem, approximately 38 kilometres downstream of the Huangfuchuan/Yellow River confluence. The location of the Toudaoguai and Fugu stations provides an excellent opportunity to quantify the sediment contributions of the Huangfuchuan watershed.

The downstream Fugu gauging station, sediment levels are much higher than at the upstream Toudaoguai gauging station (Table 1). It is thought that the Huangfuchuan River watershed contributes sediment to the Yellow River main stem and accounts for the higher sediment concentration downstream. Also, in contradiction to what is expected, the downstream Fugu station has a lower long term annual discharge in comparison to the upstream Toudaoguai station (Table 1). This may be a result of the high amount of water withdrawals from the Yellow River for agricultural, domestic and industrial use.

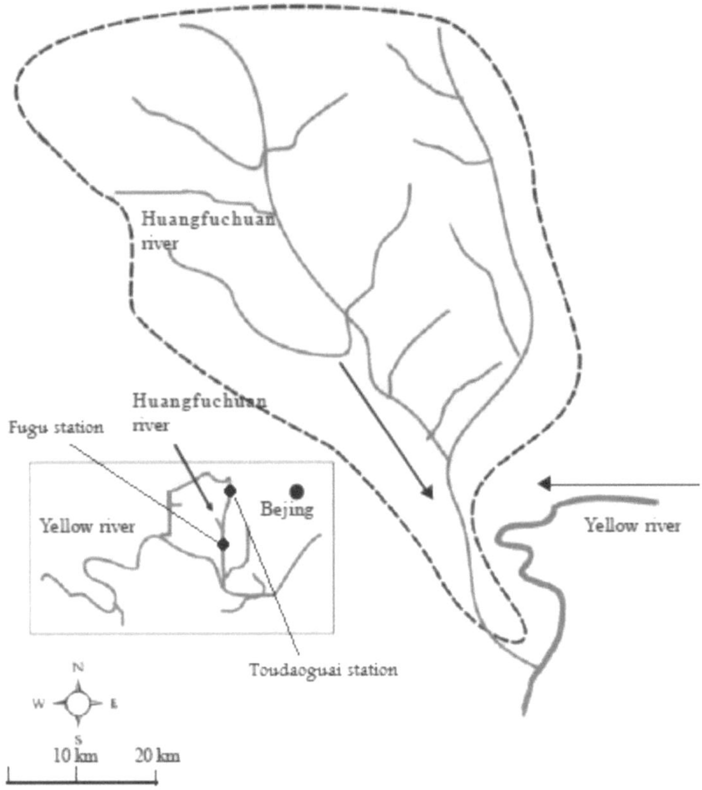

Fig. 5. Huangfuchuan watershed, located in the middle reach of the Yellow River (modified from Sui et al., 2008).

Long term annual discharge (m^3/s)			Long term annual sediment concentration (kg/m^3)		
Huangfuchuan	Toudaoguai	Fugu	Huangfuchuan	Toudaoguai	Fugu
5.23	802.47	789.18	29.47	2.98	5.24

Table 2. Long term annual discharge and sediment concentrations, calculated from Sui et al (2008).

Average monthly precipitation depth was calculated by using data from the Hequ, Yulin and Dongsheng climate stations. As shown in Figure 7, over the last 50 years all three gauging stations exhibit a downwards trend in precipitation. Approximately 80 percent of the total precipitation falls during the months of June to September. Little precipitation falls in this region between the months of November and May. During this time, the soil moisture content is low and it is though that the loess sediment is eroded through wind processes. During the rainy months, June through August, moisture seeps into the soil and less surface runoff occurs.

The runoff depths at the Toudaoguai (upstream) and Fugu (downstream) gauging stations is similar; however the sediment concentration is much higher at the downstream station. This is mainly due to sediment inputs from the Huangfuchuan sub-watershed, located between the upstream and downstream gauging stations. The sediment transport is higher at the Huangfuchuan gauging station in comparison to the Toudaoguai and Fugu gauging stations. Approximately 80 percent of the precipitation and runoff in the Huangfuchuan sub-watershed occurs during the months from June to September; the largest sediment grain size also occurs in the summer when the kinetic energy is highest.

Typically, sediment concentration in rivers increases with discharge due to the increased kinetic energy of the water (Yang et al., 2005). Interestingly, the above relationship between discharge and sediment transport is not consistent in the Huangfuchuan River during the summer months. As shown in Figure 6, the highest monthly average sediment concentration occurring in July is 130 kilograms per cubic metres, while the average monthly July discharge is 16 cubic metres per second. It is thought that the desynchronization of sediment concentration and discharge is related to the antecedent soil moisture of the area. Specifically, little rain falls during the spring months and it is thought that erosion of loess soils in the Huangfuchuan watershed occur through wind transport. By the time summer rain falls in the watershed, the precipitation is absorbed by the loess rather than contributing to surface runoff.

Overall, it appears that the Huangfuchuan watershed is an important contributor of sediment to the Yellow River. Over the past 50 years, runoff and sediment transport from the Huangfuchuan watershed have been decreasing. Climate stations in the area indicate that the average annual precipitation depth has also been decreasing; this may account for the decrease in annual runoff. However, the region has also been subject to various soil conservation initiatives in order to manage erosion. This may also be a contributing factor in the decrease of sediment transport from the Huangfuchuan watershed.

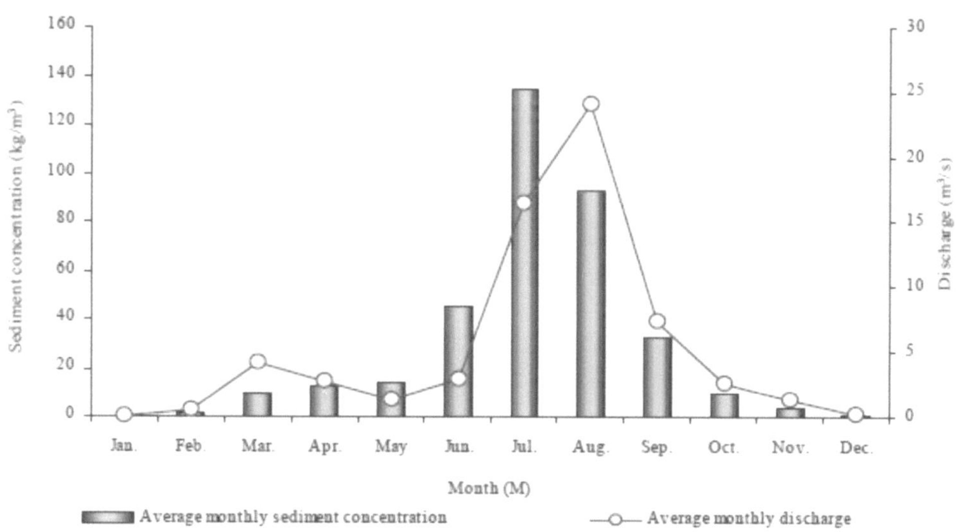

Fig. 6. Sediment concentration and discharge for the Huangfuchuan River station, 1954-1989 (from Sui et al., 2008).

3.3 Sediment transport in the Kuye River watershed, Loess Plateau

Similar to the Huangfuchuan watershed, the Kuye River watershed is important tributary located in the Yellow River's Middle Reach. The following examination of sediment transport in the Kuye watershed is based upon Sui et al (2009). The Kuye River watershed is

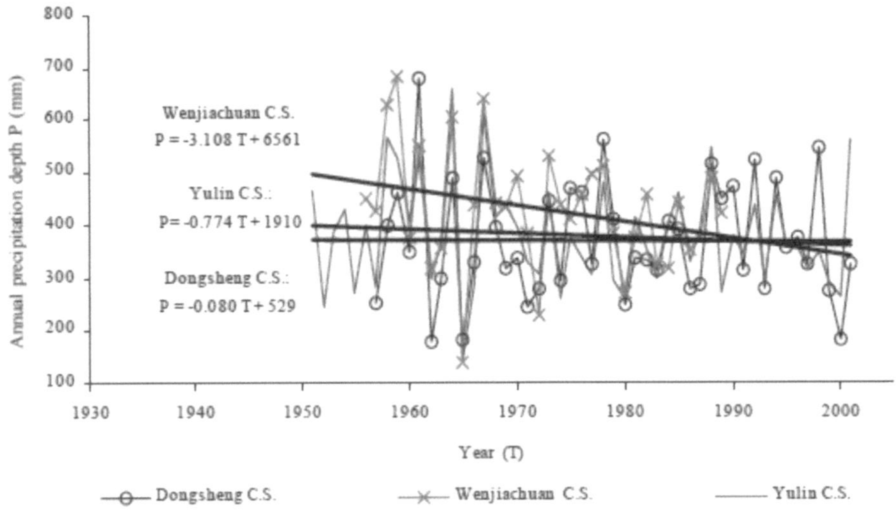

Fig. 7. Annual change in precipitation depth of 3 climate stations around the Huangfuchuan watershed (from Sui et al., 2008).

located downstream of the Huangfuchuan watershed as shown in Figure 8. The Kuye main channel length is approximately 242 kilometres in length and the watershed covers an area of 8,706 square kilometres. The Kuye watershed topography consists of hills, gullies and exposed bedrock; vegetation exists in only 6 percent of the entire watershed.

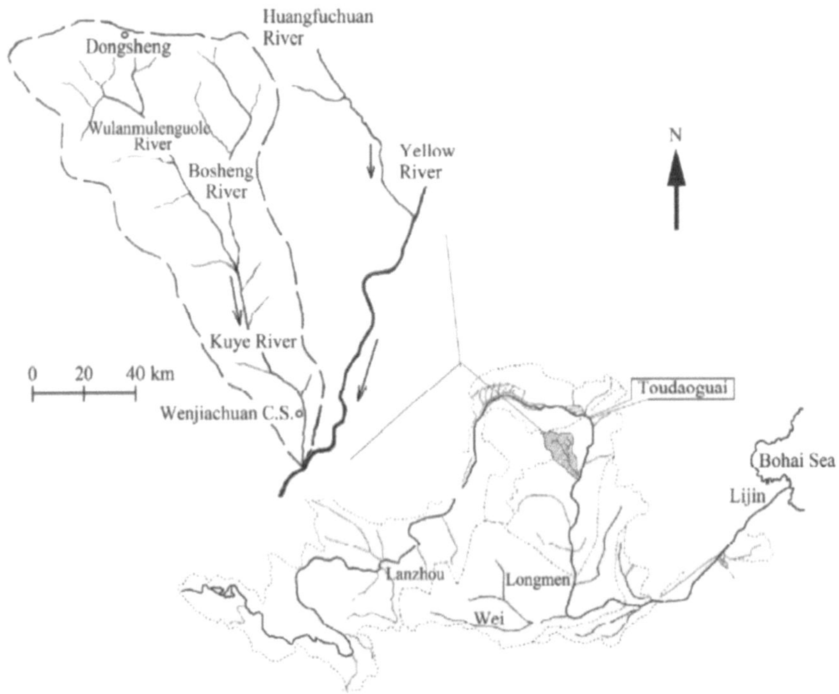

Fig. 8. Kuye River watershed (from Sui et al., 2009).

The loess ranges in depth from 20 to 80 metres. In comparison to the Huangfuchuan watershed, the Kuye watershed is almost twice the size, has little vegetation and the loess layer is much deeper. There is one gauging station on the Kuye River located approximately 7 kilometres upstream of the confluence with the Yellow River's Main Stem. Two gauging stations on the Yellow River Main Stem were also examined in order to study the sediment contributions of the Kuye watershed. The Toudaoguai gauging station is located approximately 302 kilometres upstream of the Kuye/Yellow River confluence (Figure 8). The Longmen gauging station is located 431 kilometres downstream of the Kuye/Yellow River confluence (Figure 8). At all gauging stations, annual discharge and sediment concentration were examined for the period from 1955 to 2006. Precipitation records from the Dongsheng and Wenjiachuan climate stations (Figure 8) were also used in order to calculate runoff depth.

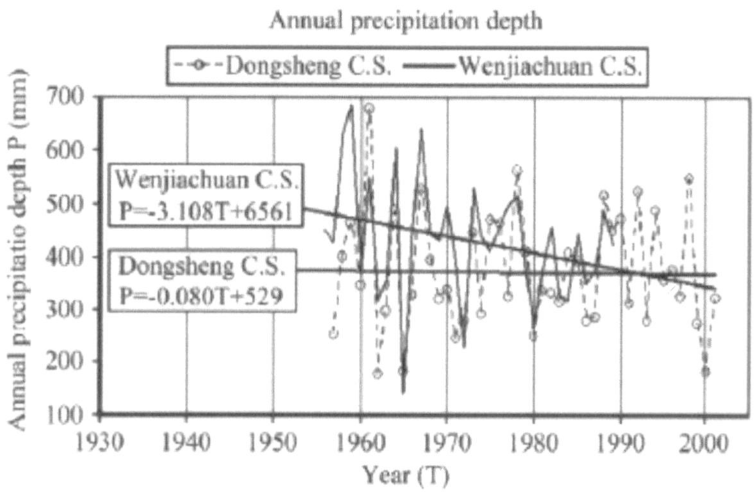

Fig. 9. Precipitation depth at the Wenjiachuan and Dongshen climate stations in the Kuye watershed (from Sui et al. 2009).

The majority of the precipitation in the Kuye watershed falls during the months of June through September. Over the past 50 years the annual precipitation in the Kuye watershed has been decreasing; the Dongsheng station has a slight downward trend (Figure 9). The highest precipitation months are generally July and August, which corresponds to the months with the highest discharge (Figure 10 & 11). Interestingly, similar to the Huangfuchuan watershed, the highest sediment concentration does not occur during the month of highest discharge (Figure 11). In the Kuye watershed the highest average annual sediment concentration of 125 kilogram per cubic metre, occurs in July; the average discharge in July is 43 cubic metres per second. The month of August has the highest average annual discharge, 66 cubic metres per second; however the sediment concentration is 90 kilograms per cubic metre. In the Kuye watershed, this desychronization of sediment and discharge measurements is thought to be related to the intermittent tributaries present throughout the watershed. It is thought that during times of low precipitation the intermittent tributary streams in the Kuye watershed do not have connectivity. The spring season is generally quite windy and high amounts of loess are transported into the tributary reaches; then during months of high precipitation (July and August, Figure 10), tributary channels form connectivity. The month of July is typically dominated by flash flood events where large amounts of stored sediment can be transported from the tributaries to the Kuye main stem and eventually the Yellow River. Even though precipitation and discharge are high in August, it is thought that the majority of the stored tributary sediments are transported during July flash flood events.

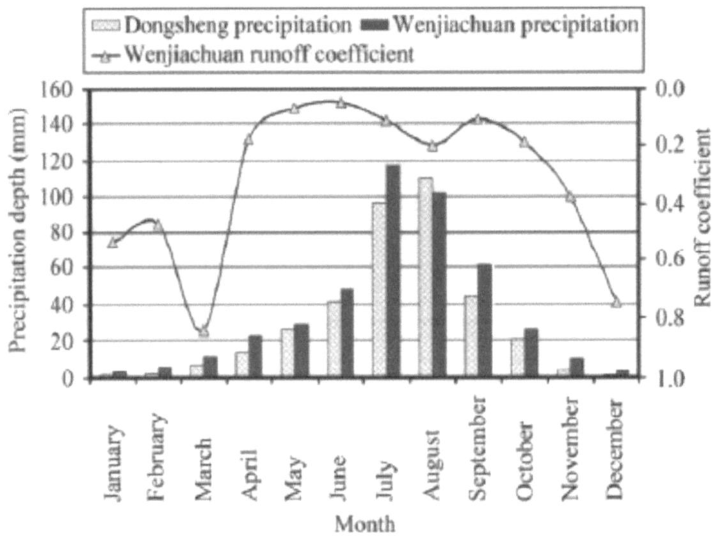

Fig. 10. Average annual precipitation depth and runoff coefficient for the Dongsheng and Wenjiachuan climate stations (from Sui et al. 2009).

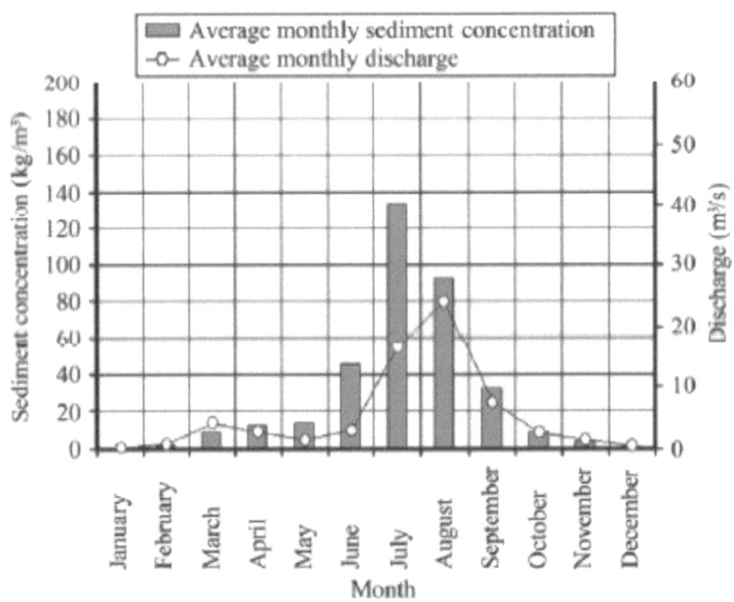

Fig. 11. Sediment concentration and discharge in the Wenjiachuan watershed from 1955 to 2006 (from Sui et al. 2009).

3.4 Sediment transport characteristics of a reservoir in the Middle Reach

In general when examining sediment transport and deposition within a reservoir, it is understood that finer, lighter particles will remain in suspension longest and be transported the farthest distance. When a reservoir is located within a river reach, the channel velocity decreases substantially in the transition zone from the channel to the reservoir. This decrease in channel velocity is due to the widening of the river channel into a reservoir. Conversely, larger, heavier particles will be deposited sooner, in the upper reach of the reservoir. Over time, constant sediment deposition will reduce the storage capacity of the reservoir and may eventually fill the area. Sediment removal practices must be undertaken in order to adequately manage a reservoir. Reservoir flushing and sediment dredging are common ways to mitigate sediment deposition.

Typically there are 3 main areas of deposition within a reservoir, backwater deposits, bottom deposits and delta deposits (Figure 12). The backwater region of a reservoir is the region that forms the transition zone between the river channel and the reservoir. Backwater deposits are usually coarse grained sediments. The delta deposition region occurs within the reservoir basin, immediately after the backwater zone. Sediment in the delta region is usually a mix of course and fine grained particles. Sediment deposition in the delta region can occur both across and into the reservoir. Lastly, the bottom deposition zone in a reservoir occurs at the bottom of the deepest part of the reservoir. The sediment located in the bottom deposition zone is fine grained, consisting of silts and clays (Batuca and Jordaan, 2000).

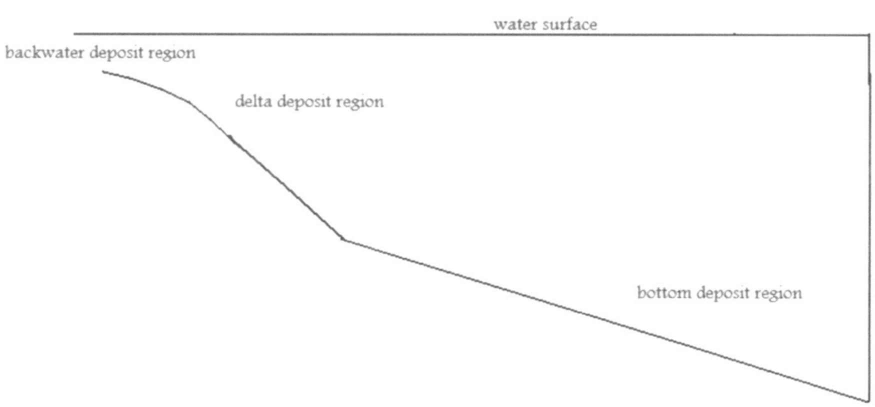

Fig. 12. Typical deposition regions within a reservoir.

Over the past 50 years, a number of hydropower stations have been developed in the main reach of the Yellow River watershed. One of the reservoirs present on the Yellow River main stem is the Tianqiao Reservoir, located just upstream of the Fugu gauging station in the Loess Plateau (Figure 13). The Tianqiao hydropower station was the first power station ever built on the Yellow River and has been in successful operation for more than 30 years. The Tianqiao reservoir consists of two reaches, an upper and lower that form a total reach length of 21 kilometres. The lower reach is 300 metres wide and 12.5 kilometres in length; the upper reach is 600 metres wide and 8.5 kilometres in length.

The water level in the Tianqiao reservoir varies depending on the hydropower operations. Sediment deposition in the Tianqiao backwater region retreats and advances depending on operational water levels. As previously discussed, the upstream Huangfuchuan sub-watershed of the Yellow River is a major source area of loess sediment. Shoals and sandbars development is common in the backwater region of the Tianqiao Reservoir and is likely due to the influx of sediment from the Huangfuchuan watershed (Sui et al., 2005). The relationship between sediment deposition in the Tianqiao Reservoir and the sediment source area upstream is evident when examining historical flash flood events. During 1982 the Huangfuchuan watershed experienced flash flooding during the end of July as indicated by the reported gauging station data (Table 3). Sediment deposition in the Tianqiao reservoir was also high during 1982 (Figure 14). Sediments are primarily deposited in the backwater region of the reservoir as the relatively low flow velocity is unable to transport all the incoming sediment. Over the season, as the backwater region experiences more sediment deposition, the water level of the area rises, which in turn would increase the kinetic flow energy. During the fall and winter seasons, when sediment supply to the reservoir is low, the high flow energy is then able to transport sediments (scouring) from the backwater region into the reservoir (Sui, et al., 2005).

Date	Discharge (m^3/s)	Sediment concentration (kg/m^3)
29 July 1982	1,100	1,250
30 July 1982	2,580	1,190

Table 3. Measured discharge and sediment concentration in the Huangfuchuan watershed, 1982 (as reported by Sui et al. 2005).

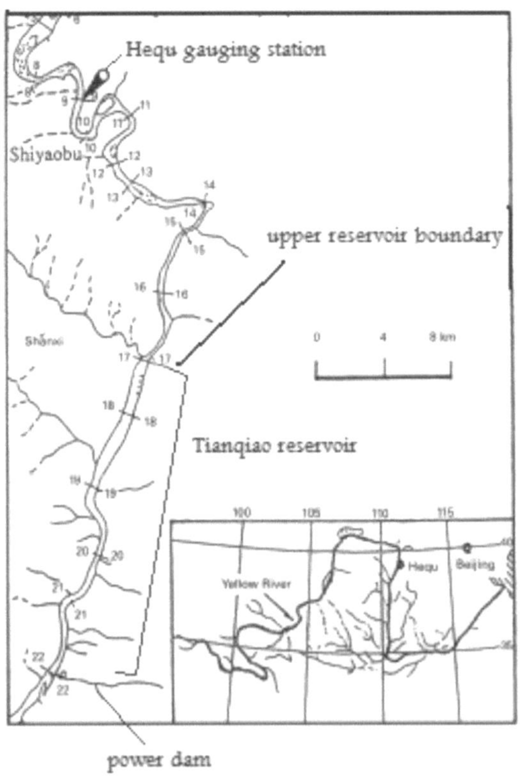

Fig. 13. Tianqiao hydropower station and reservoir along the Yellow River.

Generally, during the summer season when precipitation is high and flooding occurs, the Tianqiao reservoir is in deposition mode. For the remainder of the year, the reservoir is in a scouring mode. Over time the majority of the sediment, approximately 85 percent, becomes deposited in the lower reservoir reach (Sui et al., 2005). Understanding the dynamic relationship between climate, sediment transport and sediment deposition is important in reservoir management. Knowledge of sediment influx is key in development of sediment removal programs, budget allocation and hydropower management.

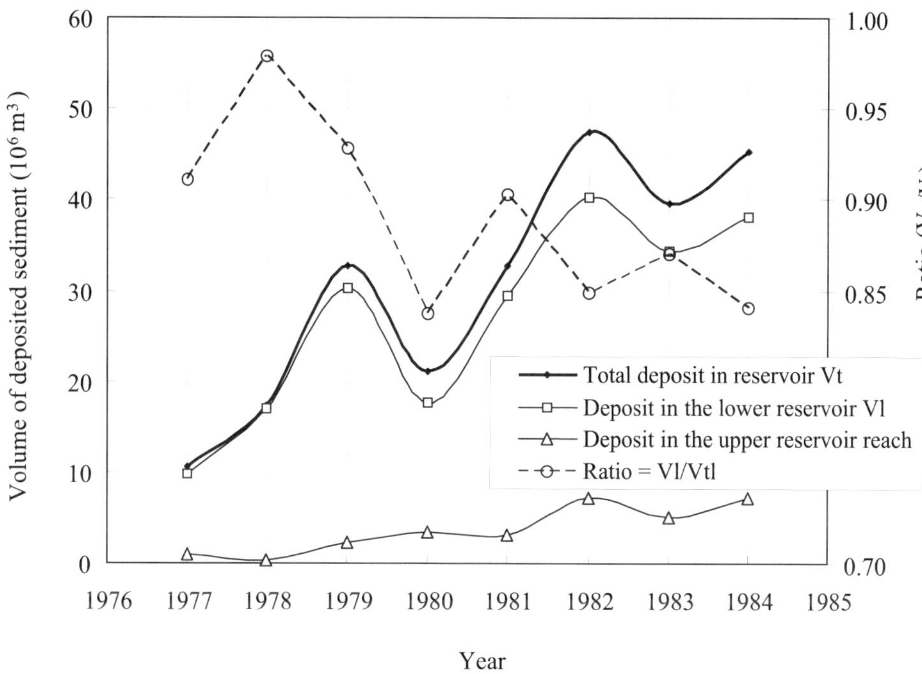

Fig. 14. Volume of sediment deposited in the upper and lower reaches of the Tianqiao Reservoir (from Sui et al., 2005).

3.4 Sediment transport characteristics during winter ice cover

While sediment transport in the Loess plateau under open channel flow has been discussed, sediment transport during winter conditions must also be addressed in order to gain a full understanding of the seasonal fluctuations in sediment. Ice cover and jams are frequent along the Yellow River's Middle Reach during the winter season. At the beginning of the Middle Reach is a channel area often referred to as the Hequ Reach; the Hequ reach spans from the Longkou Gorge to the Tianqiao Power Dam as shown in Figure 15. The Hequ reach is approximately 70 kilometres in length with a broad and shallow morphology. Upstream of the Longkou Gorge area, the Yellow River channel is a long, open reach of water; in this area a large amount of frazil ice forms during the winter. The frazil ice is transported downstream, and numerous ice jam events occur in the Hequ Reach. The following is a brief description of the relationship between frazil ice and sediment entrapment as examined in the Hequ Reach of the Yellow River.

In general for open channel flow the sediment concentration of a river depends on the current velocity and the characteristics of the river bed material. The higher the river discharge, the higher the sediment concentration. This relationship becomes much more complicated with the introduction of ice cover and frazil ice. During ice cover, sediment is transported not only through the water but also through attachment to river ice (frazil ice) (Sui et al., 2000). During frazil ice formation, supercooled water is transported throughout the water column promoting ice nucleation. During this process, ice crystals can become attached to pebbles or other sediments on the river bed. In areas of the Hequ Reach in the Loess Plateau, pebbles ranging in size from 0.2 top 0.5 kilograms have been observed in frazil jams (Sui et al., 2000). The upper layer of frazil ice generally has a higher sediment concentration than the lower frazil layers. The highest sediment concentration recorded was 25 kilograms per cubic metres of ice; this is much larger than the highest sediment in water concentration during a jam (7 kilograms per cubic metre) (Sui et al., 2000). During periods of ice jam formation or ice jam breakup, sediment concentrations were found to be higher than during times of stable jamming.

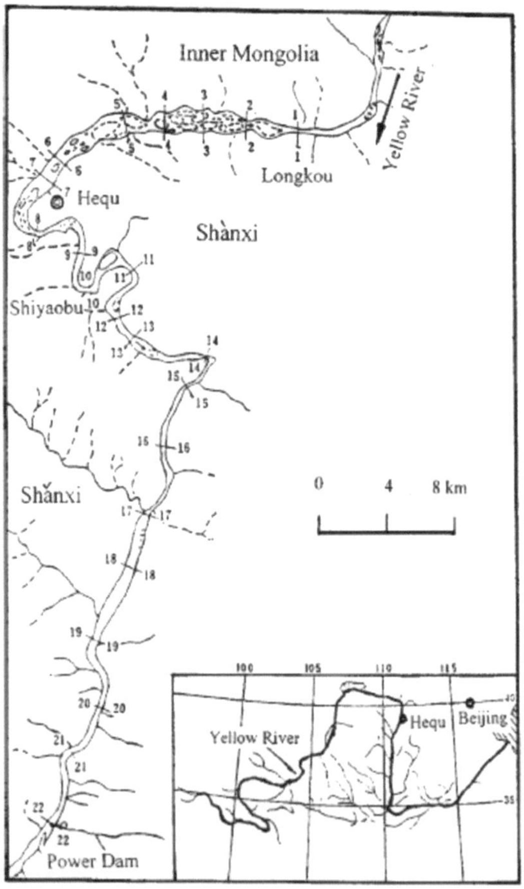

Fig. 15. The Hequ reach along the Loess Plateau region of the Yellow River.

While it has been discussed that the majority of the sediment transport in the Loess Plateau occurs during the summer season, the sediment transport during winter is unique. If ice jamming and frazil ice formation prevail, sediments of large size can be transported down the river reach.

4. Success of soil conservation initiatives in the Loess Plateau

In general over the past 50 years, while accounting for human water use and impact, runoff in all reaches of the Yellow River watershed has decreased over the past 50 years (Fu et al., 2004). Changes to land use, global warming, ground water exploitation and human activities (among others) were reported as factors influencing the basin runoff (Fu, et al., 2004). Specific to the Loess Plateau, soil conservation initiatives appear to have had a great impact on the sediment transport to the Yellow River. Large areas of once bare land have been planted with trees and grass and cattle grazing has been forbidden. By the year 2000, the seven provinces that intersect with the Loess Plateau had improved a total of 177,000 square kilometres of eroded land (CERN, 2001). Reclamation to grazing areas appears to have made the largest impact on sediment transport in the Loess Plateau. Local farmers have found drylot feeding more profitable and local vegetation is now protected (CERN, 2001). The implementation of sediment dams also appears to have altered the amount of soil erosion. Sediment dams constructed vary from 2 to 5 metres in height and work to effectively trap eroded topsoil. One a substantial area of topsoil is formed behind the dam, the area can be used as high-yield agriculture plots. The largest reduction in sediment transport can be seen in the Kuye watershed. The long term average sediment transport modulus for the Wenjiachuan gauging station is 10,591 tonnes per square kilometre per year. During the last ten years this values has significantly decreased to 2,805 tonnes per square kilometre per year (Sui et al., 2009)

5. Summary and conclusion

This chapter discussed sediment transport within the Loess Plateau region of the Yellow River. The erosion of loess sediments has long been a problem for people residing in the region. Soil erosion has been responsible for loss of agricultural land and significantly altered the lives of many communities residing in the Loess Plateau. There are a number of tributaries in the loess plateau region that contribute significant amount of sediment to the Yellow River's Middle Reach. The Huangfuchuan and Kuye watersheds were shown as sediment source areas with a large amount of sediment transport occurring during the summer flood months. Sediment transport from these watersheds also has large implications for the management of the Tianqiao reservoir. During the 1990's the Government of China implemented a number of soil conservation initiatives as part of the Loess Plateau watershed rehabilitation project. Terraces were constructed throughout the loess plateau, grazing was prohibited and sediment dams were constructed. Sediment concentration in both the main stem of the Middle Reach and in both the Huangfuchuan and Kuye rivers has decreased during the last 10 years. Overall the decrease in sediment levels in the Middle Reach are attributed to implementation of soil conservation initiatives.

6. References

Batuca, D. & Jordaan, J. (2000). *Silting and desilting of reservoirs,* A.A.Balkema, ISBN 9054104775, Netherlands.

Cai, Q. (2001). Soil erosion and management on the Loess Plateau. *Journal of Geographical Sciences.* Vol. 11, No. 1, 53-70.

China Education and Research Network, CERN. (2001). Cleaning the Yellow River. Special Report, http://www.edu.cn/special_1506/20060323/t20060323_4751.shtml.

Fu, B., Chen, L., Keming, M., Huafeng, Z. & Wang, J. (2000). The relationship between land use and soil condition in the hilly area of the loess plateau in northern Shaanxi, China. *Catena*, Vol. 39, 69-78.

Fu, G., Chen, S., Liu, C. & Shepard, D. (2004). Hydro-climatic trends of the Yellow River Basin for the last 50 years. *Climatic Change*, Vol. 75, 149-178.

Gu, W. (1994). On the reduction of water and sediment yield of the Yellow River in late years. International Journal of Sediment Research, Vol. 9, No. 1, 1-12.

Gu, W. (2002). Report on changes in water and sediment levels of the Yellow River, In: Researches on changes in water and sediment loads of the Yellow River, Wang, G. & Fan, Z (Ed.), 1-45, Yellow River Conservancy Press, Zhengzhou.

Kleine, D. (1997). "Who will feed China?" *Journal of Soil and Water Conservation*. Vol. 52, No. 6, 398-399.

Liu, T. (1985). Loess and the environment (ed. by Liu). Science Press, Beijing, China (in Chinese), 62-81.

Liu, C., Sui, J., & Wang, Z. (2008). Changes in runoff and sediment yield along the Yellow River during the period from 1950 to 2006. Journal of Environmental Informatics, Vol. 12, No. 2, 129-139.

Ludwig, W., Probst, J. & Kempe, S. (1996). Predicting the oceanic input of organic carbon by continental erosion. Global Biogeochemical Cycles, Vol. 10, No. 1, 23-41.

Pye, K. (1987). Aeolian dust and dust deposits. Academic Press, London, ISBN 0125686919, Orlando, 334 p.

Shi, C & Ye, Q. (1997). Bank breach hazards in the lower Yellow River. Destructive Water: Water-caused natural disasters, their Abatement and Control, Proceedings of the International conference on destructive water, Anaheim, California, 24-28 June 1996.

Sui, J., Wang, D. & Karney, B. (2000). Suspended sediment concentration and deformation of riverbed in a frazil jammed reach. Canadian Journal of Civil Engineering, Vol 27, 1120-1129.

Sui, J., Jackson, P., Liu, C., Fang, D., Wang, J. (2005). Characteristics of sediment transport along a river reach with a reservoir. *International Journal of Sediment Research*, 20(2), 89-102.

Sui, J., He, Y., & Karney, B. (2008). Flow and high sediment yield from the Huangfuchuan watershed. *International Journal of Science and Technology*, Vol. 5, No. 2, 149-160.

Sui, J., He, Y., & Liu, C. (2009). Changes in sediment transport in the Kuye River in the Loess Plateau in China. *International Journal of Sediment Research*, Vol. 24, No. 2, 201-213.

Wischmeier, W. (1976). Use and misuse of the universal soil loss equation. *Journal of soil and water conservation*. Vol. 31, No. 1, 5-9.

World Bank. (2003). China – Loess Plateau watershed rehabilitation project. Implementation completion and results report. 1-45 pp.

Xu, J.X. (2003). Sedimentation rates in the lower Yellow River over the past 2300 years as influenced by human activities and climate change, *Hydrological Processes*, Vol. 17, No. 16, 3359-3371.

Yang, S., Gao, A., Hotz, H., Zhu, J., Dai, B. & Li, M. (2005). Trends in annual discharge from the Yangtze River to the sea (1865-2004). *Journal of Hydrologic Sciences*, Vol. 50, No. 5, 825-836.

Zhu, Z., Zhou, I., An, Z. & Liu, T. (2004). A river erosion estimate on the Loess Plateau: a case study from Luohe River, a second order tributary of the Yellow River. *Global and Planetary Change*, Vol. 41, 215-220.

11

Modeling of Developed Meanders of an Alluvial Channel

L. Yilmaz
Civil Engineering Faculty, Hydraulic Division, Technical University of Istanbul,
80626, Maslak,
Istanbul,
Turkey

1. Introduction

The meandering planforms of alluvial rivers pose hazardous impacts on the human life, for a major portion of inhabitation lies on the banks of such rivers for the reason of easy water availability, land fertility and food grain productivity, and economic navigation and transportation of goods. Meandering is a self-induced plan deformation of a stream that is (ideally) periodic and anti-symmetrical with respect to an axis, which may or may not be exactly straight. According to this definition, which will be used in the following, an alluvial stream which deforms its initially straight channel into one of the periodic and anti-symmetrical plan forms is meandering; whereas a stream flowing in a tortuous rocky terrain or in a rigid sinuous flume, whose curvilinear plan pattern has not been created by that stream itself, is not meandering. Meandering can be classified as regular, irregular, or skewed, depending on the form of the meandering bend migration. The thalweg and tortuosity describe the meandering characteristics.

The most important parameter defining meander geometry is the tortuosity or sinuosity of the curved channel. Tortuosity was defined by Leopold and Wolman (1957) as the ratio of thalweg length (length along the line of maximum depth) to valley length, by Friedkin (1945) as the ratio of thalweg length to air line distance, and by Leopold and Wolman in their subsequent contribution (1960) as ratio of arc distance to wave length in a single meander. The first definition appears to be preferable because of its simplicity, requiring less judgement in measurement. To help visualize the degree of meandering associated with a particular tortuosity ratio, available concepts were examined. The concept which appeared most satisfactory was that presented by Langbein and Leopold (1966), who postulated that the planimetric geometry of a meander is that of a random walk, whose most frequent form is that which minimizes the sum of squares of the changes in direction, in each successive unit length. The direction angles are then sine functions of channel distance. This yields a meander shape typically present in meandering rivers.

Every phase of meandering represents a changing relationship between three closely related variables: the flow and the hydraulic properties of the channel, the amount of sand moving along the bed, and the rate of bank erosion. These three variables constantly strive to reach a balance, but never do even with a constant rate of flow. The bends of a meandering river have limited widths and lengths. The flow and the hydraulic properties of the meandering river, the amount of sand moving along the bed, and the rate of bank erosion determine these limits. When a bend reaches this width, a chute forms and a new bend develops farther downstream. Distorted bends and natural cut-offs are caused by local changes in the character of bank materials.

Several hypotheses have been brought forward to explain meandering. Meandering has been attributed to the earth's rotation; to the excessive slope and energy of a river (Anderson, 1967; Raudkivi, 1966, 1967; Fredsoe, 1982; McLean and Smith, 1986; Gust, 1988; Nelson, 1989; Yilmaz, 1990; and to changes in stage of sediment formation at the boundary (Exner, 1919). It is often considered that an irregularity in the bankline or another causing factors will disturb the flow and cause meandering. Alternate bars give rise to a sinuous migrating thalweg within initially straight banks, and might somehow evolve into meanders, provided channel banks are also erodible. The formation of alternate bars imply incipient meandering (Tubino and Seminara, 1990). The coexistence of free or migrating and forced or fixed bars in a meandering channel has been investigated through laboratory experiments by Gottlieb (1976) and Fujita and Muramoto (1982).

Rivers with a sinuosity, defined as the ratio of valley slope to channel slope, of 1.5 or greater are considered as meandering (Leopold and Wolman, 1957). In a simple of 50 rivers differing in size as well as in physiography, Leopold et al., 1964, found that two-thirds of the ratios were in the range 1.5-4.3, with a median value of 2.7. In view of this striking geometric regularity of winding rivers, they suggested that meanders are no accident and they appear to be in the form in which a river does the least work in turning. A river is the author of its own geometry. It is adjusted, in the long term, so that its ability to transport balances the water discharge and sediment load supplied from the watershed.The adjustments, which may include channel geometry, slope, meandering pattern, roughness, etc., reflect in part changes in the river's resistance; that is, in energy expenditure. It has previously been suggested that the basic reason for meandering is related to the rate of energy, or power, expenditure (Leopold and Wolman, 1960; Yang and Song, 1971). Meander geometry is obtained such that the inflow quantities of water and sediment are carried with minimum power expenditure per unit channel length as well as minimum power for the river reach. Chang (1979) applied the concept of minimum stream power per unit channel length together with relations of continuity, bed load, flow resistance, bank stability, etc., to obtain the regime geometry of alluvial streams under uniform flow conditions.

For a given discharge, meandering occurs on smaller slopes (Lane, 1957; Leopold and Wolman, 1957; Schumm, 1977). At steeper slopes, rivers are often braided in multiple channels separate by interlaced islands. In addition to the smaller slope and sinuous pattern, meandering rivers are characterized by a nearly uniform width along the channel. For purposes of river meander analysis (Chang, 1984), variables for regime conditions have been identified as independent variables, dependent variables, and constraints (Kennedy, and Brooks, 1963; Leopold, Wolman and Miller, 1964). Those which are imposed on the river from its watershed are independent variables or controlling variables and those which result when equilibrium is reached are dependent variables. Water discharge and sediment inflow and their respective properties which are determined by the watershed are independent variables for the river. Dependent variables include the flow velocity, channel width, flow depth, channel slope, and radius of curvature. The channel roughness and transverse bed slope in the curved channel are not additional dependent variables as they may be computed based on other variables. The valley slope is treated as another independent variable since the time scale for its formation is much greater than for regime channel geometry. The bank slope is another dependent variable.

The increased concern with riverbank erosion has increased the demand for theoretical models that can predict flow and bed features in a meandering alluvial channel. The most significant meander-flow characteristics are the spiraling of the mean flow due to channel curvature and nonuniformity of the velocity profile, the point bar, and deep pool bed topography near the apex of each bend. In order to plan, design, construct, or maintain bank-erosion control structures and river-basin projects in general, the meander characteristics must be quantified. Most of the studies have been concerned with the fully developed flow in a constant-radius; singular bend with uniform approach flow. A partial summary of these studies was given previously (Odgaard, 1981). The natural bend neither has a uniform approach flow; nor has a constant radius of curvature. The flow, generally, is in a state of either development or decay, or both. The purpose is to present an analytical approach to describe the flow and bed topography in such a channel (Odgaard, 1986). A change in channel curvature is as important as the curvature itself to the behavior of the bed profile. The model predicts that the secondary-flow component and the transverse bed slope react to the curvature changes like a damped oscillating system subjected to a driving force. The driving force can be any conceivable input function (an abrupt change in curvature, a harmonically oscillating curvature, or any other curvature variation).

Several formulas are theoretically and empirically proposed for the alternate bar wavelength (Ikeda, 1984). Due to dense population, most Japanese rivers are channelized, and many meandering rivers have been straightened. The emergence of alternate bars in these rivers destabilizes the channels, and induces subsequent side bank erosion.

Kinoshita (1987) found that the alternate bars are formed even in a straight laboratory flume with fixed side walls, and his subsequent field work revealed that the formation of alternate bars in straight rivers results in the development of meandering. Hayashi (1970) analyzed the flow in straight flumes with alternate bars with a potential flow model, and examined the stability of alternate bars. Hayashi and Ozaki (1980) treated the conditions of alternate-bar occurrence, and obtained the bar wavelength. Sukegawa (1971, 1972), Ikeda (1973), Kuroki et. al. (1975), Tamai et. al. (1978), and Muramoto and Fujita (1978) investigated the conditions of bar occurrence and proposed various empirical stability diagrams. Among others, Hansen (1967); Callander (1969); Engelund and Skovgaard (1973); Parker (1976); Parker and Anderson (1975); and Fredsoe (1978) presented theoretical studies on alternate bars. Field surveys are also documented abundantly, and much useful information is presented.

The objective of this paper is to describe qualitatively experimental and theoretical observations of meander evolution. Using dimensional analysis, Van Rijn (1984) concluded that H/λ, should be dependent on D^* (dimensionless particle parameter that reflects the effect of viscosity), T ($= \tau_*' / \tau_{*cr} - 1$) and ds/h.For most of the data, no appreciable influence of D^* on the steepness was detected. Several rules may be stated from the observation of Yalin's (1977) arguments: Ripples may only occur at the lowest sediment transport rates close to the initiation of grain motion, since those transport rates of grain sizes for which ripples are possible (ds<0.7 mm) (Raudkivi, 1976) are inevitably associated with small Re^*. For a given grain size, a sediment transport rate exists for which dunes with superimposed ripples begin to form. As the transport rate increases above this value, so does the size of dunes, while the ripples steepness decrease and finally comes to zero. At this range of dual bedforms, H/λ, ought to be a certain combination of both form dimensions.

1.1 Investigation of bed formations at the meandering channels

In order to investigate the reason of bed formations at the meandering channel theoretically, the knowledge of the oscillatory flow field near a wavy sandy bottom is needed. To that end, in a meandering stream model of a wavy sandy bottom, experimental observations of meander evolution are described qualitatively. The second part of experimental procedure consists of a description of the shear stress distribution changes of a straight channel with artificially wavy bottom during the meander development.

The form of the steady streamings set up in addition to the oscillatory motion by the bed profile is analyzed for different values of the parameters (Vittori, 1989). The non-linear viscous oscillatory flow over a wavy wall of small amplitudes is determined. The solution holds for arbitrary values of a^*/L^* (a^* is the amplitude of fluid oscillations near the wall and L^* is the wavelength of wall perturbation); previous work by Lyne (1971) and Kaneko and Honji (1979) are thus extended. An independent analysis for small values of a^*/L^* is performed by Vittori (1989), and the relevance of the results to the study of ripples formation at the bottom of sea waves is discussed.

In recent years many investigations have been devoted to the study of viscous oscillatory flow over a wavy wall in the study of the interaction between the meandering flow and meandering channel boundary layer in the nearbank region. Indeed it is well-known (Sleath, 1984) that a slight sinusoidal perturbation of the bottom in an oscillatory flow modifies the flow field so that steady streamings of the type described by Stuart (1966) can be observed. These steady drifts consist of recirculating cells whose form, intensity and direction depend on the values assumed by relevant parameters (Vittori, 1989). As the sediment of the bed is driven by the stress field associated with the fluid motion, a steady drift directed from the troughs towards the crests of the perturbation, may lead to the growth of the latter and thus to a pattern of sand waves (rolling grain ripples). Once formed, ripples will not continue to grow indefinitely; the steady drift is modified by non-linear effects and as ripples get steeper an equilibrium configuration is reached for which the gravity force acting down the slope balances the stress field associated with the fluid motion. Moreover when the ratio between the height and length of the ripples exceeds a value ranging about 0.1 the flow separates behind the crests and leads to vortex shedding which modifies the flow field and the ripple pattern. The latter was named vortex ripples by Bagnold (1946), who first introduced a distinction between sea-bedforms for which flow separation is absent (rolling-grain-ripples) and those which cause flow separation (vortex ripples). Lyne (1971) studied the flow induced by fluid oscillations near a wavy wall of amplitude (a dimensional quantity) much smaller than the characteristic viscous length. However, Lyne (1971) restricted his attention to the cases of small or large values of the ratio between the amplitude a^* of fluid oscillations and the wall wavelength L^*. Furthermore, Kaneko (1981) proposed a numerical solution in the same range of values of the parameters and a^*/L^*. As pointed out by Sleath (1984), the limited applicability of the solutions for small and large values of a^*/L^* lies in the fact that ripples usually form for values of a^*/L^* of order one. According to Vittori (1989), a second harmonic with amplitude of second order is introduced in the bed profile to describe the flow over a rippled bed with sharp peaks and flat troughs.

Every phase of meandering represents a changing relationship between three closely related variables: the flow and the hydraulic properties of the channel, the amount of sand moving along the bed, and the rate of bank erosion. These three variables constantly strive to reach a balance, but never do even with a constant rate of flow.

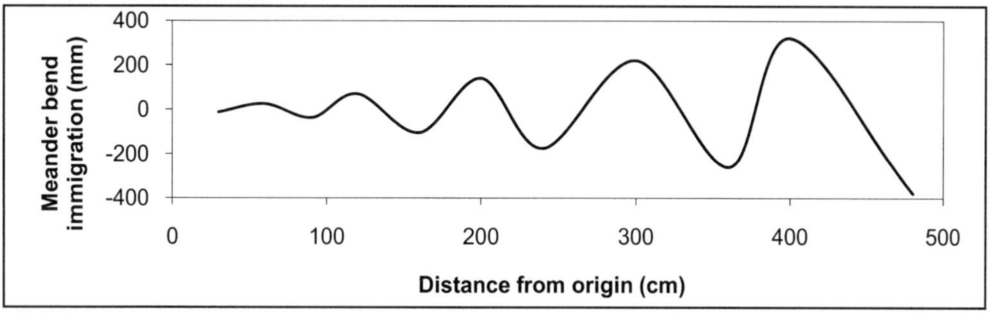

(a)

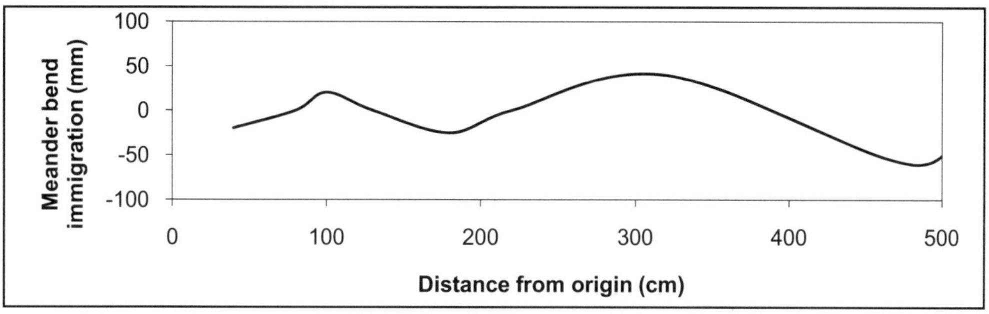

(b)

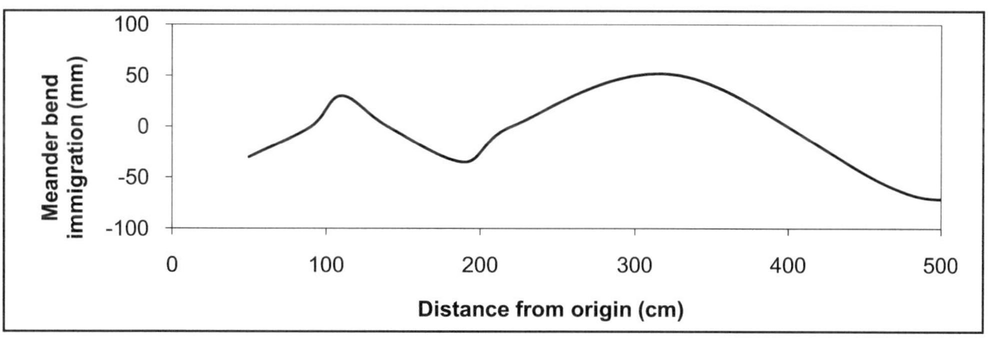

(c)

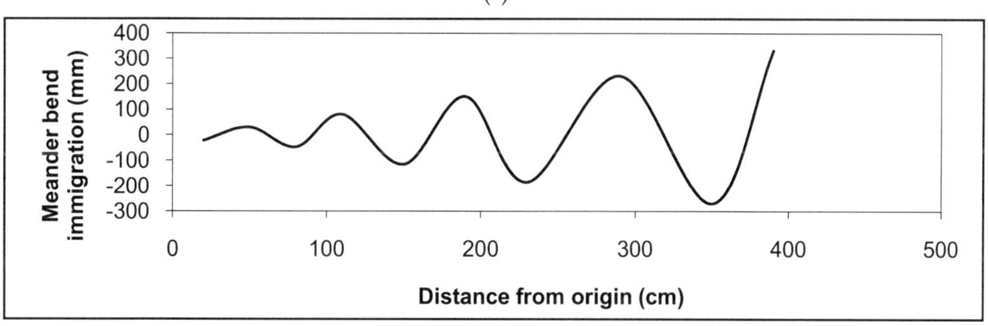

(d)

Fig. 1. Fully developed meandering channel patterns for various of discharge (Q) and bed slopes (So) of the channel after 72 hrs.

The bends of a meandering river have limited widths and lengths. The flow and the hydraulic properties of the meandering river, the amount of sand moving along the bed, and the rate of bank erosion determine these limits. When a bend reaches this width, a chute forms and a new bend develops farther downstream. Distorted bends and natural cut-offs are caused by local changes in the character of bank materials.

2. Experimental study

A movable laboratory channel with varying slope was installed for an experimental study of the meander evolution, for measuring the shear stress at the meandering channel, and for studying the sediment-water interaction at meandering channel. The main channel, as shown in Figure 1, was 10.00 meter long, 1.60 meter wide and 0.42 meter deep. There was movable carriage on the side rails, 8.80 meter long, which was situated for measurement of laboratory stream characteristics. Water coming from the head tank passed through the water tranquilizer which was 1.60 meter wide and 0.26 meter long and 0.65 meter deep. On the bottom of the channel an artificially wavy bottom was put in the middle of the straight initial channel. While the water flow through the straight canal, experimental observations of discharge, slope of the main channel, and shear stress distribution on the wavy bottom were made. An analytical model of free-surface flow over an erodible bed is developed and used to investigate the stability of the fluid-bed interface on an artificial wavy bottom, which gives the same characteristics of the bed features by measuring the oscillatory shear stress distribution with the hot-film sensors on the artificially wavy bottom in a laboratory canal . The downstream reservoir for outflow with the end gate was 1.10 meter long, 1.60 meter wide and 0.65 meter deep.
It is often considered that an irregularity in the bankline or another causing factors will disturb the flow and cause meandering. In an effort to ascertain the fundamental causes of meandering, a test was conducted in which a constant rate of flow was passed through a straight channel with artificial wavy bottom and at the oscillatory boundary layer the shear stress distribution in form of shear velocity is given in Table 1. and Figure 1.
As a result the shear stress distribution at x-coordinate is given as

$$\tau_{ox} = \tau_o (1.0 - 0.5 \cos (kx - \sigma)$$

from the experimental investigation.
The movable carriage on the side rails was very helpful for setting the profile indicator instrument and the velocity measurement instrument for obtaining the geomorphological and physical characteristics of meandering channels. For every run, it was easy to change the slope, the discharge and other parameters using the laboratory channel with variable slope.
Other test was conducted in the initial straight channel, artificially carved in uniform sandy material, through which a constant rate of flow was passed which had a velocity of 20 cm/s and which was sufficient to move sand along the bed and to erode the banks (Figure 2). The sediment supplier for the feedback system lays at the entrance after the tranquilizer. The sand collector , 0.42 mx1.60mx0.65m, was at the end of the channel. The flume channel on the laboratory alluvial boundary layer was 8.80 meter long and had a trapezoidal cross section, which was carved in the uniform sand of 1.35 mm of median diameter, with a bottom width of 0.10 meter and water width of 0.20 meter and 0.10 meter depth. The end gate from the end water tank to the sediment collector part of the channel was 0.15 meter

x/L (Distance)	Elevation from bottom (mm)	Run.1 (τ_{max}= 100mN/m²)	Run.2 (τ_{max}= 239mN/m²)	Run.3 (τ_{max}= 300mN/m²)	Run.4 (τ_{max}= 390mN/m²)
0	-5.00	0.63	1.06	1.27	1.5
0.05	-4.00	0.64	1.17	1.32	1.58
0.1	-3.00	0.655	1.30	1.40	1.77
0.15	-2.00	0.84	1.40	1.50	1.89
0.20	-1.00	0.92	1.5	1.67	1.96
0.25	0.00	0.97	1.58	1.78	2.06
0.30	1.00	1.09	1.70	1.93	2.2
0.35	2.00	1.19	1.80	2.00	2.3
0.40	3.00	1.25	1.85	2.08	2.3
0.45	4.00	1.25	1.92	2.1	2.4
0.50	5.00	1.27	1.94	2.1	2.4
0.55	4.00	1.27	1.95	2.1	2.4
0.60	3.00	1.14	1.87	2.0	2.4
0.65	2.00	1.00	1.76	2.0	2.36
0.70	1.00	0.79	1.67	1.97	2.29
0.75	0.00	0.71	1.50	1.87	2.2
0.80	-1.00	1.0	1.58	1.80	2.1
0.85	-2.00	1.0	1.40	1.58	1.9
0.90	-3.00	0.72	1.25	1.48	1.7
0.95	-4.00	0.62	1.10	1.20	1.5
1.00	-5.00	0.556	1.10	1.19	1.4

Table 1. Shear velocity distribotion on the artifically wavy boundary layer (x/L versus u_*) (Re= 15 000 and v_{mean}=0.80 m/s)

wide and 0.15 meter high. No sand was fed at the entrance of the stream. Beginning with these initial conditions, the photographs show that the stream developed naturally. An initial straight channel was formed into the meandering channel on the uniform sand bottom of the main channel. The modification of the initial straight channel into the meandering channel took place after 320 hours of the flowing of the sand-water mixture at the initial straight channel. This time was very long for experimentation. After every experimental run, it was difficult to carve the initial straight channel on the meandering deformations and to change the initial hydraulic parameters. An equilibrium condition was reached after 32 hours from the beginning of measurements at the laboratory meandering channel. Like an original prototype, the feed-back system of sand took place at the entrance of the sand reservoir of the main channel. With a developing meandering channel, bed profiles and velocities were obtained by laboratory profile indicator instruments and velocity-measurement instruments. However, only secondary velocities were obtained. The characteristic along the shifting sinuous channel of a meandering river (Figure 2.b and c) which is result of the oscillatory boundary layer. The indentations along the boundaries of the meander belt are the banklines of earlier courses. Throughout the duration of the test the stream constantly shifted its path of flow in the mid and lower sections. It is noted in the photographs that the degree of meandering increased downstream.

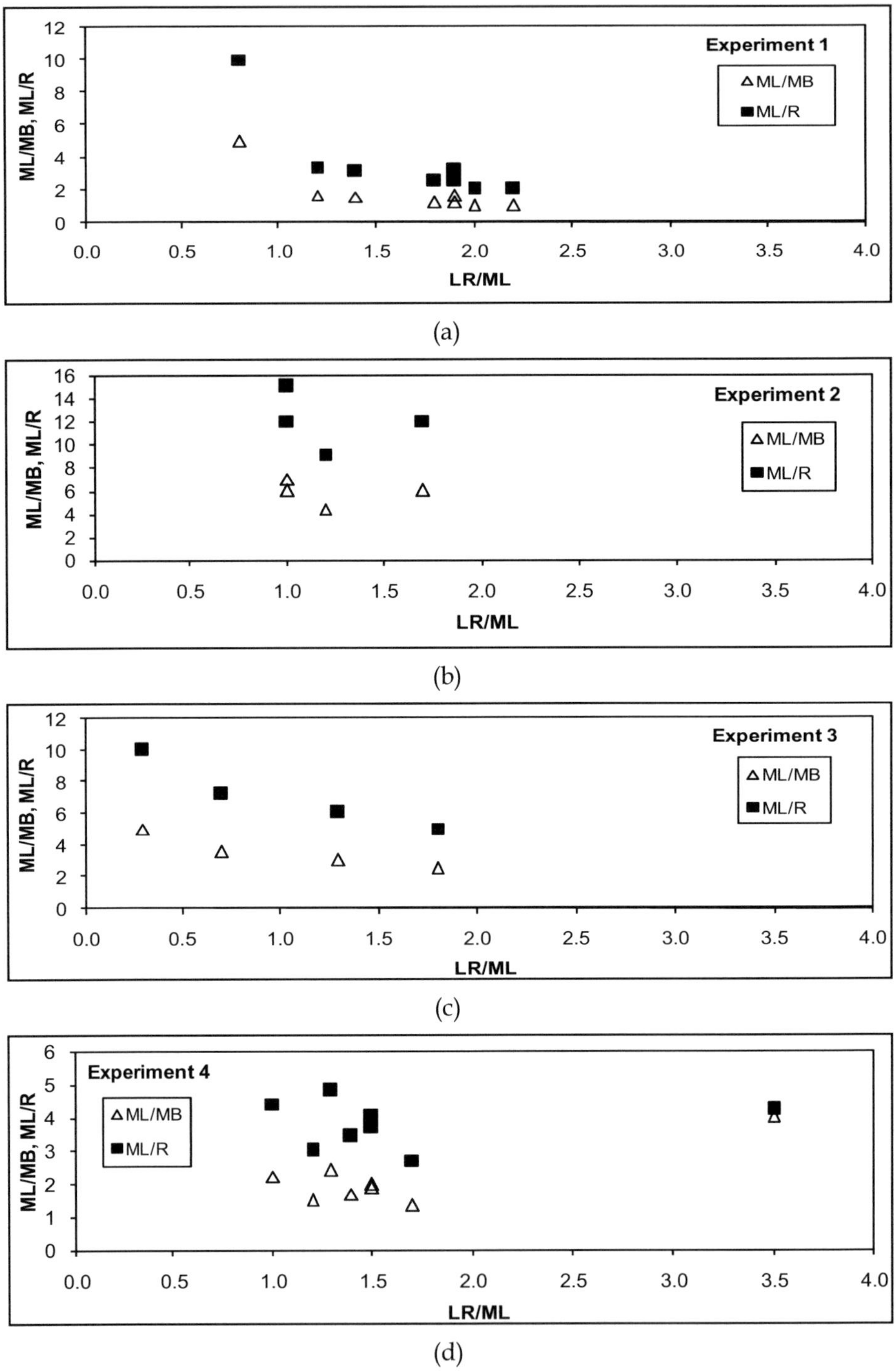

Fig. 2. Relation between tortuosity ratio (ML/LR) and aspect ratios of tortuosity, ML/MB and ML/R. ML = meander wave length, LR = meander bend length, R = meander bend, and MB = meander bend migration.

It was observed in the laboratory that the slope of the channel increased as it developed from the straight from through a shoaled condition to the meandering pattern, and that these three-dimensional channel forms were associated with significantly higher sediment transport rates than was the straight form. Due to the increasing gradient, the experimental channel would overflow its banks at the head of the system unless steps were taken to build up the banks to maintain a constant small freeboard. This was the standard procedure employed in the experimental program.

The laboratory observations provided a firm evidence of the association between the cause of meandering and an instability of sediment transport that was evidenced by the appearance of alternate shoals. Whether that instability aroses due to fluid or was caused by the bed sediment transport was not established by observations. Nevertheless, it was clear that the spacing of shoals was closely related to the wavelength of the meanders. It then follows that the spatial instability that generates alternate shoals also determines the wavelength of meanders. If an aggrading channel is permitted to overflow its banks, braiding occurs. Such a circumstance corresponds to a valley slope which is insufficient for the development of the hydraulic gradient necessary to transport the discharge of sediment and water.As the channel attempts to steepen its gradient, aggradation through sediment deposition results in a braided system (Dietrich, et al., 1979).

The experimental set-up seems to be a good model of the prototype of a meandering river. It reflects the development of a meandering channel on the cohesionless bottom boundary layer, exhibiting every phase from the beginning to the end of the erosion event at the bank and the bottom boundary layer. To determine the average sediment transport rate in weight per unit time took much time after every experimental run, because amount of sand was taken from the sand collector for it and one had to wait for it to dry in otoklav (drying oven). To estimate every measured value at the sand transport curve took too much time.

The experimental model in the present study consists of a very simple straight channel, which showed, after some time, meander planforms on the sandy bottom of the main channel. It was difficult to take some measurements at the meander bends. The sediment-transport rate took place before the meandering planform occurred. From experimental observations, it was seen that there was no sediment transport after meandering and the stability of the flume was observed with the meandering planform. One of the limitations of the experimental set-up was not observing the braiding planforms.

The measurement of sediment transport by weighting the dry sand amount at the end of the flume was error prone. In the prototype there is no influence of the dried sand transportation in the river erosion. The experimental procedure is used to find a relationship between the sediment transport in laboratory flume and that in the main channel. The sand transport distribution for every short time duration was not measured, although sediment transport equations may need a detailed sand transportation curve. Also, flume erodibility is not the only limiting factor for cross-section widening. When a cross-section becomes very wide and shallow, its cross-sectional shape may become unstable and develop into a number of separate, narrower channels, thus transforming into a braided or an anabranching river. In the experimental procedure, limited runs were taken, only for achieving the meandering planforms, but for braided or anabranching river cross-sections, the experiment was not continued. The main channel slopes varied between 0.04% - 0.5%. The main channel discharge varied between 0.07 l/s –0.73 l/s.Outside of this range, measurements could not be made for experimental limitations. The transported bed material weight in time was given for Q = 0.1536 l/s and J=0.08%. These values were restricted one for the given bed-load equation and for the meander-bend equation because of laboratory measurements difficulties. Out of this range of the main canal slope, the flow in the flume of sandy bottom varied from subcritical form into the supercritical form. That is why, one must provide more universal equations for bed load transport and for meandering bend planform equations.

3. Analysis of data and discussion of results

Following the preliminary experimental results, experiments with proper discharge and slope combinations were planned. Also, additional experiments were arranged to check the repeatability of the meander tests and to examine the meander performance at the extremes of steep and flat slopes. For each test, the flow had to be prepared at the beginning. This was done according to the following procedures. (1) The bed material throughout the whole flume was loosened. It was necessary because otherwise the sand could be cohesive and resistant to erosion. (2) The initial straight channel and the flood plain were formed. This was done by moving the carriage along the flume. On the carriage, a blade was fixed, which scraped the sediment to form a smooth flood plain. Meanwhile, a properly designed template was fixed on the blade cut into the sediment bed to form the initial channel along the center line of the flume. The cross-section shape and area of the initial channel along the center line of the flume were decided by the geometry of the template. The initial channel and the flood plain created in this way assumed the same slope, declining in the downstream direction. (3) The sediment was compacted. A certain compactness of the sediment bed was necessary in order to prevent the sediment from being easily washed away just as the flow entered the flume. (4) Both end-boundaries of the channel were shaped. At both upstream and downstream ends, the initial channel boundaries had to be carefully prepared in order to provide gradually changing sections for the entrance and exit of water flow.

The experimental results are presented in two parts. The first part consists of a description of what the shear stress distribution changes during the meander development. In the second part, experimental observations of meander evolution are described qualitatively. Generally, in the laboratory channel two different types of the meander development were observed, with and without sediment feeding at the entrance of the channel. At the beginning of the test when the flow front just reached the downstream end of the initial channel, sediment transportation took place immediately in terms of bed load. The initial flow velocity was about 40 cm/s. The eroded sediment was mostly from channel banks rather than from the bed. About 25 minutes later, the channel width increased from the initial width 10 cm, to 24 cm; the channel still remained straight. After another 10 minutes ripples (length = 40 cm) could just be seen on the channel bed, and they were moving in the flow direction with a speed of about 16 cm/s. Meanwhile, the ripple length became longer and longer and tended to be alternate bars. When elapsed time was about 30 minutes the ripples developed into alternate bars with an average bar length, approximately equal to 90 cm. Starting from that moment, the bed topography was dominated by the regular alternate bars, which seemed to begin to exert some influence over the main flow. The flow direction was changed from its initial course and diverted according to the shape of the bar

boundaries. Intensive bank erosion began to occur at the places where the flow tended to join the bank, and deposition occurred where water flowed away from the bank.
Although the main course of the channel was still essentially straight, a sinuous thalweg was noticeable. As the erosion at one bank and deposition at another continued a meandering channel was created, with a regular zigzag form. Later, a meander channel became noticeable, the meander bends started to expand in a transverse direction. Meanwhile the bend apexes started to move in the downstream direction. The lateral bend movement, known as bend expansion, initially progressed at a speed higher than that of the bend's movement in the down-valley direction which is known as the bend migration (Figure 2 b and c). However, after about one hour, the bend expansion stopped, while the bend migration not only continued but showed no sign of a decrease in speed. At this stage, despite the bend migration, the meander bends assumed substantially the same size and pattern. Hence, the meander plan form geometry became more or less constant. This condition of a meandering river is defined as the stabilized meander.

3.1 Analysis of data in the shear stress distribution on the wavy bed

In this part of experimental procedure, a given discharge was allowed to flow in the flume with a wavy bed. The uniformity of flow was ensured by adjusting the depth of flow constant near the entrance and exit reaches of the flume. The oscillating shear stress distribution on the wavy bed of the channel was measured with a hot-film-WTG-50-sensor. A technique developed by Gust (1988) measured skin friction in flume boundary layers, with and without suspended particulate matter, by constant temperature anemometry. The experimental setup consisted of a 1.5 m wide and 1.5 m deep rectangular flume (10 m). The straight length of channel had a bottom covered with artificial waves of iron plate, as the same of the walls with wave length of 120 cm, and amplitude a=2 cm. The clear water supply was obtained from a 5 m high overhead tank. In order to study the effect of oscillating shear stress distribution the bottom slope was changed. In the second part of experiments, skin friction measurements were made with an array of flush-mounted-hot-films at 24 points on the different slopes of one of a field of two-dimensional immobile artifically made dunes in the straight channel. The total boundary shear stress was also measured.
With sensor scales of 3 mm and frequency responses of 20 Hz flush-mounted, epoxy-coated hot films yielded mean and fluctuating components of wall shearing stress at different water depths. Fluctuating components of the skin friction are present in all turbulent flows, including those of hydrodynamically smooth walls (Eckelmann, 1974), and require consideration in turbulent flow of calibration techniques. In experimental runs mean friction velocities are obtained by multiplying recorded mean voltage signals by calibration coefficients obtained through either polynomial or exponential least squares fits of the calibration data. In turbulent flow the calibration curve of instantaneous (equal to mean) skin friction can be expressed for constant temperature anemometry as

$$\tau^{1/2} = A_t E^2 + B_t \qquad (1)$$

where the coefficients A, B are determined by a least squares fit from the shape of the calibration curve (Hanratty and Campbell (1983)).

4. Comparison with previous studies

Paola (1983) used similar hot film anemometry technique to measure skin friction and close-to-bed velocity profiles over fixed ripples of median grain size of 0.2 mm in a flume experiment. The ripples were 1 cm high and 10 cm long. His results thus should be comparable to those of the present study. His measurements were obtained under a range of roughness Reynolds number $Re_*=Hu_*/\nu$. He obtained the total shear velocity u_* from the energy gradient method. Estimates of this shear velocity from available velocity profiles agreed well with the values from the energy gradient method. Skin friction shear velocity u_{*s} was measured at four locations on the stoss slope of a ripple and were then spatially averaged. The data points closely follow the diagonal line, suggesting a good agreement between the measured shear velocity and those predicted by the model. Data from Paola (1983) and Zilker et al. (1977) show that u_{*s} obtained from the inner boundary layer velocity profiles relatively far above the sand bed decreases from the ripple trough to the crest. This trend is just opposite to the u_{*s} measured using the skin friction probes, though the spatially averaged values of the two methods do not differ significantly. A recent study by Nelson and Smith (1989) also shows that logarithmic fit to the whole inner boundary layer velocity profile will underpredict u_{*s}. Based on these findings, the inner boundary layer ripple crest velocity profiles of Paola (1983) have been fitted to the van Karman-Prandtl equation to obtain the profile-based skin friction shear velocity. The results of the present study and data from previous investigations generally support the application of the shear stress model to sand ripples for partitioning skin friction and form drag from known grain size, ripple geometry and conventional outer boundary layer velocity profiles. Though the model is relatively simple, it does require iterative calculations, and uncertainties exist in various coefficients. It is preferable, therefore, to derive some overall empirical relationships between outer boundary layer shear velocity u_* and inner boundary layer skin friction shear velocity u_{*s}. This empirical relationship should provide approximate but easily derived estimates of skin friction over sand dunes from known dune geometry and the outer boundary layer shear velocity..

The ratio of the inner boundary layer shear velocity u_{*s} and the outer boundary layer shear velocity u_* in a unidirectional flow is mainly controlled by the flow condition, grain size and dune geometry. The flow condition is represented by the total shear velocity u_* and the dune character can be represented by the dune height H since the aspect ratio H/L is roughly constant for natural current dunes (=0.05). Therefore, a universal empirical relationship between the ratio u_{*s}/u_* and a flow dune parameter u_*/H can be used to properly estimate u_{*s} from u_* and H. Due to acceleration and deceleration of near-bed flows over bed forms, flows tend to separate at the dune crest and a wake forms at the lee side of the dune (Raudkivi, 1967) . Higher ripple steepness and low to medium flow stresses have been found to favor this flow separation and wake development (Davies, 1980; Smith, 1977). At low u_*/H ratios (2.3), the smaller flow stress and or bigger ripple height determine that the flow is separated and lee side wakes are fully developed. Under this condition, u_{*s} increases relatively faster than the form drag as flow becomes stronger. Thus, u_{*s}/u_* increases systematically with u_*/H. When the flow dune parameter u_*/H is beyond the critical value of 4.2, the flow stress becomes very strong and sand bypassing occurs. Under this condition, the dune height is reduced and wakes start to break down, and flow is no longer fully separated. These will cause a dramatic increase of drag and hence increase of total shear velocity u_*.

5. Conclusions

The flow resistance in a meander bend is considerably increased due to the form resistance of the patterns about which much is not known. It depends on a number of factors including grain friction, form resistance of two- and three dimensional patterns, skin friction of the non-separated oscillatory component and the sediment transport rate. The following results from the investigation were obtained:

The measured skin-friction field is consistent with a simple model for sediment transport over bed forms, where the fluctuating skin friction is important. The data are also consistent with the drag-partition theories of Engelund (1966) and Paola(1983). Normalized skin-friction spectra vary with stream-wise position but not vary with Reynolds number.

6. Definition of terms

The tortuosity ratio determines the shape of meanders. For any one value of tortuosity ratio, there are associated single values of the ratios of:

1. Length along river bend to radius of center line of bend: (LR/R)
2. Straight length along the river valley to radius of center line of bend: (ML/R);
3. Meander belt to radius of center line of bend (MB/R);
4. Angle of bend to radius of curvature (θ/R).
5. Meander length to meander belt (ML/MB)

7. Analysis

As shown in Table 1, four experiments were conducted with discharge varying from 0.08 to 0.50 l/s and slope changing from 0.08 to 0.35%. In this table, F_o is the Froude number, which was determined using the velocity and depth measured at the extreme upstream end of the channel. In Table 1, R_e is the Reynold's number that is computed taking kinematic viscosity equal to 1.57×10^{-5} m^2/s, which corresponds to the standard temperature and pressure (STP). MBN represents the number of meander bends occurred in the initially straight channel after 72 hrs of the experimentation. L, LR, and MB represent, respectively, the straight channel length, total meander bend length of the channel, and the maximum migration of the initially straight channel. Some of these terms are defined in Fig. 1. In this figure, θ is the central angle of the meander bend and R is the radius of the bend. The resulting forms of the meandered channel in the four experiments are shown in Figs. 2a-d, which correspond to experiments no. 1 through 4, respectively.

It is apparent from Table 1 that the number of meander bends (MBN) increases with the increase in either F_o or R_e and vice versa. It implies that the number of meandering bends depends on the flow regime described by the Froude and Reynold's numbers. The ratio of the total length of the initially straight channel, L, to the total meander length of the channel, LR, describes the tortuosity of the meandering channel. It is an indicator of the meander formation; if L/LR is equal to 1, the channel does not meander at all and if LR approaches infinity, the tortuosity ratio approaches zero, implying that the tendency of a channel to meander increases with the reduction in the tortuosity ratio and vice versa. It is seen from table 1 that the tortuosity ratio decreases with the increase in either F_o or R_e and vice versa. It leads to inferring that the flow in the regime of low velocity and high depths allows an alluvial channel to meander less than in the otherwise situation.

The above results, however, deviate much from those observed in the field and shown in Table 2 for some rivers. The ratio L/LR for laboratory channel (Table 1) varies from 0.033 to 0.048 whereas it ranges from 0.035 to 0.582 in natural rivers, implying that the natural rivers will meander less than the laboratory channels even though both are in the same flow regime. It leads to inferring the prominent role of boundary conditions in meandering. Laboratory channels are usually of restricted geometry (length and width) whereas the natural rivers may adopt any size. The channel characteristics are connected with meander geometry. For example, wide shallow channels exhibit lesser tortuosity than narrow deep channels do and vice versa. Furthermore, the flow lines in the laboratory channel follow sharper curvature than in the prototype, indicating that deeper and narrower channels produce more acute bends (Rozovasky and Makkavuev, 1964). According to Rozovasky and Makkavuev (1964), the velocity of flow in transverse direction that causes river meandering is directly proportional to the depth and mean velocity of flow and inversely proportional to the radius of curvature. It is noted that the depth and the velocity of flow governs the regime of flow described by F_o and R_e. Thus, the inference of the results from laboratory channel is consistent with the results of earlier studies.

The above results, however, provide only an overall view of the meandering of the laboratory channel. A discussion of the individual meander forms of all the four experiments shown in Fig. 2 and summarized in Table 3 follows. In this table, the location describes the location of MBN and other columns describe its corresponding ML, MB, R, and LR. It is seen from the table that, in all the four experiments, the meander length (ML) and the radius of curvature (R) increase with the distance. It implies that the wave length (ML) increases in the direction of flow and vice versa. Since R also increases with ML, the increase in ML is coupled with the increase in the meander bend migration (MB), as also seen in Fig. 2. Thus, the meandering behavior of the laboratory channel expands in the direction of flow. Here, it is appropriate to mention a little about the role of the initial flow conditions that may be of significance in the process of meander development. For example, a meandering natural river at a certain location exhibiting certain meandering characteristics in a particular flow regime will differ significantly from the behavior of the laboratory channel in the same flow regime. It is because of the difference in initial and boundary conditions, which affect significantly the flow wave behavior in open channels (Mishra and Singh, 1999). Since the ratio L/LR describes local features of channel meandering, this ratio is sensitive to the considered river reach and, therefore, is larger in natural rivers than in laboratory channels.

The aspect ratio defined by ratio of the length along river bend to meander length (= LR/ML) in experiment 1 varies from 0.8 to 2.0, with most of the values near 2.0. In experiment 2, the aspect ratio varies from 1.0 to 1.7, with most values near 1.0; it ranges between 0.3 and 1.8 in experiment 3, and between 1.0 and 3.5 in experiment 4. The overall average of these values is of the order of 1.5.

The experimental meandering channel developments show the relationships between tortuosity ratio; LR/LV; and ML/R and ML/MB. Similar curves could be drawn for the other ratios. Important channel characteristics are connected with meander geometry. Wide, shallow channels are usually associated with lesser tortuosity. When distortion is used in hydraulic models, flow lines are found to follow sharper curvature than in the prototype, indicating that deeper and narrower channels produce more acute bends.

MB No:	ML (cm)	MB (cm)	R (cm)	*ML/MB	*ML/R	LR (cm)	LV (cm)	*LR/ML
Experiment no. 1								
1	50	10	5	5.00	10.0	40	50	0.8
2	25	20	10	1.25	2.5	45	25	1.8
3	25	25	13	1.00	2.0	50	25	2.0
4	50	30	15	1.67	3.3	60	50	1.2
5	35	35	18	1.00	2.0	70	35	2.0
6	50	40	20	1.25	2.5	95	50	1.9
7	70	45	23	1.56	3.1	100	70	1.4
8	50	50	25	1.00	2.0	110	50	2.2
9	80	50	25	1.60	3.2	150	80	1.9
Experiment no. 2								
1	45	8	4	6.00	12.0	75	45	1.7
2	105	15	8	7.00	15.0	100	105	1.0
3	150	25	13	6.00	12.0	150	150	1.0
4	165	38	18	4.40	9.1	200	165	1.2
Experiment no. 3								
1	50	20	10	2.50	5.0	90	50	1.8
2	75	25	13	3.00	6.0	100	75	1.3
3	175	35	18	5.0	10.0	50	175	0.3
4	180	50	25	3.60	7.2	125	180	0.7
Experiment no. 4								
1	10	3	2	4.00	4.2	35	10	3.5
2	30	20	10	1.50	3.0	35	30	1.2
3	30	13	6	2.40	4.8	40	30	1.3
4	50	23	11	2.22	4.4	50	50	1.0
5	40	20	10	2.00	4.0	60	40	1.5
6	55	33	16	1.69	3.4	75	55	1.4
7	65	35	18	1.86	3.7	95	65	1.5
8	60	45	23	1.33	2.7	100	60	1.7

Note: ML = meander wave length, MB = meander bend migration, R = radius of meander bend, LV = thalweg, superscript '*' indicates measures of tortuosity of the channel.

Table 3. Detailed features of the meanders in four experiments

From Table 1 the longest meander bend length in thalweg is given in the smallest discharge and bed slope as in the experimental run number 1; Q= 0.08 l/s; and S_o=0.08%. The meander bend number was in 500 cm length from the origin to the experimental measurements are 9. Tortuosity number = Meander wave length/ Meander bend length was

(ML/LR = 475/15100 = 0.03146) the smallest, but meander bend migration after 72 hours was 50 cm, the largest, because of the smallest bed slopes (S_o= 0.08%).

The second longest meander bend length in thalweg is given in the experimental run number 4, which has the discharge Q=0.50 l/s; S_o= 0.35%. The meander bend number was in 400 cm distance in the observations from the origin was 8. The tortuosity number has the second biggest value as: ML/LR=400/11900= 0.0336, but the meander bend migration after 72 hours was 45 cm, as the second largest migration.

The third biggest tortuosity number (ML/LR=500/11300= 0.044) belongs to the third experimental run, with Q= 0.40 l/s, and S_o= 0.20 %, which has altogether 4 meander bend numbers.

From the figure of fully developed meandering channel patterns for various discharges (Q) and bed slopes (S_o) of the channel after 72 hours are given below results:

a. Figure (a) and (d) show similar trends, and (b) and (c) have no much meander bends.
b. After 72 hours of the experimental run the boundary layer material is compacted and does not show many changes in small bed slopes like at the experimental run (a). If we change the slope from S_o= 0.10%, and S_o= 0.20% to S_o= 0.35% and the discharges from Q=0.08 l/s, and Q= 0.20 l/s, and Q= 0.40 l./s to Q=0.50 l/s, and S_o= 0.35% it is observed the planform changes like at the beginning of the experiments, because the compactness of the sand material is too much after 72 hours.

From the Table 1, meander bend migration after 72 hours is only 50 cm like in the first experimental run, but the smaller meander bend number as 4 is the same with the experimental run in number 2 with Q= 0.20 l/s, and S_o= 0.10 %, but the tortuosity number is with ML/LR= 500/10500= 0.0476 is the biggest one, because second experimental run has the shortest meander bend length in thalweg like totally 10500 cm. It means if the meander bend length in thalweg (LR) is shorter, it has the biggest tortuosity number as ML/LR=500/10500=0.0476.

From the Table 2, giving the details of all features of meanders, the biggest ratios of ML/MB and ML/R for tortuosity grade, the second experimental run shows with Q= 0.20 l/s, and slope as S_o= 0.10%, ML/MB= 6 and ML/R=12, which has the meander bend number as 4. The second largest values (ML/MB=5) and ML/R=10 value is given in the experiment number 1, by Q=0.08 l/s; and S_o= 0.08 %, which has meander bend number as 9. These comparisons show that there is no necessity for having the largest meander bend number for large tortuosity ratios.

The third tortuosity ratio belongs to the experimental run number: 4, with Q=0.50 l/s; with slope S_o= 0.35%, which has the tortuosity ratio ML/MB=4 and ML/R= 4.166 at the first meander bend. The last values at the fourth experimental run with bend number as 8, which has ML/MB=1.33 ratio and ML/R=2.66 are the smallest in comparing with the other experimental runs. The first experimental run has the biggest meander wave number as 9, number 2 has 4 and number 3 has also 4, but the last run has also 8 meander bends. It means, if the boundary layer material is compacted after the 72 hours, it has the biggest discharge as Q=0.50 l/s, and the biggest slopes as S_o= 0.35%. It means for more tortuosity ratios we need the biggest discharge and the biggest bed slope in laboratory conditions.

Comparison of the laboratory meander tortuosity with the natural river meander tortuosity: Data is taken from Langbein and Leopold (1966) for the natural river meander tortuosity and gives the below results: The prototype for the laboratory meanders is choosen as San Juan River, Utah, and has this values:

ML (miles)	MB (miles)	R (miles)
1.2	2.4	2.0
1.8	2.7	2.3
2.2	2.9	2.5

The tortuosity ratio is ML/MB = 1.2/2.4 = 0.5 and the other ratio is ML/R = 1.2/2.0 = 0.6 Comparing these values with the laboratory values, the laboratory meanders are not well developed because of the canal bottom slope, but in natural meander bends, with the whole developed bends the tortuosity ratio is too small, when comparing with the laboratory data, ML/MB = 6 and ML/R = 5 and natural data gives 10 times smaller values.

Experiment No.	L (cm)	Discharge (l/s)	Slope (%)	F_o	$^*R_e$	MBN	LR (cm)	L/LR	MB (cm)
1	500	0.08	0.08	0.040	12658	9	15100	0.033	50
2	500	0.20	0.10	0.010	3165	4	10500	0.048	38
3	500	0.40	0.20	0.020	6329	4	11300	0.044	50
4	500	0.50	0.35	0.025	7911	8	11900	0.042	45

Note: F_o = Froude number, R_e = Reynold's number, MBN = number of peaks and troughs of the channel after meandering, L = straight channel length, LR = meander bend length, MB = largest meander bend migration, and L/LR = tortuosity number. *kinematic viscosity is taken equal to 0.157×10^{-6} m²/s.

Table 1. Meandering characteristics of the laboratory channel

River	Location	Slope %	MBN	L (mile)	LR (mile)	L/LR	MB
San Juan	Utah	-	4	0.43	0.739	0.582	200 ft
Popo Agie	Wyoming	0.20	3	0.24	0.426	0.563	150 ft
Mississippi	Greenville	-	3	5.00	19.00	0.263	3.6
Potomac	West	-	3	6.63	14.25	0.465	1.5
Pole Creek	Wyoming	0.21	2	0.14	3.978	0.035	300 ft

Note: MB No. = number of peaks and troughs of the channel after meandering, L = straight channel length, LR = meander bend length, MB = largest meander bend migration, and L/LR = tortuosity number. *kinematic viscosity is taken equal to 0.157×10^{-6} m²/s. '-' implies not available

Table. 2. Meandering characteristics of some natural meandering rivers

MB No:	ML (cm)	MB (cm)	R (cm)	*ML/MB	*ML/R	LR (cm)	LV (cm)	*LR/MB
Experiment no. 1								
1	50	10	5	5.00	10.0	40	50	0.8
2	25	20	10	1.25	2.5	45	25	1.8
3	25	25	13	1.00	2.0	50	25	2.0
4	50	30	15	1.67	3.3	60	50	1.2
5	35	35	18	1.00	2.0	70	35	2.0
6	50	40	20	1.25	2.5	95	50	1.9
7	70	45	23	1.56	3.1	100	70	1.4
8	50	50	25	1.00	2.0	110	50	2.2
9	80	50	25	1.60	3.2	150	80	1.9
Experiment no. 2								
1	45	8	4	6.00	12.0	75	45	1.7
2	105	15	8	7.00	15.0	100	105	1.0
3	150	25	13	6.00	12.0	150	150	1.0
4	165	38	18	4.40	9.1	200	165	1.2
Experiment no. 3								
1	50	20	10	2.50	5.0	90	50	1.8
2	75	25	13	3.00	6.0	100	75	1.3
3	175	35	18	5.0	10.0	50	175	0.3
4	180	50	25	3.60	7.2	125	180	0.7
Experiment no. 4								
1	10	3	2	4.00	4.2	35	10	3.5
2	30	20	10	1.50	3.0	35	30	1.2
3	30	13	6	2.40	4.8	40	30	1.3
4	50	23	11	2.22	4.4	50	50	1.0
5	40	20	10	2.00	4.0	60	40	1.5
6	55	33	16	1.69	3.4	75	55	1.4
7	65	35	18	1.86	3.7	95	65	1.5
8	60	45	23	1.33	2.7	100	60	1.7

Note: ML = meander wave length, MB = meander bend migration, R = radius of meander bend, LV = thalweg, superscript '*' indicates measures of tortuosity of the channel.

Table 3. Detailed features of the meanders in four experiments

8. References

Anderson, A. G., 1967, On the development of stream meanders, Proc. XII. Congress IAHR, Fort Collins, U.S.A .

Bagnold, R. A., 1946, Motion of waves in shallow water, Interaction between waves and sand bottoms, Proc. Roy. Soc., London, A 187, pp. 1-18.

Blondeaux, P. and Seminara, G., 1985, A unified bar-bend theory of river meanders, Journal of Fluid Mechanics; Vol. 157, pp. 449-470.

Callander, R. A., 1969, Instability and River Channels, Journal of Fluid Mechanics, London, England, Vol. 36; Part.3.

Chang, H.H., 1984, Analysis of river meanders, Journal of Hydraulic Engineering, Vol. 110, No. 1.

Chang, H. H., 1983, Energy expenditure in curved open channels, Journal of Hydraulic Engineering, ASCE, Vol. 109, No. 7, pp. 1012-1022.

Chang, H. H., 1979, Geometry of Rivers in Regime, Journal of the Hydraulics Division, ASCE, Vol. 105, No. HY6, Proc. Paper 14640, pp. 691-706.

Cardoso, A. H., Graf, W. H., and Gust, G., 1989, Uniform flow in a smooth open channel, Journal of Hydraulic Research; Vol.27; No. 5; pp. 603-616.

Cardoso, A. H., Graf, W. H., and Gust, G., 1991, Steady gradually accelerating flow in a smooth open channel, Journal of Hydraulic Research, Vol. 29, No.4, pp. 525-543.

Davies, T. R. H., 1980, Bedform spacing and flow resistance, Journal of Hydraulic Division, Am. Soc. Civ. Engr., 106, pp. 423-433.

Dietrich, W. E., and Smith, J. D., 1983, Influence of the point bar on flow through curved channels, Water Resources Res., 19(5), pp. 1173-1192.

Eckelmann, H., 1974, The structure of the viscous sublayer and the adjacent wall region in a turbulent channel flow, J. of Fluid Mech., pp. 439-459.

Engelund, F., and Fredsoe, J., 1976, A sediment transport model for straight alluvial channels, Nordic Hydrol., 7, pp. 293-306.

Engelund, F, and Hansen, E., 1967, A monograph on sediment transport in alluvial streams, Technical Press, Copenhagen.

Engelund, H., and Skovgaard, O., 1973, On the origin of meandering and braiding in alluvial streams; Journal of Fluid Mechanics, London, England, Vol. 76, Part. 3, pp. 457-480.

Exner, F. M., 1919, Zur Theorie der Flussmaender, (in German), Sitzungs-Berichte der Akademie der Wissenschaften, Wien, Oesterreich.

Fredsoe, J., 1978, Meandering and Braiding of Rivers, Journal of Fluid Mechanics, London, England, Vol. 84, Part.4, pp. 609-624.

Fredsoe, J., 1982, Shape and dimensions of stationary dunes in rivers, J. of Hydraulic Div., Am. Soc. Civ. Eng., 108, pp. 932-947.

Gottlieb, L., 1976, Three-dimensional flow pattern and bed topography in meandering channels, ISVA, Tech. Univ. Denmark, Series Paper 11.

Gust, G., 1988, Skin friction probes for field applications, Journal of Geophysical Research, Vol. 93, No. Cii, pp. 14121-14132.

Hansen, E., 1967, On the formation of meanders as a stability problem. Basic Research Progress Report No.13, Hydraulic Laboratory, Technical University of Denmark, pp. 9-13.

Hanratty, T. J., and J. A. Campbell, 1983, Measurement of wall shear stress, in Fluid Mechanics Measurements, edited by R. J. Goldstein, Hemisphere, Washington D. C., pp. 559-615.

Hayashi, T., 1970, The formation of meanders in rivers, Proceedings, Japan Society of Civil Engineers, No. 180, pp. 61-70.

Hayashi, T., and Ozaki, Y., 1980, Alluvial Bed Form Analysis-Formation of Alternating Bars and Braids; Application of Stochastic Process in Sediment Transport, H. W. Shen and H. Kikkawa, Water Resources Publications, Littleton, Colo.

Ikeda, H., 1973, A study on the formation of sand bars in an experimental flume, Geographical Review of Japan, Vol. 46, No.7, pp. 435-450.

Ikeda, S., 1984, Prediction of alternate bar wavelength and height; Journal of Hyd. Engineering, ASCE, Vol. 110, No.4.

Kaneko, A., 1981, Oscillation sand ripples in viscous fluids, Proc. J. S. C. E., 307, pp.113-124.

Kaneko, A., and Honji, H., 1979, Double structures of steady streaming in the oscillatory viscous flow over a wavy wall, J. of Fluid Mech., 93, pp. 727-736.

Kennedy, J.F., and Brooks, N. H., 1963, Laboratory Study of Alluvial Streams at Constant Discharge, Proceedings, Federal Inter-Agency Sedimentation Conference, Miscellaneous Publication No. 970, Agricultural Research Service, pp. 320-330.

Kinoshita, R., 1987, Formation of Dunes on River Bed, Transactions, Japan Society of Civil Engineers, N. 42, pp. 1-21.

Kuroki, M., Kishi, T., and Itakura, T., 1975, Hydraulic Characteristics of alternate bars, Report for National Science Foundation, Department of Civil Engineering, Hokkaido University, Hokkaido, pp. 80-88.

Leopold, L.B., and Wolman, M. G., 1960, River Meanders, Bulletin of the Geological Society of America, Vol. 71, pp. 769-794.

Leopold, L. B., Wolman, M.G., and Miller, J. P., 1964, Fluvial processes in Geomorphology, W. H. Freeman and Co., San Francisco, Calif., p. 522.

Leopold, L. B., and Wolman, M. G., 1957, River channel patterns; Braided, Meandering and Straight, USGS Professional Paper 282-B, pp. 45-62.

Lane, E. W., 1957, A study of the shape of channels formed by natural streams flowing in erodible material, U. S. Army Eng. Division, Missouri river, Corps. of Engineers, M. R. D. Sediment Series, No. 9, Omaha, Neb.

Lyne, W. H., 1971, Unsteady viscous flow over a wavy wall, J. Fluid Mech., 50, pp. 33-48.

McLean, S. R., and J. D. Smith, 1986, A model for flow over two-dimensional bedforms, J. Hydraulic Eng., 112, pp. 300-317.

Muramoto, N., and M. Fujita, 1978, The classification of meso-scale river bed configuration and the criterion of its formation, Proceedings of the 22nd Japanese Conference on Hydraulics, Japan Society of Civil Engineers.

Nelson, J. M., and J. D. Smith, 1989, Mechanics of flow over ripples and dunes, J. Geophys. Res., 94, pp. 8146-8162.

Nezu, I., and Rodi, W., 1986, Open-channel flow measurements with a laser doppler anemometer, Journal of Hydraulic Engineering; Proceedings of the ASCE; Vol. 112, No. 50, pp. 335-355.

Odgaard, A. J., 1981, Transverse bed slope in alluvial channel bends, J. Hydraulic Division, ASCE, 107 (HY12), pp. 1677-1694.

Odgaard, A. J., 1986, Meander Flow Model. I: Development, J. of Hydraulic Engineering; ASCE, Vol. 112; No. 12.

Odgaard, A. J., 1982, Bed characteristics in alluvial channel bends, J. of Hydraulic Division, ASCE, 108 (HY11), pp. 1268-1281.

Parker, G., 1984, Discussion of lateral bed load transport on side slopes by S. Ikeda, Journal of Hydraulic Engineering; Vol. 110; pp. 197-199.

Parker, G., 1976, On the cause and characteristic scales of meandering and braiding in rivers; Journal of Fluid Mechanics, London, England, Vol. 76, Part. 3, pp. 457-480.

Parker, G., and Peterson, A. W., 1980, Bar resistance of gravel-bed streams, Journal of the Hydraulics Division, ASCE, Vol. 106, No. HY10, pp. 1559-1576.

Paola, C., 1983, Flow and skin friction over natural rough beds, Doctoral Dissertation; Joint Program in Oceanography and Ocean Engineering; Woods Hole Oceanographic Institution/ Massachusetts Institute of Technology: WHOI- 83-18.

Raudkivi, A. J., 1976, Bed forms in alluvial channels, J. of Fluid Mechanics, 26, pp. 507-514.

Raudkivi, A. J., 1967, Loose Boundary Hydraulics, pp. 175-221, Pergamon, New York.

Schreider, M. I. and, Amsler, M. L., 1992, Bedforms steepness in alluvial streams, Journal of Hydraulic Research, Vol. 30, No.6.

Schumm, S.A., 1977, The fluvial system, John Wiley and Sons, New York, N. Y., pp. 338.

Shen, H. W., 1971, Stability of alluvial channels, Chapter 16, River Mechanics, Vol.I, edited and published by Shen, H. W., Fort Collins, Colo.,

Shimizu, Y, and Itakura, T., 1989, Calculation of Bed Variation in Alluvial Channels, Journal of Hydraulic Engineering, Vol. 115, No. 3, pp. 367-384.

Sedimentation Engineering, 1975, Manual No. 54, ASCE, New York, N. Y.

Simons, D. B., and Senturk, F., 1977, Sediment Transport Technology, Water Resources Publications, Fort Collins, Colo.

Smith, J. D., 1977, Modelling of sediment transport on continental shelves, in The Sea, Vol.6, edited by E. D. Goldberg et. al., Wiley-Interscience, pp. 539-577, New York.

Smith, J. D., and McLean, S. R., 1977, Spatially averaged flow over a wavy surface, J. Geophys. Res., 82, pp. 1735-1746.

Stuart, T., 1966, Double boundary layers in oscillatory viscous flow, J. Fluid Mech., 24, pp. 673-687.

Struiksma, N., Olesen, K. W., Flokstra, C., and DeVriend, H. J., Bed deformation in curved alluvial channels; J. of Hydraulic Res., 23(1), pp. 57-79.

Sukewaga, N., 1971, Study on meandering of streams in straight channels, Report of Bureau of Resources, Department of Science and Technology.

Sukewaga, N., 1972, Criterion for alternate bar formation in experimental flumes; Proceedings, Japan Society of Civil Engineers, No. 207.

Tamai, N. and Nago, T., and Mikuni, F., 1978, On the large-scale bar patterns in a straight channel, Proceedings of the 22nd Japanese Conference on Hydraulics, Japan Society of Civil Engineers, pp. 265-273.

Tubino, M., and Seminara, G., 1990, Free-forced interactions in developing meanders and suppression of free bars, Journal of Fluid Mechanics, Vol. 214, pp. 131-159.

Van Rijn, L. C., 1984, Sediment transport, Part III, Bed forms and alluvial roughness, J. of Hydraulic Engineering, ASCE, Vol. 110, No. 12, Dec., pp. 1733-1754.

Vittori, G., 1989, Non-linear viscous oscillatory flow over a small amplitude wavy wall, Journal of Hydraulic research Vol. 27, No.2.

Yalin, M. S., 1977, Mechanics of Sediment Transport, 2nd. Ed., Pergamon Press Inc., London, England.

Yang, C. T., and Song, C. S., 1979, Theory of Minimum Rate of Energy Dissipation, Journal of the Hydraulics Division, ASCE, Vol. 105, No. HY7, Proc. Paper 14677, July, pp. 769-784.

Yilmaz, L., 1990, Meanderbildung, (in German) Beitrag in Turkisches Wissenschafts-Technologiezentrum e. V. Berlin, IWAWI- Technische Universitaet Berlin.

Yen, B. C., 1965, Characteristics of subcritical flow in a meandering channel, Institute of Hydraulic Research, The University of Iowa, Iowa City, Iowa, pp. 155.

Wiberg, P.L., and Rubin, D. M., 1989, Bed roughness produced by saltating sediment, J. Geophys. Res., 94, pp. 5011-5016.

Zilker, D. P., Cook, G. W., and Hanratty, T. J., 1977, Influence of the amplitude of a solid wavy wall on a turbulent flow; Part.1. Non-separated flows, J. Fluid Mech. 82; pp. 29-51.

Zilker, D. P., and Hanratty, T. J., 1977, Influence of the amplitude of a solid wavy wall on a turbulent flow. Part.2. Separated flows, J. Fluid Mech. 90: pp. 257-271.

Zimmermann, C., and Kennedy, J. F., 1978, Transverse bed slopes in curved alluvial streams, J. of Hydraulic Division, ASCE, 104 (HY1), pp. 33-48.

Zimmermann, C., and Kennedy, J. F., 1978, Transverse bed slopes in curved alluvial streams; J. of Hydraulic Division; ASCE, 104 (HY1), pp. 33-48.

12

Evolution of Ripple Field Architecture during Sediment Transport, as Measured by CT Scanning

Long Bernard and Montreuil Stéphane
Centre Eau Terre Environnement
Institut national de la recherche scientifique
Université du Québec
Canada

1. Introduction

Despite the advent of ever more sophisticated technologies, the study of sediment transport remains highly relevant. Ongoing climate change adds urgency to the need to understand phenomena governing sediment transport. During sediment movement, the architecture of the sediment bed evolves (Nummedal et al., 1993; Allen and Posamentier, 1994; Van Wagoner, 1995). This chapter focuses on the fundamental processes involved in sediment transport and aims to define and characterize the evolution of the water-sediment mixture in a migrating sand ripple field under a stationary flow regime, from the flow surface to the base of the sediment bed. In an experimental sediment flume with attached computed tomography (CT)-scanner, the sediment density of a ripple field was studied at different phases of its architectural evolution using three- and four-dimensional X-ray scanning, under several flow regimes and with different grain sizes. This approach yielded a detailed understanding of the parameters involved in transport of sand-grade sediment.

2. Methods

This study used two hydraulic flumes. One was a vertical, closed-loop flume 3048 mm long and 1219 mm high designed by Teeter and Pankow (1989). The inner rectangular section of the flume (152 mm by 304 mm) passed through the mobile gantry of a computed tomography (CT)-scanner. The lower part of the tunnel was partly filled with a 30 mm-thick layer of C-109 Ottawa sand (0.37 mm grain size, with bulk dry density of 1.56 $g{\bullet}cm^{-3}$). The flow regime was adjusted by an electrical motor to 25 $cm{\bullet}s^{-1}$ and controlled by a Doppler flow meter (Controlotron 1010N). For more details of this apparatus, see Montreuil (2006). In this flume experiment, the volume element (voxel) chosen was 0.39 mm by 0.60 mm by 1.00 mm. Typical density profiles along the ripple bed were extracted from the CT-scanner measurement matrix of 512 x 512 x 512 voxels and a 305 x 305 x 200 mm volume.
The second flume was an open acrylic flume (300 x 300 x 7000 mm) looped by a pump with 1.51 $m^3{\bullet}min^{-1}$ capacity. The rectangular section of the flume (300 x 300 mm), passed through the mobile gantry of the CT-scanner (Fig. 1). The steady flow, measured by a Controlotron

1010 Doppler flow meter, was controlled by a PID regulator using a pneumatic valve installed in the pump looping circuit. Three flow velocities and two grain sizes were used to confirm the existence of several points and zones and to identify their respective locations over a ripple. To create a natural sand column, the flume base was sand-blasted, as recommended by Madsen and Durham (2007), and was filled with a 50 mm-thick sand layer. The CT-scanner was able to accommodate a large open-channel flume scanner (300 x 300 x 7000 mm; Fig. 2) and allowed continuous measurement of both suspended and bedload sediment, without hydraulic interference on the water or sediment, for the 512 x 512 x 1500 mm voxel matrix. In the open-flume experiment, the voxel dimensions were 0.6 x 0.6 x 0.6 mm.

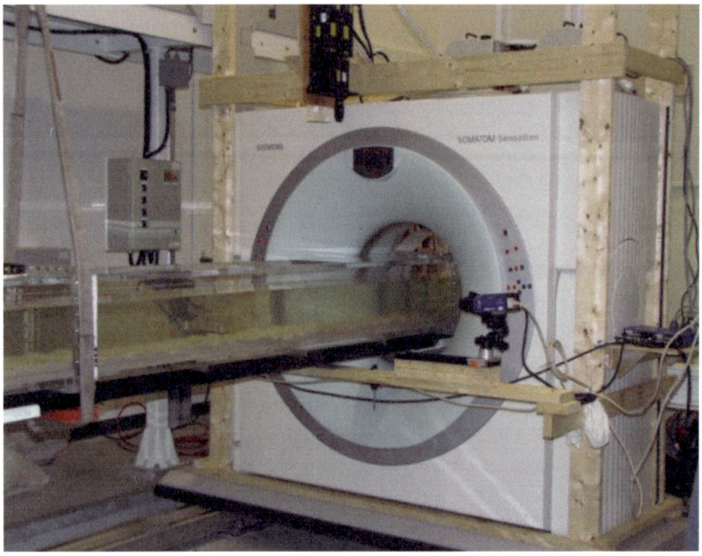

Fig. 1. CT-scanner gantry moving along the flume on four parallel rails and open-channel flume fixed on the examination table.

The CT-scanner imagery corresponds to a 3D linear X-ray attenuation voxel image (Fig. 2). The instantaneous slice group measured forms a volumetric 3D matrix (Fig. 3). X-ray attenuation is a function of the material density, its effective atomic number, and the analyzed sample thickness. Although Duliu (1999) indicated that four effects characterize X-ray interaction with a material, medical CT-scanners, which use energy below 150 keV, exhibit only two of these effects: the photoelectric and Compton effects. In this type of study, the photoelectric effect dominates (Moore, 2004).

In this study, two different techniques were used to undertake pseudo-instantaneous measurement (Lagrangian) and time-evolution measurement (Eulerian). The global volume technique (GVT) uses the CT-scanner to measure a large volume, giving a picture at a specific moment (Lagrangian) (Fig. 2). This technique, used by Montreuil (2006) and Montreuil and Long (2007), gives a 3D image of the entire sand ripple with a maximum available length of 1500 mm. The global volume technique is useful in comparing all densimetric profiles over a sand ripple within 90 s for a 1500 mm-long bedform. To measure an entire ripple, the CT-scanner needs to move along the flume. The simultaneous motion of the CT-scanner gantry and the fluid inside the flume reveals an uncertainty due to moving sediment during X-ray integration. The global single-volume technique is, therefore, a pseudo-instantaneous measurement technique.

The second technique, the periodic single volume technique (PSVT), provides an Eulerian representation of the internal bedload architecture constancy during sand ripple migration.

Fig. 2. Three-dimensional X-ray image of a sediment bed 300 mm wide by 300 mm high by 900 mm long, formed from 1500 slices of 512 x 512 pixels.

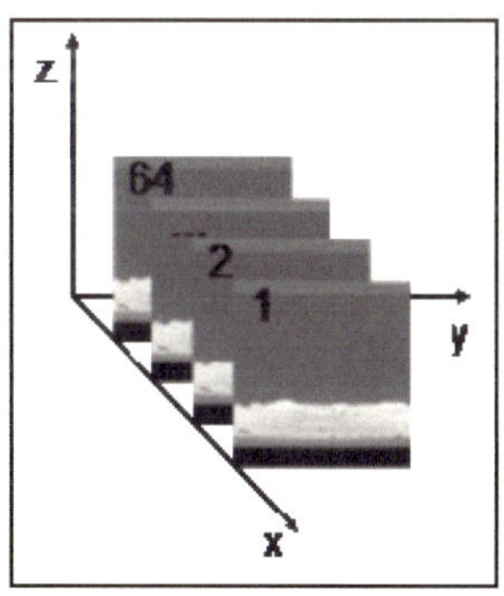

Fig. 3. CT-scanner measurement volume formed from 64 slices 0.6 mm thick at 0.6 mm intervals. Each slice has a resolution of 512 x 512 pixels and the set forms a volume of 300 x 300 x 18 mm.

To perform this measurement, a volume formed from thirty slices (each 300 x 300 x 0.6 mm) was measured, keeping the same flume position (i.e., the CT-scanner gantry remained immobile). The volume, including the sand ripple bed and the water column, was measured instantaneously by the 64 CT-scanner line of sensors. Each volume was measured with a time-space Δt sample in seconds or minutes depending on ripple migration velocity, yielding the density of the ripple moving across a spatially fixed volume (Fig. 4).

The flume axes were defined as follows: x for the flow direction, y for the horizontal cross-section, and z for the flume height (including both sediment and water column).

Various other methods have been used to measure bedform propagation, including stereophotogrammetry (Sternberg, 1967), direct measurement (Langhorne, 1981), high-resolution echo-sounding (McLean, 1983), and ripple-shadowing (Wilkinson et al., 1985).

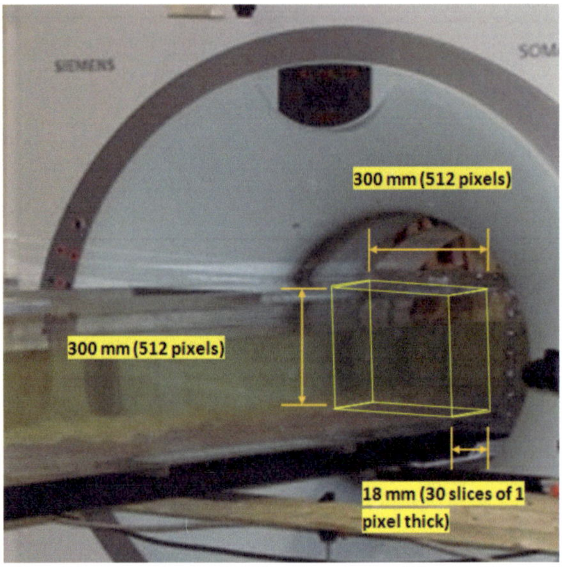

Fig. 4. Acrylic flume passing through the immobile CT-scanner gantry yielding a measurement volume of 300 x 300 x 18 mm.

Medical CT-scanners can be used to measure relative density of materials such as sand layers in sedimentological studies (Duliu, 1999; Ketcham and Carlson, 2001; Mees et al., 2003; Duchesne et al., 2006). Relative density values measured vary from -1024 to +3071 Hounsfield units (HU; Kak and Slaney, 1999). Air and water values are -1000 HU and 0 HU, respectively (Ketcham and Carlson, 2001). The quartz sand used has a density similar to that of human bone (1477 HU) and spinal core (1609 HU; Schneider et al., 1996). Density measurements were converted to $g\bullet cm^{-3}$. The following equation was used to convert and correct the beam-hardening and Beer's law effects of X-ray measurements (ASTM, 1992):

$$\rho_s = 0{,}0257\left(\frac{HU+1024}{1000}\right)^2 + 1{,}0209\left(\frac{HU+1024}{1000}\right) - 0{,}0254 \quad (1)$$

where ρ_s is the density of the fluid-sediment mixture inside the flume and HU is the relative density measured by the CT-scanner.

To validate the confidence level of the data, a standard deviation was calculated for each section of the ripple using a group of profiles. The results indicate a standard deviation of 0.018 $g\bullet cm^{-3}$ for the water column, which reflects turbulence and Brownian motion (Brown, 1828) for zero-velocity measurement. For the suspended-transport zone, the standard deviation of 0.022 $g\bullet cm^{-3}$ is due to the turbulence, bed proximity, and heterogeneous concentration of suspended sediment. At the interface between suspended load and the upper bedload transport zone, the standard deviation reaches a maximum of 0.023 $g\bullet cm^{-3}$. This high value indicates the presence of air bubbles trapped in the sand and a difference in the degree of compaction between the different laminations.

For the water column, the concentration of sediment in water was calculated using the Bridge and Demicco (2008) equation,

$$C = \frac{\rho_s - \rho}{\rho_{mix} - \rho} \tag{2}$$

where ρ_{mix} is the density of the sediment-water mixture, ρ_s is the sediment density, and ρ is the water density at the experimentation temperature.

The suspended load, located between the reattachment point and the separation point, is calculated using dynamic viscosity profiles defined in the Bridge and Demicco (2008) equation:

$$\mu_s = \mu(1 + 2.5C + 6.25C^2 + 15.26C^3) \tag{3}$$

where μ is the pure fluid viscosity and C is the sediment concentration.

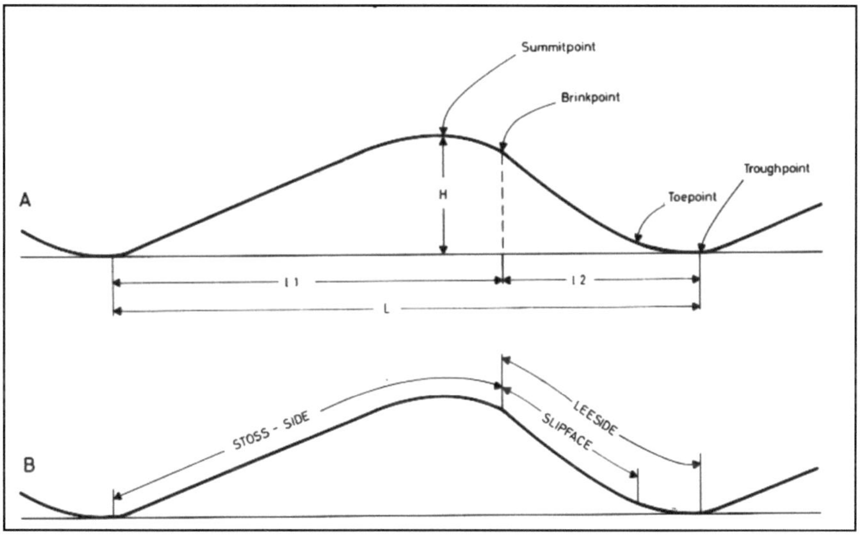

Fig. 5. Definition of sand ripple "sections," after Reineck and Singh (1975). (A) Sand ripple form is defined by a succession of deterministic points. (B) Sand ripple is divided into erosional and depositional zones.

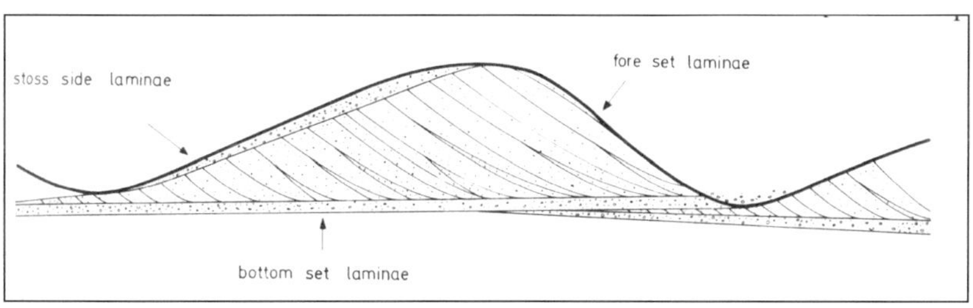

Fig. 6. Definition of sand ripple "sections," after Reineck and Singh (1975). The internal structure of the sand ripple consists of three lamina sets.

This study used the sedimentological definitions of Reineck and Singh (1975; Figs. 5 and 6). Ripple zones are detailed in Figure 7. Reineck and Singh (1975) defined sediment parameters by delineating and naming distinct zones and points in the formation of a sand ripple. Values for parameters were assigned based on measurable or observable attributes at a given time. Ripple "zones" are defined perpendicular to layering, whereas the ripple "section" is defined according to horizontal location.

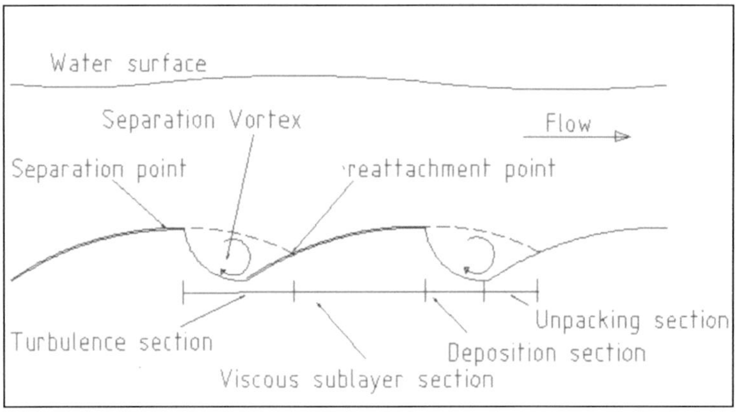

Fig. 7. Definition of sand ripple "sections": turbulent section (contains vortices), viscous sublayer section (also called the pure transport section), depositional section, and, within the turbulent section, a sediment unpacking section (also called the erosion section). Adapted from Bridge and Demicco (2008) and based on Bennett and Best (1995).

Montreuil and Long (2007) emphasized the presence of a maximum point, at which the density along the ripple reaches a maximum value. A ripple grows by deposition of successive foreset laminae that avalanche down the lee side of the ripple. Foreset laminae are constantly being eroded on the stoss side of each ripple during ripple migration; some of this remobilized sediment moves by bedload and defines stoss-side laminae that are nearly perpendicular to the foreset laminae. Reineck and Singh (1975) identified a static foreset lamina field that is constantly eroded and stoss-side laminae which consist of sediment in continuous motion.

Other definitions, including crest and trough concepts, are from Tucker (2001) and Bridge and Demicco (2008), who described the formation of current ripples (Figs. 7 and 8). The formulation of fluvial-type processes in a flume (Yalin, 1992) and the sediment transport mechanics associated with current ripples (Yalin, 1977) have been clearly stated and are the foundation for the definitions in this study. Kastaschuk and Villard (2001) proposed a conceptual model for suspended load flow, in which the internal boundary layer meets the viscous sublayer and the wake region is not characterized by vortical activity. Middleton and Southard (1977) proposed a schematic definition of shear stress and pressure relative to the flow velocity profile and profile position. Finally, Allen (1968, 1984) provided a general picture of sandy sediment-depositional systems that is the starting point of this work.

In general, the internal geometry of a ripple is a consequence of the forces acting at the beginning on stationary and moving fluids, and corresponds to the relationship between velocity and pressure in a moving fluid, as described by the Bernoulli equation. Flow over a ripple field, however, is not laminar, but turbulent. The transition between laminar shear flow and turbulent shear flow is produced by turbulent eddies. Such eddies are critical to bedform formation and evolution. Turbulent shear flow is due to the addition of a viscous shear stress like that defined in the Reynolds equation.

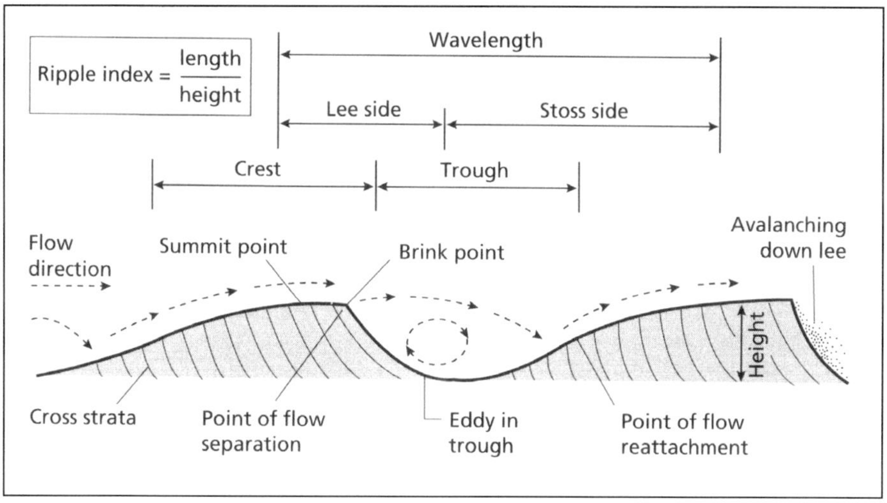

Fig. 8. Complementary general sand ripple definition, after Tucker (2001).

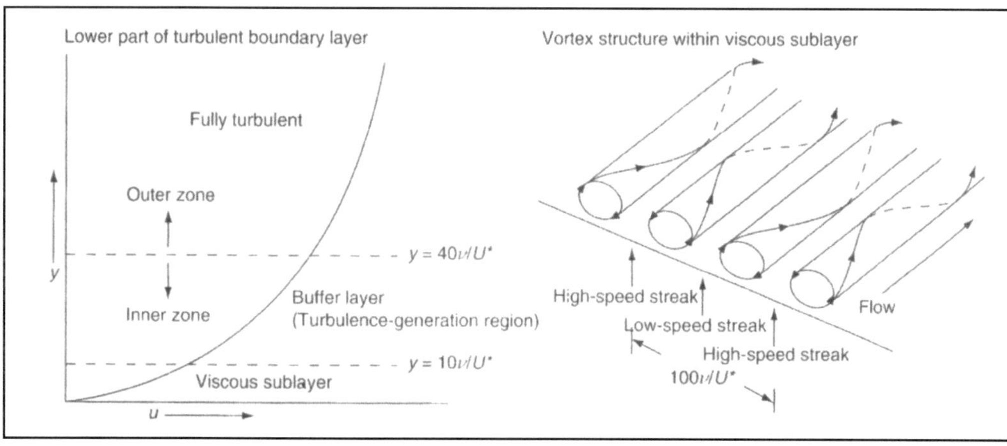

Fig. 9. Main structural features of turbulent boundary layers (after Bridge and Demicco, 2008, and modified from Bridge, 2003).

The turbulent boundary layer in steady, uniform, turbulent flow is a function of bottom friction. Friction at the solid boundary of the water flow (Fig. 9) causes a reduction in flow velocity near the boundary. The main structural features of the turbulent boundary layer define three zones (Bridge, 2003). Very close to the sediment is a viscous sublayer a few millimeters thick. This basal layer is overlain by a buffer zone called the inner zone, which consists of a turbulence-generating region. The uppermost layer is the full-turbulence zone, which is referred to as the outer zone. The boundary between the viscous sublayer and the inner zone is defined by

$$\delta y = \frac{10\vartheta}{U_*} \tag{4}$$

The boundary between the two turbulent zones is defined by

$$\delta y = 40\vartheta / U_* \tag{5}$$

where υ is the kinematic viscosity and U_* is the shear velocity, defined as

$$U_* = \sqrt{(\tau_0/\rho)} \tag{6}$$

where τ_0 is the mean fluid shear stress at the bed and ρ is the fluid density.
The three zones are not uniformly distributed above the ripple bed during sediment transport because U_* is not constant. The present work illustrates this variation empirically.

3. Definitions

3.1 Three-dimensional sedimentological architecture of a sand ripple field

Density profiles were examined using the classical approach of separating and isolating the suspended load from the bedload, as done by Carmenen and Larson (2004). The density profiles provided by the 3D matrix were analyzed continuously from the water surface to the bedload base. A typical density profile (Fig. 10) depicts transport zones that are delimited by characteristic density points. More details are available in Montreuil and Long (2011b).

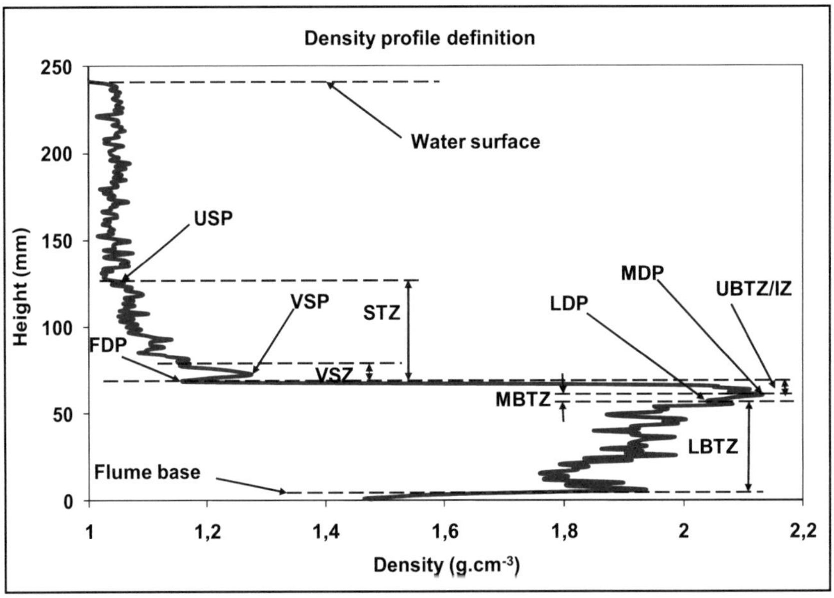

Fig. 10. Typical density profile of a sand ripple, as measured by the CT-scanner.

Starting in the upper water column, a typical curve of sediment column density encounters the upper suspension point (USP), where clear water (washload) is distinguished from suspended load by the latter's higher density (above 1.05 $g{\bullet}cm^{-3}$; Fig. 10). The upper suspension point may be at the water surface if turbulent flow has generated a high sediment transport rate.
The suspended transport zone (STZ) contains a significant amount of suspended sediment and may include a viscous sublayer.
The point separating the suspended load and the bedload is the frontier density point (FDP). This point is used as a reference concentration, but is not synonymous with the ripple

surface. Instead, it is at the maximum density point (MDP) at the base of the suspended zone, as defined later in the text.

The suspended transport zone includes the viscous sublayer zone (VSZ) (section 3.2), which is defined by its low density, high viscosity, and obvious distinction from the bedform. The viscous sublayer zone contains a viscous sublayer point, defined as the point of highest density and viscosity. When a viscous sublayer zone is present, the viscous sublayer point (VSP) can be used as a reference concentration instead of the frontier density point. The upper bedload transport zone (UBTZ) is delimited by the frontier density point at the top and the maximum density point at its base. Sediment in the upper bedload transport zone moves as a thin lower-density layer around low density points, and is characterized at its base by the maximum density point, an abnormally high-density point at which hydraulic pressure is linked to bed density. The maximum density point is the top of the bed *sensu stricto* and is the upper limit of shear stress lamination of the inner bed. Evolution of the maximum density point affects the lee-side avalanching process and stoss-side erosion because of fluid-related effects.

Below the maximum density point is a lower density point (LDP), characterized by water-rich sediment (less compaction) in which fluid pressure favors sediment mobility. The lower density point forms part of the inner bed shear stress lamination structure (Madsen and Durham, 2007).

Below the upper bedload transport zone is the middle bedload transport zone (MBTZ), with an upper limit at the maximum density point and a lower boundary at the lower density point. The middle bedload transport zone and upper bedload transport zone move on the low-density layer that forms around the lower density point.

Within the upper bedload transport zone is an inflection zone (IZ) that increases the thickness of the upper bedload transport zone across a hinge point or inflection density point close to the maximum density point. The inflection zone is due to lee-side avalanching and to water penetration caused by vortices located over the ripple unpacking section.

The last transport zone, the lower bedload transport zone (LBTZ) consists of a succession of layers (Wheatcroft, 2002) that move under fluid pressure induced deep in the bed (Madsen and Durham, 2007), and by an inner bedload shear stress, which causes high-density sheet sliding on low-density layers produced by water-filled porosity.

3.2 New definitions and locations of boundary layers

Bridge and Demicco (2008) explained attributes of the boundary layer using a solid ripple surface. In the present study, however, measurements show that upper bedload transport zone thickness and bedload density variation indicate pseudo-liquefaction and water ejection along part of the ripple (Figs. 21 and 22 (section 5.4)).

Water penetration through the boundary requires that the y-axis be modified such that motion below the boundary is included (Madsen and Durham, 2007); this also demands a different definition of the viscous sublayer environment.

Based on the sedimentological architecture previously defined in section 3.1, the upper bedload transport zone thickness is compared with the maximum density point forming the bedload *sensu stricto*. The results (Fig. 11) indicate low-density erosion (1.75 to 2.25 $g \bullet cm^{-3}$) for 8 to 12 mm of thickness, where the water penetrates deeply into the bedform (Montreuil and Long, 2011a). This transportation, on the upper part of the ripple's stoss-side, is near the surface (4 to 8 mm) and is characterized by higher density (2.00 to 2.25 $g \bullet cm^{-3}$) where and when the water leaves the bedform.

On the lee side of ripples, results indicate phase recovery of the water rather than deposition. Within the sediment column, which is 4 to 14 mm thick depending on location, density was 2.00 to 2.35 g•cm^{-3} with an average of 2.20 g•cm^{-3}, indicating expulsion of water.

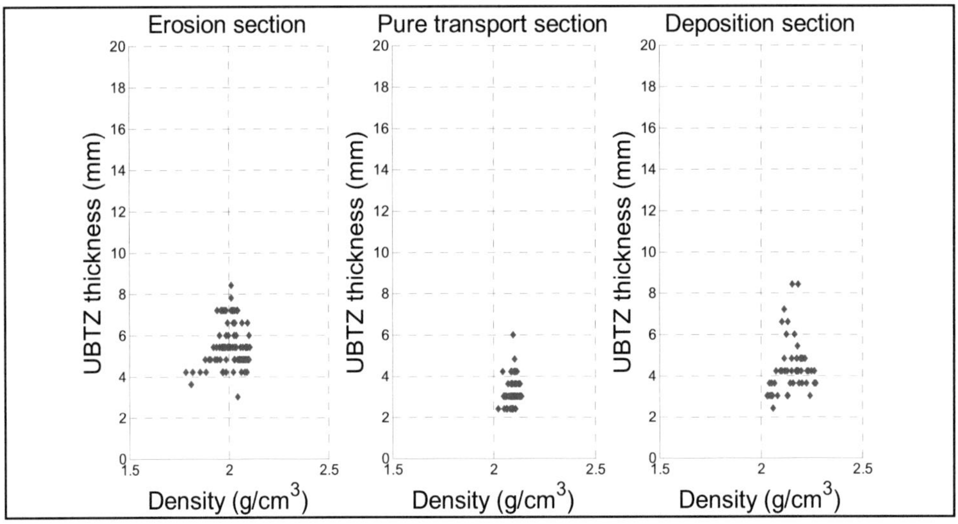

Fig. 11. Upper bedload transport zone (UBTZ) thickness as a function of the maximum density point (MDP) value, for three ripples.

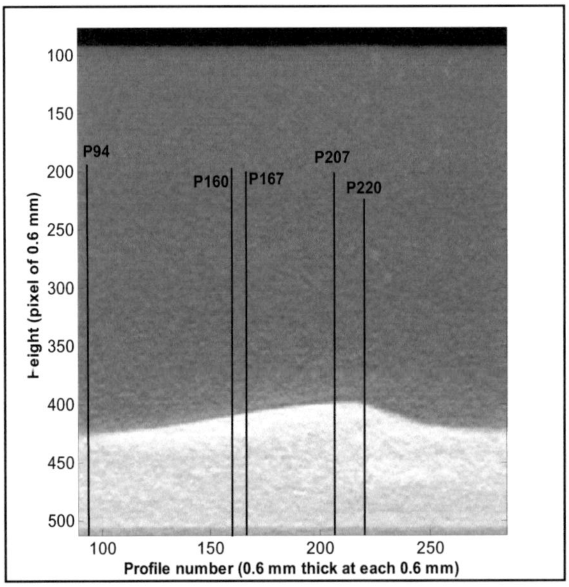

Fig. 12. X-ray image of the sand ripple (2008-No27-G), pixel size: 0.6 mm, d_{50}: 124 µm, and flow: 19 cm•s^{-1}. The partial gray-scale density image measured using the global volume technique provides a general view for 195 of the 512 density profiles (P94 is the 94th of 512 profiles).

4. Influence of flow on the vertical density profile definition along the ripple length.

4.1 Hydraulic effect on the density profile

Hydraulic phenomena over sand ripples govern the density profiles of sediment-water mixtures (Fig. 12). Measurements were conducted using the global volume technique to obtain a pseudo-instantaneous image.

An upper suspension point with density above 1.05 g•cm^{-3} indicates washload in the water column or suspended load in the suspended transport zone (Fig. 10). The difference between the upper suspension zone and the frontier density point gives the suspended load height, the concentration, and the viscosity profile measured by the CT-scanner in each voxel. The frontier density point, located near the ripple surface, is the boundary between bedload and suspended load and has a density between 1.05 and 1.18 g•cm^{-3} depending on the sediment suspension, vortices, separation point, and re-attachment point. This point is used as a reference elevation for concentration when the viscous sublayer is absent.

4.1.1 Effect on the stoss side

Density profile P220 (Fig. 13) shows a vortex 18.4 mm in diameter. The size of this vortex in subsequent profiles indicates its shape, located between 69.5 and 87.9 mm in the turbulent section. In the turbulent section, a second suspended transport zone is present, indicating continuous transport.

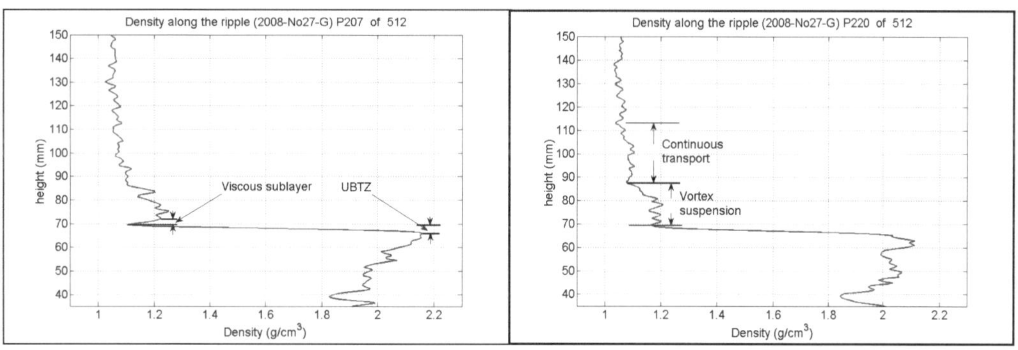

Fig. 13. Sand ripple density profile P207 shows a viscous sublayer, and density profile P220 indicates continuous transport above a vortex. (GVT; ripple ID: 2008-No27-G).

The presence of an inflection density point high on the stoss side (Fig. 13, P207-P220) indicates the presence of a saturated sediment layer that is part of the upper bedload transport zone. The maximum density point of 2.0 to 2.3 g•cm^{-3} is generated by hydraulic pressure near the surface on the lower part of the stoss side; its location is deeper within the sediment with increased height on the ripple and along the lee side of the ripple. The maximum density point indicates the point of traction within an area of minimum saturation.

Profile P94 (Fig. 14) indicates the absence of a viscous sublayer and the concentration of suspended sediment (Fig. 14, left). Separation section profiles are near the brinkpoint in association with the fluid separation point (Fig. 13).

Density profiles (Fig. 15, left) present the transition between a fully developed vortex and a viscous sublayer higher on the stoss side (Fig. 15, right) (Allen, 1982). The appearance of a viscous sublayer (Fig. 15, right) indicates re-attachment of the fluid (Fig. 8). The location of this point is influenced by current velocity and the shape of the sand ripple.

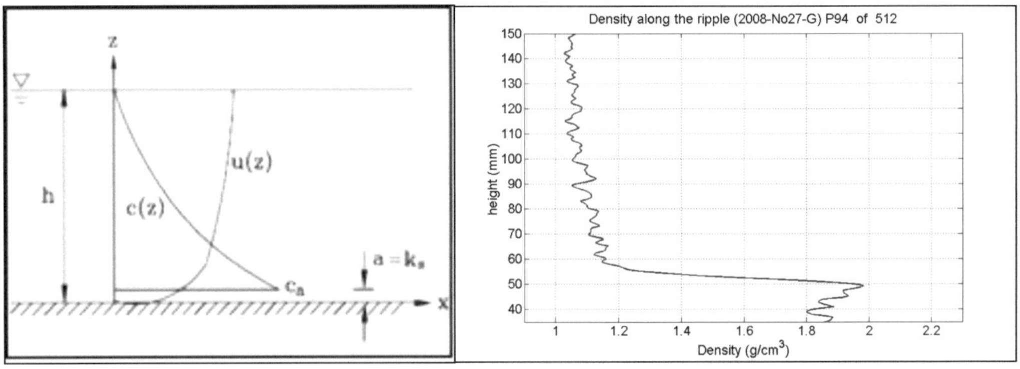

Fig. 14. Concentration and velocity profile after Liu (2001), and sand ripple density profile P94, presenting a classical log concentration profile. (GVT; ripple ID: 2008-No27-G).

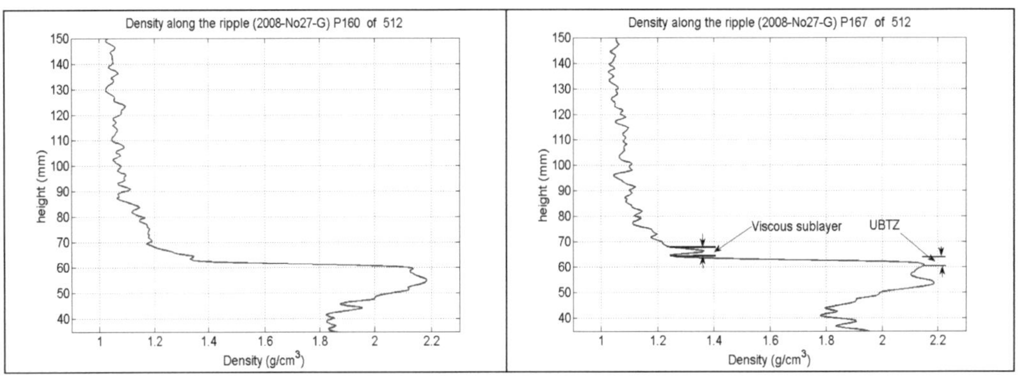

Fig. 15. Sand ripple density profiles P160 and P167, including the transition from a vortical section to a fully developed viscous sublayer and associated water ejection. (GVT; ripple ID: 2008-No27-G).

On profile P167 (Fig. 15, right), the viscous sublayer zone and upper bedload transport zone are about 1.8 mm and 3.6 mm thick, respectively. The viscous sublayer is thinner than the upper bedload transport zone, but its water-sediment density variations are different, ranging from 1.37 to 1.24 g•cm^{-3} for the viscous sublayer zone and 2.15 to 1.24 g•cm^{-3} for the upper bedload transport zone, indicating mainly suspended transport due to fluid pressure and velocity influencing the bedload erodibility.

Comparison of density profiles P207 and P220 (Fig. 13) indicates modification of fluid behavior representing a separation position over the ripple. Profile P207 shows a viscous sublayer, but profile P220 does not (Fig. 13). Density at the frontier density point retreats near the water density (1.024 g•cm^{-3}) and becomes 1.168 g•cm^{-3}. This thin layer (Fig. 13, left) represents a viscous sublayer between the suspended transport zone, whereas the upper bedload transport zone represents a thicker layer containing vortical perturbations that affect the lee sides of the ripples (Fig. 13, right).

4.1.2 Effect on the lee side

From the ripple crest to the top of the lee side, the thickness of the upper bedload transport zone increases from 6.5 to 13.7 mm, a change that is linked to the avalanching process (Fig. 13, P220). Density regression of the lower density point indicates the upper erosion limit of the previous foreset on the stoss side and deposition on the lee side. The presence of an inflection density point on the lee side (Fig. 10, P94) shows the presence of a saturated layer caused by avalanching. On the lee side, density changes from 2.3 to 1.9 g•cm^{-3}.

4.2 Thickness variation in sand ripple transport zones

Each of the transport zones presented in section 3.1 has its own characteristic thickness and evolution. These typical characteristics are described below, beginning in the water column (Fig. 16).

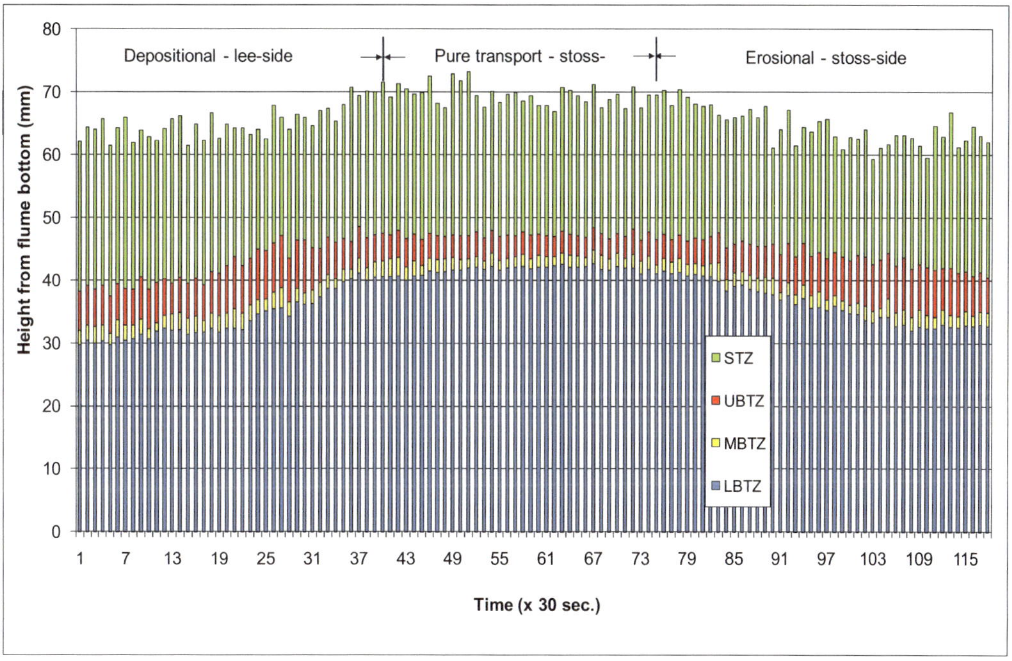

Fig. 16. Composition of the sand ripple transport zone. (PSVT; ripple ID: 2010-No5-C).

The suspended transport zone thickness along the ripple remains relatively stable at 21 mm, when the upper suspension point is chosen at 1.18 g•cm^{-3}. Below the suspended transport zone, the upper bedload transport zone is at about 6.1 mm on the lee side. This thicker zone is associated with an inflection zone correlated with a thicker middle bedload transport zone (approximately 3 mm). Inside the upper bedload transport zone, the shear-stress decreases vertically in inverse proportion to the water concentration, which governs the viscosity and consequently the degree of sediment transport or deposition.

In the pure transport area on the stoss side (Montreuil and Long, 2011a), the middle bedload transport zone and upper bedload transport zone are thinner (2 and 4 mm, respectively). These thicknesses indicate a weak fluid influence due to protection by the viscous sublayer zone.

During erosion under vortical action, thickness of the upper bedload transport zone reaches 10 mm, and that of the middle bedload transport zone is 2 mm, with a standard deviation of 2 mm. These values are caused by strong vortices active on the lower stoss side and by micro-turbulence around the reattachment point. The transport zone thickness varies along the ripple as a function of hydrodynamic variations, indicating that variable compaction is taking place deep in the ripple interior.

5. Bedload density parameters and their relation to hydrodynamics.

5.1 Effect over the lower stoss-side section

In erosional areas P90 to P160 and P150 to P180 (Fig. 17), vortex pressure causes an augmentation of suspension just above the bedload and a reduction of density, or liquefaction, which helps the sediment unpack, resulting in suspended load. When leaving the erosional section (P140), as compared to the previous profile (P150), the slope of the upper bedload transport zone transition approaches the horizontal, indicating an absence of sediment passing from bedload to suspended load. Profile P140 is approximately in a pure transport section.

In the part of the vortex in which the current is directed downward, the suspended sediment concentration is low. Sediment concentration increases in the lower, horizontal part of the vortex, between the downward and upward movement. The highest sediment concentration is in the part of the vortex where water movement is upward (Montreuil and Long, 2011a).

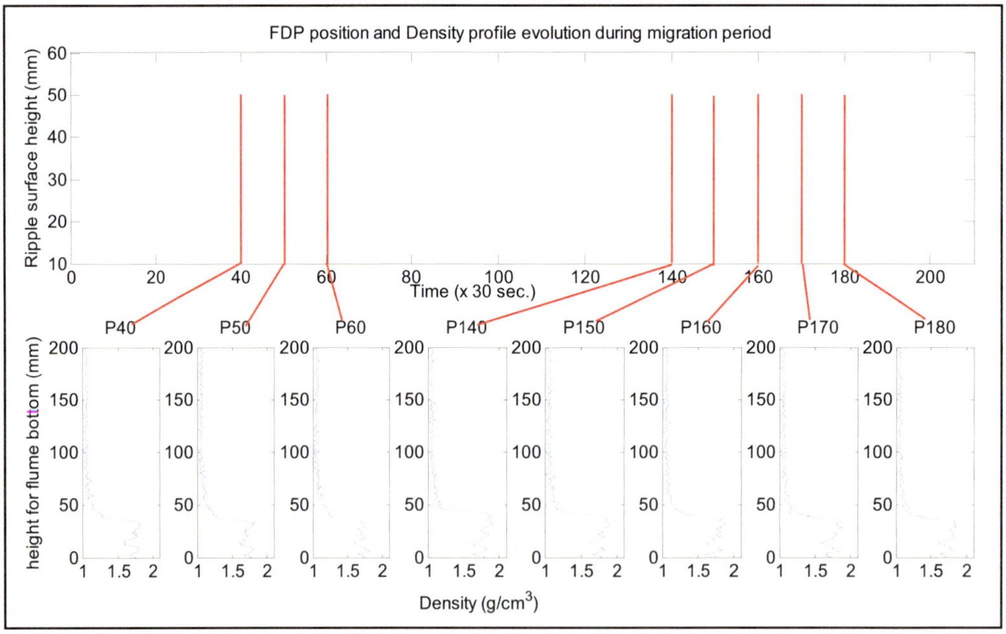

Fig. 17. Frontier density point, upper part of upper bedload transport zone (above), and sand ripple density profiles measured with periodic single volume technique (below). Profiles P40 to P60 and P140 to P180 are located in the erosional section of the lower stoss side. (PSVT; ripple ID: 2010-No5-C).

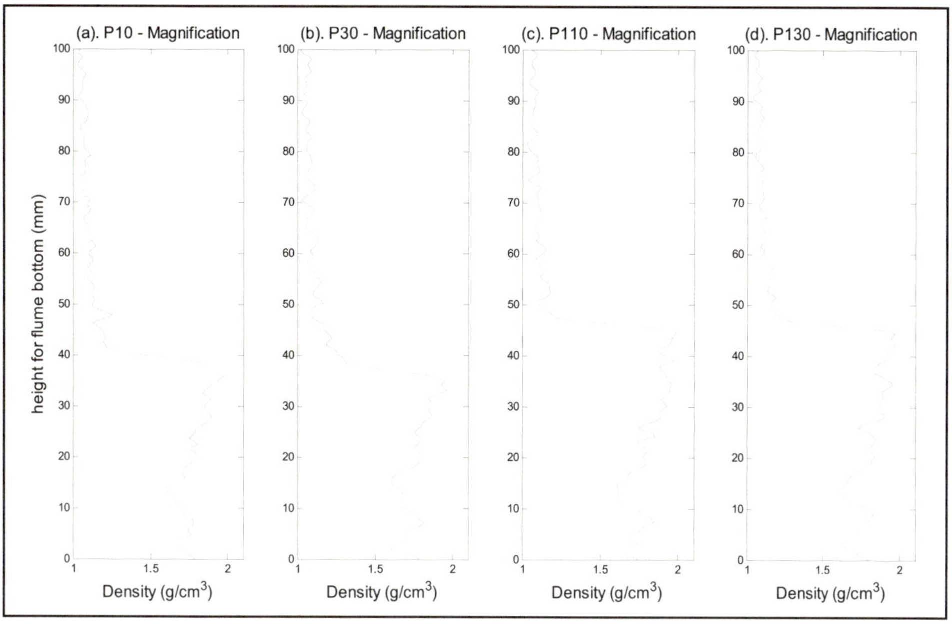

Fig. 18. Magnification of density profiles P10 (viscous sublayer section), P30 (vortex section), P110 (pure transport section with viscous sublayer), and P130 (between vortex and viscous sublayer section). (PSVT; ripple ID: 2010-No5-C).

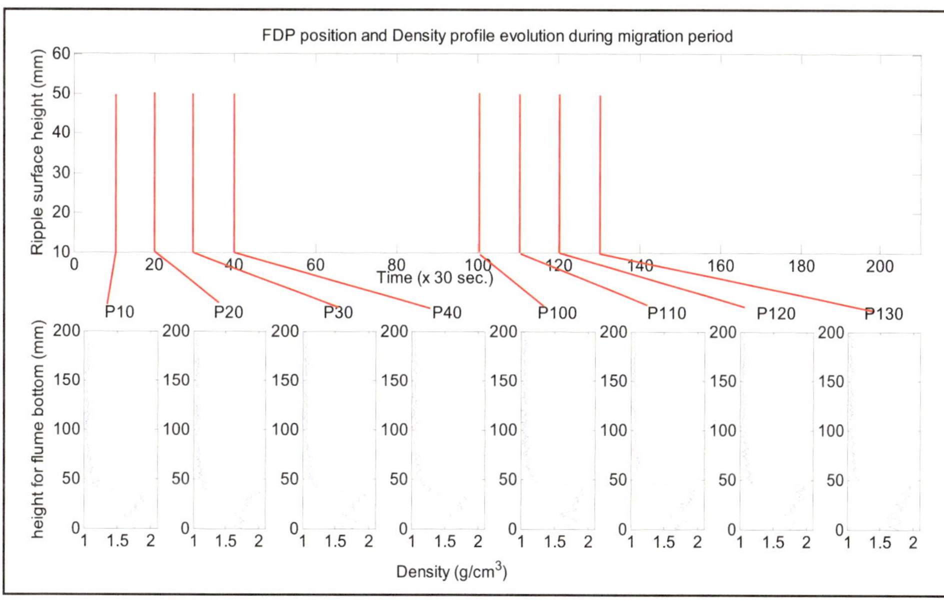

Fig. 19. Frontier density point and upper part of the upper bedload transport zone (above) and sand ripple density profiles measured with the periodic single volume technique (below). Profiles P10 to P40 are located in the erosional section (middle stoss side) and profiles P100 to P130 are located in the pure transport section (upper stoss side). (PSVT; ripple ID: 2010-No5-C).

5.2 Effect over the upper stoss-side section

The transition from erosional profile P40 (Fig. 18) to pure transport profile P10 (Fig. 19) is marked by density variation in the upper bedload transport zone (0.56 g•cm^{-3} at P40 to 0.78 g•cm^{-3} at P10). This increased upper bedload transport zone density transition, indicated by a frontier density point reduction (0.06 g•cm^{-3}) and a maximum density point increase (0.16 g•cm^{-3}), signifies a reduction in liquefaction, followed by unpacking. This transition corresponds to the disappearance of the vortex and change to a pure transport section with a thin upper bedload transport zone. Profiles P100 (4.72 mm) to P130 (8.26 mm) contain a viscous sublayer zone (Fig. 19, P110) and a maximum density point that decreases from 1.97 to 1.93 g•cm^{-3} as fluid pressure is decreased.

5.3 Effect over the lee-side section

Profiles P180 to P200 (Fig. 20) confirm a density evolution that is related to avalanching, and profiles P60 to P100 depict the transition from suspended load to bedload settling. This transition is characterized by an increase in bedload density (1.73 to 1.93 g•cm^{-3}) and decrease in reference concentration *Ca* (1.25 to 1.14 g•cm^{-3}). Bedload density near the flume bottom remains roughly constant, with slight variation due to compaction.

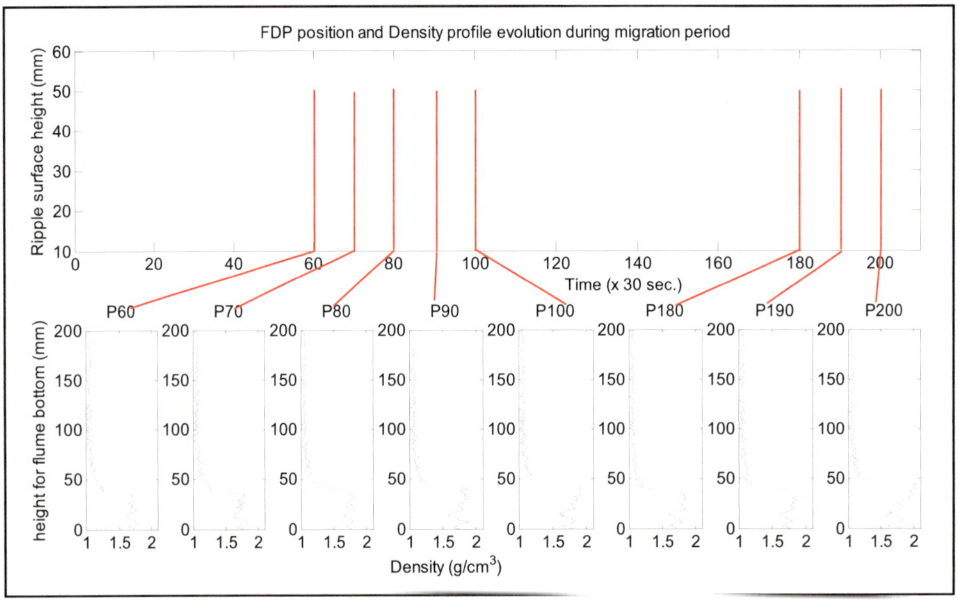

Fig. 20. Frontier density point for the upper part of the upper bedload transport zone (above) and sand ripple density profiles measured with the periodic single volume technique (below). Profiles P60 to P100 and P180 to P200 are from the depositional sides (lee sides) of ripples and record evolution of suspended load and bedload density in the lee side (PSVT; ripple ID: 2010-No5-C).

The lower part of the lee side differs from the upper part in having a lower bedload density (1.71 g•cm^{-3}) indicating weak compaction at the base. The upper lee side is dominated by a high frontier density point value of 1.16 g•cm^{-3}, as compared to 1.25 g•cm^{-3} for the base, and has a higher maximum density point, which resulted from compaction during the 900 s (15 min) time interval that separated P60 from P90. Through this time interval (periodic single volume technique), the internal structure of the bedload remains similar, but the bedload density value, from the base to the top, varies as a function of sediment compaction in the lower position and in response to fluid pressure near the surface. The sediment matrix is, therefore, under the influence of both phenomena.

5.4 Effect over the three main sections

The surface of a sand ripple has three main sections: an erosional section, a pure transport section, and a depositional section (Montreuil and Long, 2011a). The most important difference among those sections is water penetration depth (Fig. 21). Unlike the pure transport zone, in which the upper bedload transport zone is 5 mm thick, the erosional section has an upper bedload transport zone 12 mm thick, indicating deeper water penetration and lower viscosity, both of which favor sediment erosion and initiation of transport (Fig. 22). The density at ripple surfaces is lower (1.85 g•cm^{-3}) under the vortical erosional effect than it is in the pure transport zone (1.90 g•cm^{-3}). This difference is due to water penetration and pressure applied by surface fluid, as demonstrated by Middleton and Southard (1977).

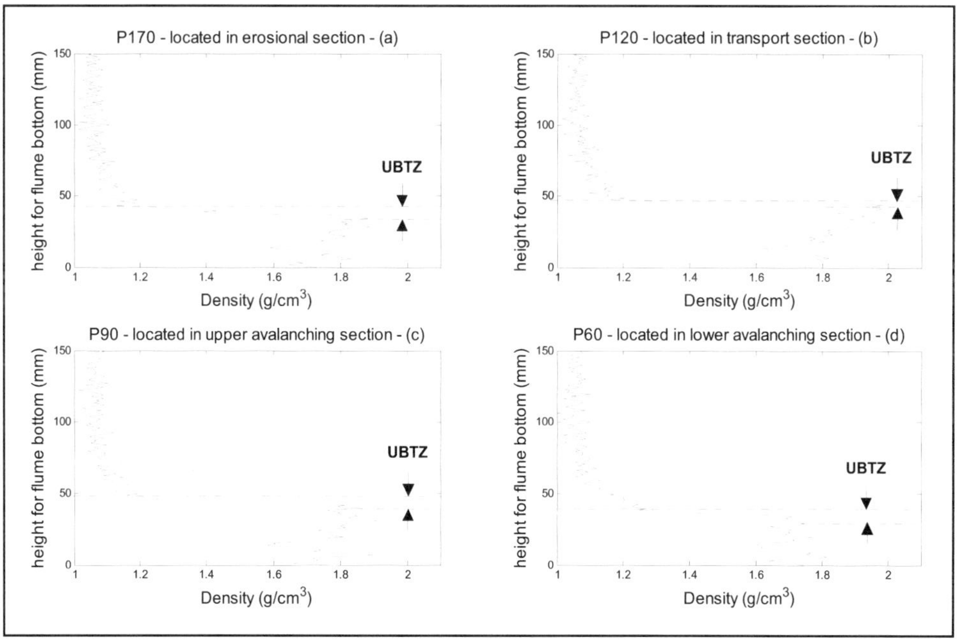

Fig. 21. Density profile of sand ripples and thickness of upper bedload transport zone in the erosional section, pure transport section, upper lee side, and lower lee side, as measured using the periodic single volume technique (PSVT; ripple ID: 2010-No5-C).

When ripples migrate, the stoss side of each ripple is covered by sediment deposited on the lee side of the next, approaching ripple. Thickness of the upper bedload transport zone decreases because of water expulsion caused by forward movement of the vortex and sediment compaction (Fig. 22). Water ejection is recorded in the density change in the upper bedload transport zone, showing a decrease from 2.06 to 1.84 g•cm^{-3}. The maximum density point exhibits a small variation in density (0.028 g•cm^{-3}) between 28.32 and 41.89 mm above the bed bottom. Deeper in the bed, in the lower bedload transport zone, subtle density variations are due to compaction and water migration.

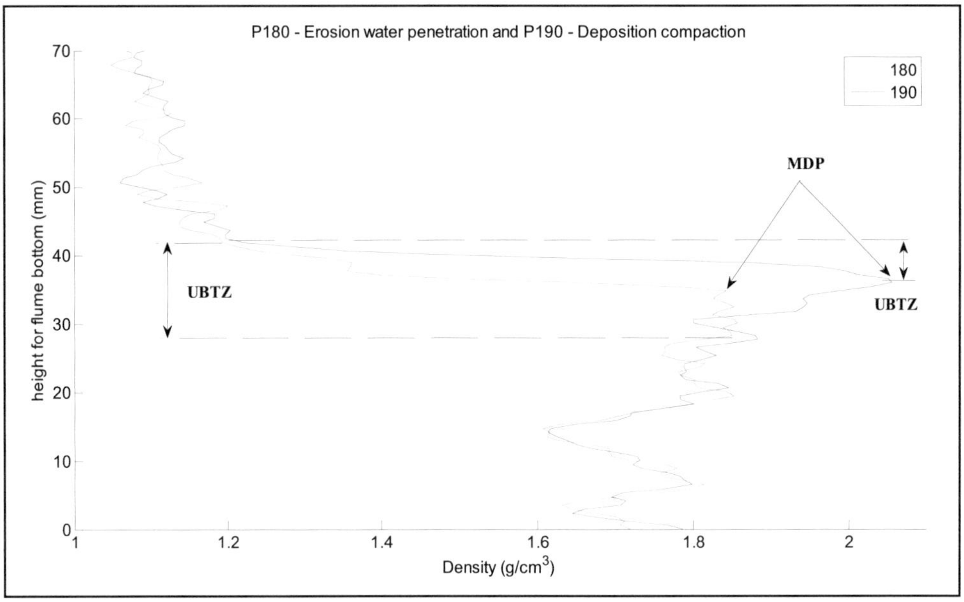

Fig. 22. Density profile P180 (blue), from the vortex section, has a high concentration caused by fluctuation in the location of the transition between the frontier density point and the maximum density point. Density profile P190 (red), from the lower lee-side avalanching section, shows an elevated maximum density point, an increased density in the middle bedload transport zone, and a similar density in the lower bedload transport zone, an area that had already been affected by migration of the sediment mass of the preceding ripple. (PSVT; ripple ID: 2010-No5-C).

Bedload transport occurs in two regimes: initial bedload mobilization and sheet-flow bedload (King, 2005). The density of the lower part of the sheet flow, just above the stationary (flat) bed, is defined by the maximum density point of the upper bedload transport zone. In this study, sheet-flow refers to the entire bedload transport zone and is used to define the boundary with suspended transport, but maximum density is located between the upper bedload transport zone and the middle bedload transport zone. The upper bedload transport zone, which rises and then collapses along the sand ripple under moving fluid, can be conceptualized as a moving sheet of material. It works well until the thickness of the layer decreases to the order of a grain diameter, at which point the viscous sublayer breaks down.

Any sediment bed has a characteristic surface roughness (Fig. 14; Liu, 2001). In this study, profiles along a ripple indicate changes in this roughness around the frontier density point and a more complex transition between suspended load and bedload. This observation emphasizes the need for CT-scanner measurements (Fig. 14) for further investigation of the transition between suspended load and bedload.

According to Bridge and Demicco (2008), viscous (surface) sublayer drag is dominant on both hydraulically smooth and hydraulically rough boundaries. Pressure (form) drag is so strong that viscous forces can be ignored. In this study, the smooth boundary is in the pure transport zone and contains the viscous sublayer zone, which moves on a well-defined bedload with a thin upper bedload transport zone with a high density at the maximum density point. In the erosional ripple zone, which has a rough boundary, the viscous layer is replaced by a thicker upper bedload transport zone caused by a more intense and larger diameter vortex.

Bed roughness thickness is commonly considered to be equivalent to ripple height (Bijker, 1992; Liu, 2001), but in this study, concentration profiles and porosity along the ripples were not constant, and this general statement is accurate only for a macroscopic view of a small ripple (~20–30 mm long). For a ripple longer than 30 mm, bed roughness can be micro-analyzed in subsections. Such ripple micro-analysis is necessary because of the fluctuating conditions along its length. The main zones involved in bed roughness are a function of flow: suspended transport zone, viscous sublayer zone, upper bedload transport zone, and middle bedload transport zone.

6. Vortex analysis

6.1 Effect of vortex on bedload architecture of the lower stoss side

The contribution of vortices to sediment erosion is undisputed, but their effect deep inside the bedload has been unknown until now. CT-scanner imaging of ripples reveals the vortex sediment content through density profiles and the effect of vortices on the structural integrity of the bedload.

Vortices have three parts. The first (Fig. 23, P147) is where the sediment-poor fluid (lowest concentration) is dragged downward and begins to apply pressure that deepens the critical shear stress (discussed in section 6.2). The second part of the vortex is the longitudinal part of the vortex (Fig. 23, P153). This part is characterized by a higher average density in the suspended load (1.11 $g \bullet cm^{-3}$ around 90 mm from the bottom, as opposed to 1.06 $g \bullet cm^{-3}$ for surrounding volumes). This suspended sediment is entrained by the upward rotating part of the flow (left-pointing arrow in Fig. 23, P153). The average density appears lower because of sediment expelled by centripetal force. These density values are not entirely accurate, however, because the X-ray integration of stationary rotating sediment overestimates the value, but the relative values remain significant. At the bottom of the vortex, flow is inverted (right-pointing arrow). This backward flow initiates sediment unpacking, causing the thickness of the upper bedload transport zone to increase to 8.85 mm, and conditioning the sediment to leave the bedload. The density transition from the frontier density point to the maximum density point is disturbed by this pseudo-liquefaction. A maximum density point forms ~35 mm from the bottom (Fig. 23, P153). This deep point indicates the formation of a temporary ripple surface *sensu stricto*, due to the fluid velocity trying to move the bedform backward. The ripple surface resists backward motion by increasing shear stress, causing more friction among grains.

The third part of the vortex is its upward motion. In this area, the deep pressure of the second vortex zone is released. Sediment in the upper bedload transport zone is waterlogged and has minimal intra-granular friction. The sediment leaves the bedload as water previously contained in the bedload matrix is released. Density profile P162 (Fig. 23) shows an average suspended sediment concentration (equation 2) of 0.56, which contrasts with the values of 0.53 and 0.54 for P147 and P153, respectively.

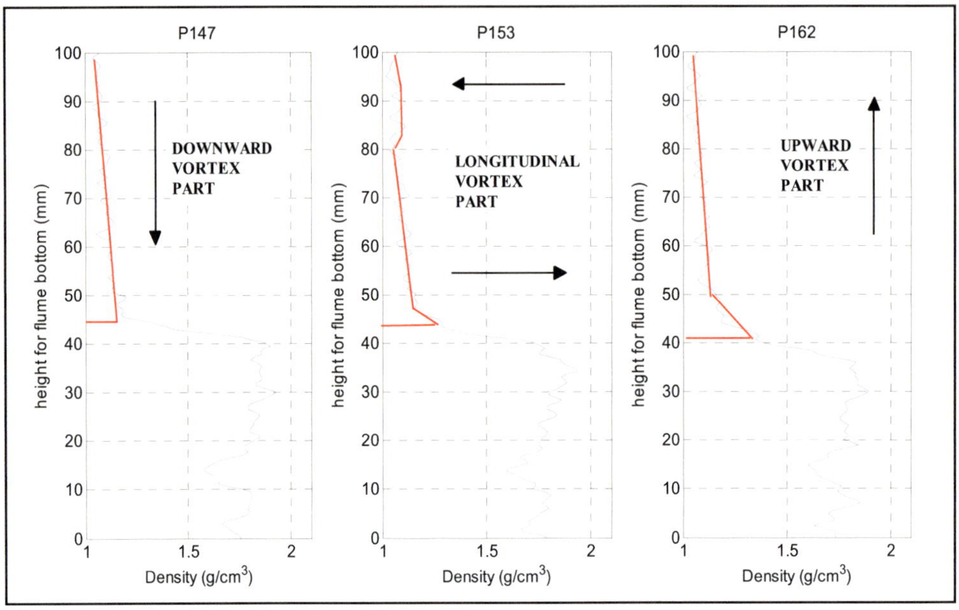

Fig. 23. Density profiles measured in three parts of a stoss-side vortex. Profile P147 is in the downward-moving part of the vortex, P153 is located in the longitudinal part, and P162 is located in the upward part.

This increased concentration originates from sediment liquefaction and uplifting. The last influence of the upward-moving part of the vortex is marked by the elevation of the maximum density point, which rebuilds a higher ripple surface *sensu stricto*. The maximum density point passes from 1.78 $g{\bullet}cm^{-3}$ in P153 to 1.83 $g{\bullet}cm^{-3}$ in P162, producing a surface elevation of ~2–4 mm.

The vortex is characterized in general by an average maximum density point density of approximately 1.83 $g{\bullet}cm^{-3}$, in contrast to 2.04 $g{\bullet}cm^{-3}$ outside the vortex part of the ripple. In the upper bedload transport zone, the transition from the frontier density point to the maximum density point is thick, but, unlike the lee-side avalanche transition zone, it is neither continuous nor smooth. It is perturbed by the turbulence surrounding the vortex.

6.2 Shear stress effect on bedload architecture

The shear stress in a viscous fluid composed of water and sediment and moving along a solid boundary induces the shear stress on the boundary (Yalin, 1992). The no-slip condition dictates that the speed of the fluid at the boundary (relative to that boundary) is zero but, at some height from the boundary, the flow speed must increase to equal the fluid speed. The region between these two points is aptly named the boundary layer. Shear stress is produced in this boundary layer as a result of this difference in velocity and can be expressed as

$$\tau_w = \mu \frac{\partial u}{\partial y} \tag{7}$$

where μ is the dynamic viscosity of the fluid, u is the velocity of the fluid along the boundary, and y is the thickness of the boundary layer. As demonstrated in section 4, the sand ripple bedload (more specifically the upper bedload transport zone) is not a solid boundary, and the dynamic viscosity of the fluid μ is replaced by the viscosity of the inner bedload μ_I, which pushes the no-slip condition deep into the bed. The boundary thickness is now taken from the real ripple surface, which corresponds to the maximum density point, located at the lowest point in the upper bedload transport zone (Fig. 24).

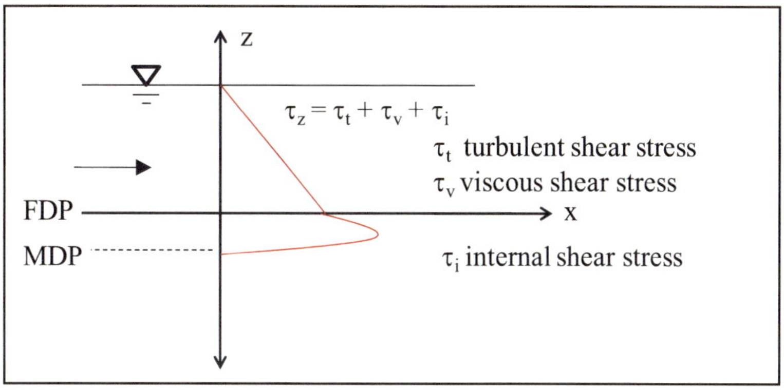

Fig. 24. New shear stress definition, which includes water penetrating into the sediment and increasing the depth at which sediment transport is initiated.

7. Conclusions

Computed axial tomography (CT-scanning) is a non-destructive technique for generating images that map the numerical density of a sand-water mixture. The data produced depict the evolution of sedimentological parameters by variation in density, concentration, porosity, and viscosity of the water-sediment mixture. Results can be obtained for every section of a ripple, using zones and inflection points in density patterns. These transport zones evolve longitudinally in the form of layers that vary in their properties along the length of each ripple.

This study demonstrates that understanding of the hydraulic phenomena behind sediment transport can be improved by reference to density points and zones. The value of CT-scanning lies in its ability to provide numerical parameters associated with the sediment-water interface and with features associated with shear stress deep inside the bed. CT-imaging highlights the importance of water penetration deep into the erosional area on the stoss side of ripples. Accurate measurements of the viscous sublayer reveal the intimate connection between suspended load and bedload. Density measurements of the lee sides of ripples highlight compaction phenomena associated with water ejection.

The density profile of a sand ripple consists of the upper bedload transport zone, the middle bedload transport zone, and the lower bedload transport zone, which form successive layers that shear against each other. These bedload transport layers move on a thin, low-density layer that forms at the low-density point. As indicated by Raudkivi (1963), Vanoni and Hwang (1967), and Middleton and Southard (1977), hydraulic pressure over the bed induces slow motion deep inside the bed (Madsen and Durham, 2007). The horizontal pressure gradient persists deep inside the bed, creating a deep shear stress that produces a laminated transport layer between the lower density point and a higher density layer (maximum density point). Thicknesses of these layers have been correlated to sediment grain size

(Montreuil and Long, 2011b), but the velocity of particles within the bed has not been quantified.

The viscous sublayer exists only in the middle of the ripple stoss side, down-flow from the fluid reattachment point, where interstitial water is ejected from the sediment. The viscous sublayer is absent in locations where vortices erode the ripple and where the sediment is deposed on the lee sides (avalanche face) of ripples.

Stoss-side laminae contain two zones, the upper bedload transport zone and the middle bedload transport zone (Reineck and Singh, 1975). The static foreset laminae, at the bottom of the stoss side, form the lower bedload transport zone.

This study confirms the existence and location of major hydraulic points and zones under various environments, and also reveals new parameters related to flow and sediment properties. The results also provide a new shear stress profile that includes activity deep inside the bedform, particularly in areas affected by vortices. Pervasive small-scale variability is present in the density of near-surface sands, which may have important implications for the measurement of sediment transport using high-frequency acoustics.

Related work, already in progress, will include an exhaustive analysis of velocity and grain-size effects on the dynamics of sedimentological parameters. The link between shear stress within the bed and the upper bedload transport zone will be examined. Bed motion was not covered in this study but will be measured using X-ray sediment tracking.

8. Acknowledgments

Special acknowledgment to the Canadian Foundation for Innovation (CFI) and the GEOIDE project for their financial contributions and to L. F. Daigle and M. Des Roches for technical support and work.

9. References

Allen, J.R.L. (1968). *Current Ripples: Their Relation to Patterns of Water and Sediment Motion.* Amsterdam, North-Holland.

Allen, J.R.L. (1982). *Sedimentary Structures: Their Character and Physical Basis* (Two volumes) Elsevier, New York, 593 p. (v. 1) and 663 p. (v. 2).

Allen, J.R.L. (1984). *Sedimentary Structures-Their Character and Physical Basis.* Elsevier Publishers, Amsterdam, 1256 pp.

Allen, G.P. and Posamentier, H.W. (1994). Transgressive facies and sequence architecture in mixed tide- and wave-dominated incised valleys: example from the Gironde estuary, France. In: Dalrymple, R. W., Zaitlin, B. A. & Boyd, R. (eds.) *Incised valley systems: origin and sedimentary sequences*, Society of Economic Paleontologists and Mineralogists, Special Publications, 51, 225–240.

Bennett, S.J. and Best, J.L. (1995). Mean flow and turbulence structure over fixed, two-dimensional dunes: implication for sediment transport and dune stability. *Sedimentology* 42, 491-514

Bijker E.W. (1992). Mechanics of sediment transport by the combination of waves and current, *23rd Int. Conf. on Coastal Engineering*, Venice, 147-173.

Bridge J.S. (2003) *Rivers and Floodplains: Forms, Processes, and Sedimentary Record.* Oxford: Blackwell.

Bridge J. and Demicco R. (2008). *Earth Surface Processes, Landforms and Sediment Deposits.* Cambridge University Press, New York, 830pp.

Brown, R. (1828). A brief account of microscopical observations made in the months of June, July and August, 1827, on the particles contained in the pollen of plants; and on the general existence of active molecules in organic and inorganic bodies, *Phil. Mag. 4,* 161-173.

Carmenen B. and Larson M. (2004). A general formula for non-cohesive bed load sediment transport, *Estuarine, Coastal and Shelf Science 63 (2005)* 249–260.

Duchesne, M. J., Long, B. F., Labrie, J., and Simpkin, P. G. (2006). On the use of computerized tomography scan analysis to determine the genesis of very high resolution seismic reflection facies, *J. Geophys. Res.,* 111, B10103, doi: 10.1029/2006JB004279.

Duliu, O.G. (1999). Computer axial tomography in geosciences: an overview. *Earth-Science Reviews 48,* 265–281.

Kak A. C. and Slaney M. (1999). *Principles of computerized tomographic imaging,* IEEE PRESS, New York, 329 pp.

Kastaschuk R. and Villard P. (2001). *Turbulent sand suspension over dunes. Fluvial Sedimentology VI.* Blackwell Science. 478pp.

Ketcham R.A. and Carlson W.D. (2001). Acquisition, optimization and interpretation of X-ray computed tomographic imagery: applications to the geosciences, *Computers & Geosciences 27 (2001)* p.381- 400.

King D. B. Jr. (2005). Influence of grain size on sediment transport rates with emphasis on the total longshore rate, *US Army Corps of Engineers, ERDC/CHL CHETN-II-48,* 24pp

Langhorne, D. N. (1981) An evaluation of Bagnold's dimensionless coefficient of proportionality using measurements of sandwave movement. (*Mar. Geol*). 43: 49-64.

Liu Z. (2001). *Sediment transport, Laboratoriet for Hydraulik og Havnebygning, Instituttet for Vand, Jord og Miljøteknik.* Aalborg Universitet 71 pp.

Madsen, O.S. and Durham, W.M. (2007). Pressure-Induced subsurface sediment transport in the surf zone. *Proceedings Coastal Sediments '07 Conference, ASCE,* New Orleans, LA, 1:82-95.

McLean, S. R. (1983). *Turbulence and sediment transport measurements in a North Sea tidal inlet (The Jade). In North Sea Dynamics,* ed. J. Sundermann, W. Lang, pp. 436-52. Berlin: Springer-Verlag.

Mees, F. Swennen, R. Van Geet, M. and Jacobs, P. (2003). Applications of X-ray computed Tomography in the Geosciences. *Geological Society of London Special Publication 215.*

Middleton G.V. and Southard J. B.; (1977). *Mechanics of sediment movement,* S.E.P.M. Short Course Number 3, Binghamton.

Montreuil S. (2006). *Évolution densimétrique des structures de rides en déplacement sous un courant stationnaire mesuré sous scanographe dans un canal en boucle verticale.* MSc Thesis, INRS-ETE, Québec. 121 pp., (unpub.).

Montreuil S. and Long B. (2007). Flume experiments under CAT-scan to measure internal sedimentological parameters during sediment transport, *Proceedings Coastal Sediments '07 Conference, ASCE,* New Orleans, LA, 1:124-136.

Montreuil S. and Long B. (2011a). Relationship between vortex activity and pseudo liquefaction at a lower stoss-side ripple surface, *Proceedings Coastal Sediments '11 Conference,* ASCE, in press.

Montreuil S. and Long B. (2011b). CT-scanner measurement of sand ripple sedimentological parameters, a definition. *Sedimentology*, In Prep.

Moore, F. 2004, *Application de la scanographie à l'étude de la déposition consolidation: Modélisation physique et théorique*, M.S. thesis, 131 pp., Inst. Nat. de la Rech. Sci., Centre Eau, Terre, Environ., Québec.

Nummedal, D., Riley, G. W. and Templet, P. L. (1993). High-resolution sequence architecture: a chronostratigraphic model based on equilibrium profile studies. In: Posamentier, H. W., Summerhayes, C. P., Haq, B. U. & Allen, G. P. (eds.) *Sequence Stratigraphy and Facies Associations, International Association of Sedimentologists, Special Publications*, 18, 55–68.

Raudkivi, A. J. (1963). Study of sediment ripple formation. *ASCE J. Hydraul.* Div., 89, 15-33.

Reineck H.-E. and Singh I.B. (1975). *Depositional Sedimentary Environments*, Springer study edition, New York, 439 pp.

Schneider, U., Pedroni, E. and Lomax, A. (1996). The calibration of CT Hounsfield units for radiotherapy treatment planning, *Phys. Med. Biol.* 41 (1996) 111–124.

Sternberg, R. W. (1967). Measurements of sediment movement and ripple migration in a shallow marine environment. *Mar. Geol.* 5:195-205.

Teeter Allen M. and Pankow Walter (1989). *Deposition and erosion testing on the composite dredged material sediment sample from New Bedford harbour, Massachusetts*, Technical report HL-89-11, US Army Corps of Engineers, Washington, 61 pp.

Tucker M.E. (2001). *Sedimentary petrology* 3ed. Blackwell Science. 262pp.

Vanoni, V. A. and Hwang, L. S. (1967) Relation between bedforms and friction in streams. (Journal of the Hydraulics Division), ASCE, Vol. 93, No. HY3, pp. 121-144.

Van Wagoner, J. C. 1995. Sequence stratigraphy and marine to non-marine facies architecture of foreland basin strata, Book Cliffs, Utah, U.S.A. In: Van Wagoner, J. C. and Bertram, G. T. (eds.) Sequence Stratigraphy of Foreland Basin Deposits, *American Association of Petroleum Geologists Memoir*, 64, 137–223.

Wheatcroft R.A. (2002). In Situ Measurements of Near-Surface Porosity in Shallow-Water Marine Sands. *IEEE Journal of Oceanic Engineering*, vol. 27, no. 3, pp. 561–570.

Wilkinson, R. H., Moore, E. J., Salkield, A. P. (1985). Photogrammetry in sediment transport studies. *In: Underwater Photography and Television for Scientists*, ed. J. D, George, G. I. Lythgoe, J. N. Lythgoe, pp. 109-21. Oxford: Clarendon. 184 pp.

Yalin, M.S. (1977). *Mechanics of Sediment Transport. (2nd edition)* Pergamon Press, Oxford, 298 pp.

Yalin, M.S. (1992). *River Mechanics*. Pergamon Press, Oxford, 219 pp.

13

Sediment Transport under Ice Conditions

Faye Hirshfield and Jueyi Sui
University of Northern British Columbia
Canada

1. Introduction

Every winter, a vast majority of rivers in cold regions undergo freeze up and subsequent ice jam events. Historically, ice jamming on rivers has led to extensive flooding and damage of property. As humans tend to settle near rivers, the impact of river ice and transport of sediment under river ice is an important phenomenon in engineering design and construction. River ice can easily build up around bridge abutments and piers, impacting not only the structure itself but also the local river bed morphology. In the past 30 years much research has led to increased knowledge of ice jams and associated river hydraulics. Studies have documented changes in break up timing during spring (Prowse & Bonsal, 2004), characteristics of ice jam release surges (Jasek, 2003; Beltaos & Burrell, 2005) and the impacts of climate variability on ice jam events (Beltaos, 2002; Prowse & Conly, 1998; Prowse & Beltaos, 2002).

The confluence of rivers is a location where ice jams often form. Shallow mixing layers are generally found at river confluences. Large-scale structures are found in these flows and should play a significant role in the transverse exchange of mass and momentum. The merging of ice runs, differences in hydrodynamic pressure between channels, ice congestion at a confluence bar, and riverbed deformation should be responsible for the evolution of ice jams at river confluences. A number of studies have documented the occurrence of ice jams at river confluences: Andres (1997) reported flooding caused by an ice jam at a river confluence in Prince George, Canada; Wuebben and Gagnon (1996) described the formation of an ice jam near the confluence of the Missouri and Yellowstone Rivers in Montana; Jasek (1997) presented a case study of an ice jam formed at the confluence of the Porcupine and Bluefish Rivers in Yukon; Andres & Doyle (1984) and Prowse (1986) gave descriptions of ice jams at the confluences of the Clearwater and Athabasca Rivers and the Liard and Mackenzie Rivers, respectively; Ettema et al. (1999, 2001) conducted experimental studies regarding ice processes at river confluences.

In general, very few records can be found detailing ice jam events supported by field measurements. Sediment transport under ice cover and characteristics of ice jam events present obstacles for field examination. Access issues and safety concerns often make it difficult for ice jams to be properly documented. In addition, sediment recording equipment in rivers can often be washed out and destroyed during winter freeze up. If ice jams do not incur damage or flooding, ice events and subsequent under ice sediment transport are often not sufficiently examined or measured.

The purpose of this chapter is to examine sediment transport under ice cover. In order to bring attention to the importance of the matter, this chapter will begin with a brief account of ice related disasters. Next will be review of sediment transport characteristics under open channel and ice cover conditions. The incipient motion of frazil ice and sediment under ice cover will be discussed. Lastly, two case studies will be reviewed: riverbed deformation of the Hequ Reach of the Yellow River in China and the Nechako River in Canada will be examined.

2. Historical ice related disasters

River ice build-up and subsequent ice jams have caused bridge failure, flooding of land and damage to property. Since man has been constructing bridges and crossing structures over river channels, there have been a number of accidents involving ice build up, ice jams and sediment scour. Between 1989 and 2000 there were over 500 bridge collapses in the United States alone (Wardhana & Hadipriono, 2003). River bed scour was responsible for 165 of the failures and ice was the cause of 10 of the collapses (Wardhana & Hadipriono, 2003). In April of 1987, the Schoharie Creek Bridge in New York collapsed due to sediment scour around the bridge pier (Hains & Zabilansky, 2007). During the collapse 5 vehicles plunged into the river, killing 10 people.

Over time, bridge structures can become unstable due annual ice jams and sediment scour. A bridge over the White River in Vermont experienced a number of ice jam breakups over several years; in the winter of 1990 the White River Bridge failed. Examination into the failure revealed that the bridge foundation gradually became deteriorated due riverbed scour around the piers (Zabilansky, 1996). Eastern Canada has also experienced numerous structure failures due to ice jams and sediment transport. An ice jam flood during 1987 in New Brunswick caused river levels to rise in the Perth-Andover region. Ice build up along the bottom structures of a railroad bridge eventually caused the bridge to fail, taking with it rail cars (Beltaos & Burrel, 2002). As early as 1902, New Brunswick records indicate that ice jam flooding caused the Nashwaak River to rise, resulting in severe damage and flooding to the area. When the Nashwaak ice jam released, the flow of ice was so powerful it destroyed a 53 meter long bridge (Beltaos & Burrel, 2005). While research has focused on river ice processes and sediment scour, there still remain gaps in our knowledge about sediment transport processes due to the difficulty in studying ice jams and their impact on the river bed.

3. Flow hydraulics of open channel and ice covered conditions

Sediment transport processes and riverbed deformation is different under ice covered conditions in comparison to open channel conditions. Open channel flow conditions have been widely examined; one of the most popular equations governing open channel flow is the Manning equation,

$$V = \frac{R^{2/3} S^{1/2}}{n} \tag{1}$$

where V is the fluid velocity, R is the channel hydraulic radius, S is the hydraulic gradient and n is the roughness coefficient. The roughness coefficient represents the resistance the bed material exerts on the water flow. Many studies have explored Manning's roughness coefficient in relation to bed material size and configuration and developed sets of resistance values (Fasken, 1963; Limerinos, 1970). Generally, the larger and more angular the riverbed material, the greater the resistance coefficient.

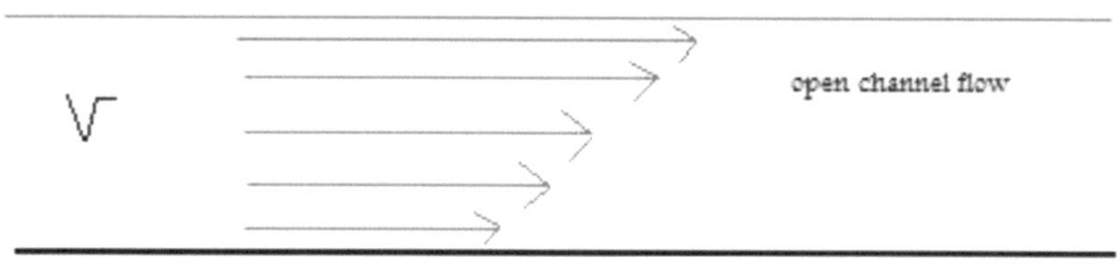

Fig. 1. Velocity profile of river under open channel flow.

For open channel flow, the flow velocity is highest near the water surface and lowest near the channel bed. The drag forces exerted on water near the river bed generally account for the decrease in flow velocity. River ice cover imposes an extra boundary on flow, altering the flow velocity and water level in comparison to open channel flow (Shen and Wang, 1995). For ice cover conditions, the portion of upper flow is mainly influenced by the ice cover resistance while the lower flow is mainly influenced by the channel bed resistance (Sui et al., 2010). The maximum flow is located between the channel bed and ice cover depending on the relative magnitudes of the ice and bed resistance coefficients. Generally, the maximum flow velocity is close to the surface with the smallest resistance coefficient. In the case of narrow river channels or near river banks, the maximum flow velocity will not occur at the surface but rather slightly below the surface due to the resistance forces of the side banks (Wang et al., 2008). The velocity profile under ice conditions also depends on the relative roughness of the ice. Wang et al., (2008) examined the location of the maximum velocity under rough ice cover, smooth ice cover and open channel flow (Figure 2). As the ice resistance increases (from smooth ice to rough ice cover), the maximum flow velocity moves closer to the channel bed.

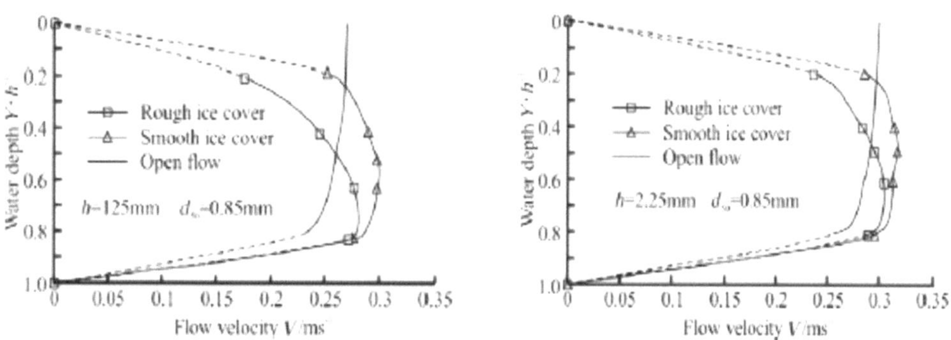

Fig. 2. Velocity profiles under ice cover with difference resistance coefficients (from Wang et al., 2008).

3.1 Incipient motion of frazil ice

Frazil is randomly oriented ice crystals that form in turbulent, supercooled water. Supercooling of river water occurs when cold air temperatures persist and cause river water to rapidly decrease in temperature below its freezing point of zero degrees Celsius. If the

decrease in temperature happens very quickly crystal nucleation can be avoided; water can be well below its freezing point without being in solid phase (forming ice). The occurrence of supercooled water is important for the development of frazil ice. Frazil ice formation occurs when turbulent flow causes supercooled river water to mix throughout the entire channel depth. The mixing of supercooled water throughout the channel depth encourages the formation of small ice crystals. Through the process of secondary nucleation, more ice crystals form and continue to grow. These growing ice crystals are termed frazil ice. Frazil ice crystals will then begin to adhere to sediments in the water and continue to grow.
The conditions for incipient motion of sediment was first introduced by Shields in 1935 and is generally studied through examining the shear stress as given by,

$$\tau_* = \frac{\rho u_{*c}^2}{g \Delta \rho d_e} \tag{1}$$

Where τ_* is the shear stress, ρ is the mass density of water, u^2_{*C} is the critical bed shear velocity, g is the gravitational acceleration, $\Delta\rho$ is the difference in the mass density of the studied sediment and water and d_e is the grain size of the sediment particle.
The conditions required for incipient motion of sediment are normally compared with the critical shear stress of the river bed. In order for a river to begin to transport sediment resting on the riverbed, the bed shear stress must be greater than the critical shear stress. The Reynolds number characterises the ratio of viscous to inertial forces for a given flow condition. The shear Reynolds number (R_{e*}), is used to study the initial motion of sediment particles,

$$\mathrm{R_e} = \frac{U*_c\, d_S}{\nu} \tag{2}$$

where u_{*C} is the critical bed shear velocity, d_s is the sediment grain size and v is the kinetic viscosity of the fluid. Water flowing under ice can exert forces that will entrain or move sediment or ice particles that are located under the ice cover. The forces that act against the flowing water depend on the particle grain size and particle size gradation. As shown in Figure 3, there are 3 main forces that act upon a non-cohesive frazil particle.
F_B is the buoyancy force of the submerged frazil particles, F_L is the hydrodynamic force acting downward perpendicular to the ice cover and F_D is the drag force parallel to the ice cover. In order for incipient motion to occur the drag force should be equal to the resistance force (Sui et al., 2010). When a frazil particle is just about to move, the resultant of the drag and resistant forces is along the direction of the friction angle as given by,

$$tan\theta = \frac{F_D - F_B\ sin\theta}{F_B\ cos\theta} \tag{3}$$

When a frazil particle is about to move, it will lose contact with the frazil particle just above it. The buoyancy force, F_B, can be defined as the following,

$$F_B = C_1 g(p - p_1) d_i^3 \tag{4}$$

where g is gravitational acceleration, p and p_i is mass density of water and frazil particles, d_i is the median grain size of frazil particles and $C_1 d_i^3$ is the volume of the frazil particle. The drag force, F_D, is defined as follows,

$$F_D = C_2 \tau_c d_i^2 \quad (5)$$

where C_2 is the particle area coefficient, τ_c is the critical shear stress for incipient motion of a frazil particle and $C_2 d_i^2$ is the effective area of contact between a studied frazil particles and other particles. By replacing F_B and F_D in Eq. 3 by their above expressions, the critical shear stress for incipient motion of frazil ice particles can be given by,

$$\tau_{cs} = \frac{c_1}{c_2} g \Delta \rho_s d_s \cos\theta \left(\tan\theta_s - \tan\theta \right) \quad (6)$$

As examined by Sui et al., (2010) the required shear stress for incipient motion of frazil particles is larger than the shear stress required for incipient motion of sediment particles. Also, the shear stress required for incipient motion of frazil particles is larger under smooth ice conditions in comparison to rough ice conditions (Sui et al., 2010). The authors also found that the coarser (or larger) the frazil particle, the larger the shear stress required for incipient motion of frazil ice.

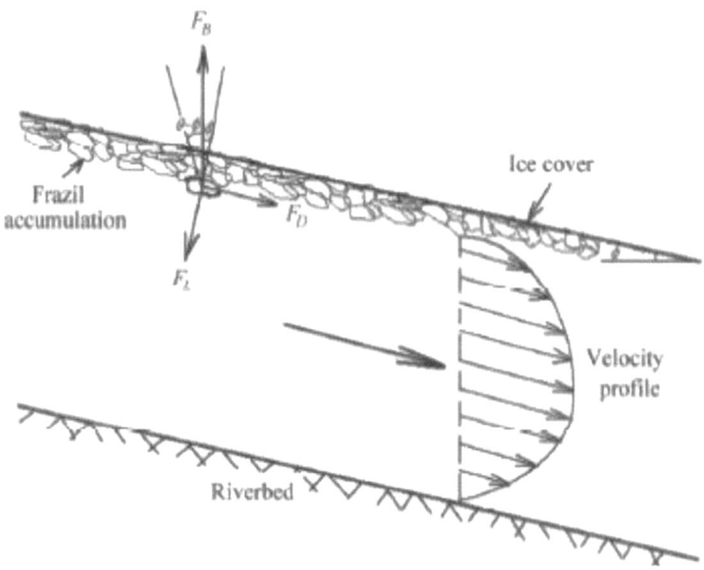

Fig. 3. Forces acting on frazil ice under a sloping ice cover (from Sui et al., 2010)

3.2 Incipient motion of sediment under ice cover

As discussed, the conditions required for incipient motion of sediment are normally compared with the critical shear stress of the river bed. Since ice cover imposes an added boundary on flow conditions, the incipient motion of sediment under ice cover is different than for open channel flow. Wang et al. (2008) conducted a number of clear water flume experiments examining the relationship between incipient motion of bed material and ice cover conditions. Since near-bed velocity is higher under ice covered conditions, a higher shear stress is exerted on the river bed under ice covered conditions (Wang et al., 2008). The threshold velocity for the incipient motion of sediment under ice cover decreases as the ice cover resistance increases. This is due to the increased kinetic energy exerted on the bed material as the near-bed velocity increases. The flow velocity required for initial movement of bed material under ice cover also increases with water depth (Wang et al., 2008). This relationship is only valid if the resistance coefficients of the ice cover and channel bed remain constant. It was also found that the larger the bed material the greater the velocity required for incipient motion (Figure 4).

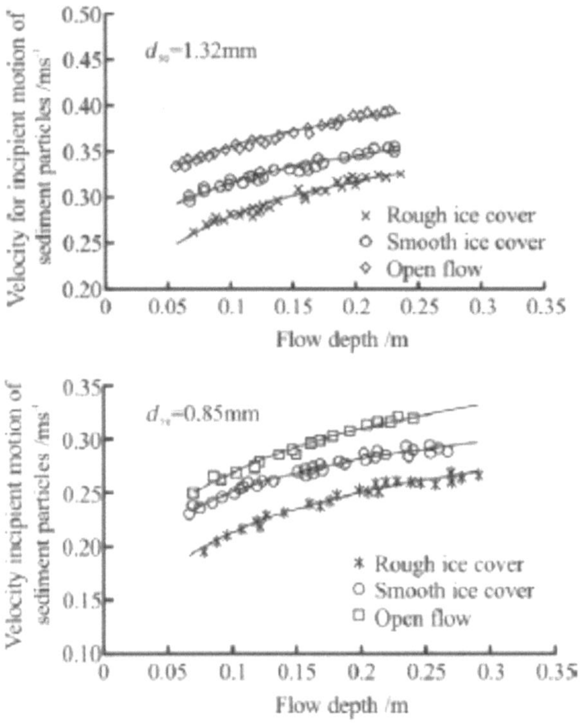

Fig. 4. The relationship between velocity and flow depth and grain size for incipient motion (from Wang et al. 2008).

The Froude number is used to determine the resistance of an object flowing through water. The ability of an object to move through water will depend on its size (object length and area) as well as the relative velocity of water. The greater the Froude number, the greater the resistance exerted on water flow by the river bed material. The incipient motion of sediment depends not only on hydraulic variables but also on the characteristics of the bed material itself. The densimetric Froude number is given by,

$$F_0 = \frac{V_1}{\sqrt{\dfrac{g d_{50}(\rho_s - \rho)}{\rho}}} \tag{7}$$

where d_{50} is the median grain size of the bed material. Generally, the larger the roughness coefficient for ice cover, the greater the near bed flow velocity. In this situation, the larger the roughness coefficient for ice cover, the smaller the densimetric Froude number for incipient motion of river bed material (Wang et al. 2008). Since the near bed velocity is relatively high, larger sediment particles can be moved. However, if the roughness coefficient of the river bed is high, the near bed velocity will decrease; this in turn will increase the densimetric Froude number required for incipient motion of bed material.

The shear velocity for near uniform flow, U_*, in Eq. 2 is given by the following equation,

$$U_* = \sqrt{gRS} \tag{8}$$

where R is the hydraulic radius and S is the hydraulic slope. When examining the incipient motion of sediment, the shear velocity is an important parameter in the Shields criterion as outlined in Eq. 1. The Shields criterion typically considers sediment to be uniform and therefore can be defined by one grain size value. In the real world, sediments are non-uniform. To account for this natural variability in sediment size, Wang et al. (2008) studied the Shields criterion under 3 different sediment types. The authors found that the larger the shear Reynolds number, the greater the shear stress for initial motion of river bed material.

4. Riverbed deformation under ice jam conditions

4.1 Hequ Reach, Yellow River, China

This section examines sediment transport and riverbed deformation in the Hequ Reach of the Yellow River in China, as outlined by Sui et al. (2000, 2006). The Hequ Reach of the Yellow River is approximately 700 kilometres in length and has a broad and shallow morphology. The channel width of the middle reach varies between 400 and 1000 metres. Upstream of the Longkou Gorge (Figure 5) is a reach of open water with numerous localized rapids. During cold winter temperatures, frazil ice forms in the channel and can lead to the formation of large ice jams further downstream in the reach. During the years from 1982 to 1992 the river experienced a large number of ice jams; some lasting well over 100 days. In 1982, a frazil jam caused water levels to rise to extremely high levels, flooding the area and causing extensive damage.

Generally, characteristics of sediment transport during ice jam formation and ice jam breakup are poorly understood. During an ice jam breakup sediment transport can increase due to higher flow velocities, increased river bank erosion and high water levels (Prowse, 1993; Beltaos, 1998).

During an ice jam event, sediment can be transported in water and ice moving down the river channel. Sediment can become attached to frazil ice if supercooled water is moved through the water column to the river bed. Through ice nucleation occurring at the river bed ice can attach to bed sediments and form anchor ice. Anchor ice can form around very large or small particles. Heat transfer from turbulent water can eventually weaken the bond anchor ice forms with the bottom of the river bed causing it to become buoyant. During frazil jam events in the Hequ Reach, pebbles ranging in size from 0.2 to 0.5 kilograms have been observed in ice cover. The largest recorded sediment concentration in frazil ice in the Hequ Reach was 25 kilograms per metre cubed of ice; this is a much larger sediment concentration than what has been recorded in water flowing under the ice (7 kilograms per cubic metre of water).

During an ice jam event from December 1986 to March 1987 Sui et al. (2000) examined riverbed deformation at a cross section in the Hequ Reach. As shown in Figure 6, over the period of approximately 3 months, the river width expanded and sediment scour expanded the deepest parts of the river channel. The following sediment transport and riverbed deformation processes were observed during frazil ice jams in the Hequ Reach:

a. once frazil ice is formed in the upstream river reach, it is moved underneath the ice in the jam head area. This process causes frazil ice to accumulate under the head region of an ice jam, and leads to a reduced cross section. As a result, the head of the ice jam becomes thicker, water levels rise and increased scouring occurs.
b. since the cross sectional area at the ice jam head is decreased, local flow velocities are increased. Increased flow velocities move frazil ice accumulated in the ice jam head

region to the toe. The ice jam toe region then becomes thicker, the river cross section in the toe location decreases, water levels rise and riverbed scour increases.

c. the latter processes will continue as the air temperature drop. Once the air temperature begins to warm, frazil ice development will decrease. Any frazil ice will be transported to the ice jam toe and deposited.

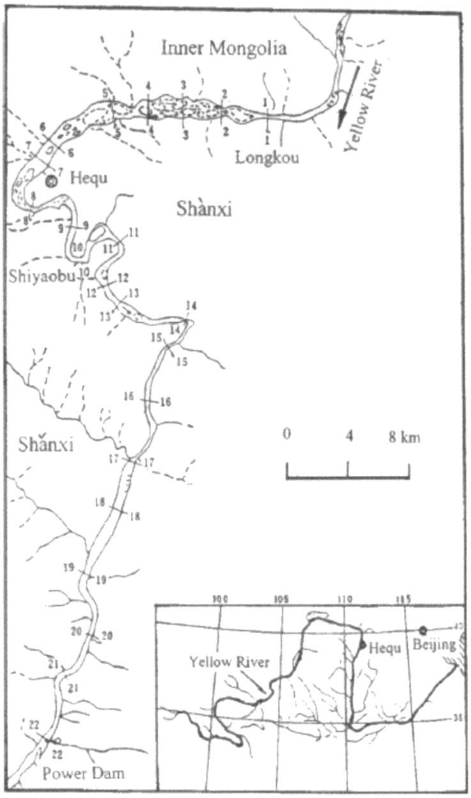

Fig. 5. Hequ Reach of the Yellow River, China.

Based upon extensive field observations of ice jam formation and frazil ice transport, Sui et al., (2000) proposed that frazil ice formation and riverbed deformation reinforce each other. The authors found:

a. as an ice jam grows, the riverbed will be scoured. Once an ice jam begins to decrease and diminish, the riverbed material will undergo deposition. During the entire life of an ice jam, patterns of scour and deposition will be repeated.

b. river flow during an ice jam becomes covered flow. The decrease in river cross section causes river flow to take a path of least energy consumption. As a result, the riverbed will become deformed.

Generally, as indicated by the study of sediment transport under ice for the Hequ Reach, it can be concluded that the larger the flow velocity under an ice cover, the larger the sediment concentration. Also, the sediment concentration during stable jamming period is smaller than the sediment concentration during ice jam breakup period. During a frazil jam it is important to note that sediment can not only be transported through suspension in current, but also through attachment to frazil ice.

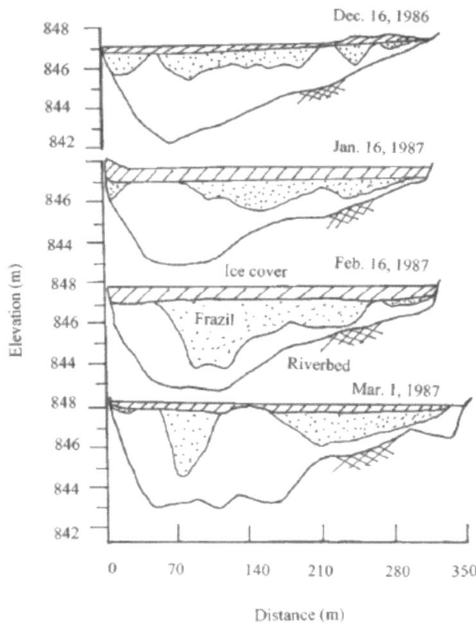

Fig. 6. Changes incurred at cross section of Hequ reach of Yellow River during ice jam period (from Sui et al., 2000).

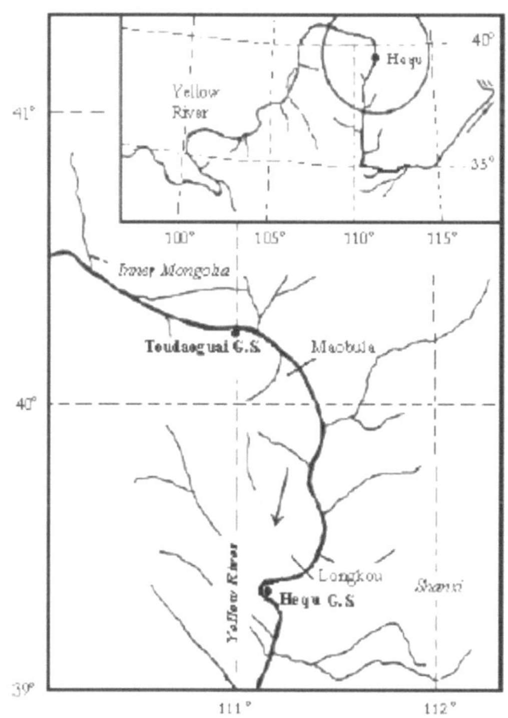

Fig. 7. Location of Hequ gauging station and depiction of river reach between Maobula and Longkou.

Frazil ice transport can also facilitate the formation of a hanging dam. Frazil ice is carried in suspension until it joins the head of an ice jam or gets carried underneath the ice cover. Frazil ice transported underneath the ice cover will become attached to the underside of ice at such a point in which the flow velocity cannot facilitate further transport. Hanging dams are often formed around the Hequ gauging station (Figure 7). The river channel between Maobula and Longkou (Figure 7) has a relatively steep slope, 1.17 percent, exhibits high flow velocity and generally remains clear of ice during the winter season. This relatively straight and open river reach is an area of frazil ice production during the winter season. Due to the river bend that occurs at the Hequ gauging station a river ice cover will often form. Frazil ice from the upstream open water reach will become entrapped on the underside of the ice cover at the Hequ gauging station and form a hanging dam.

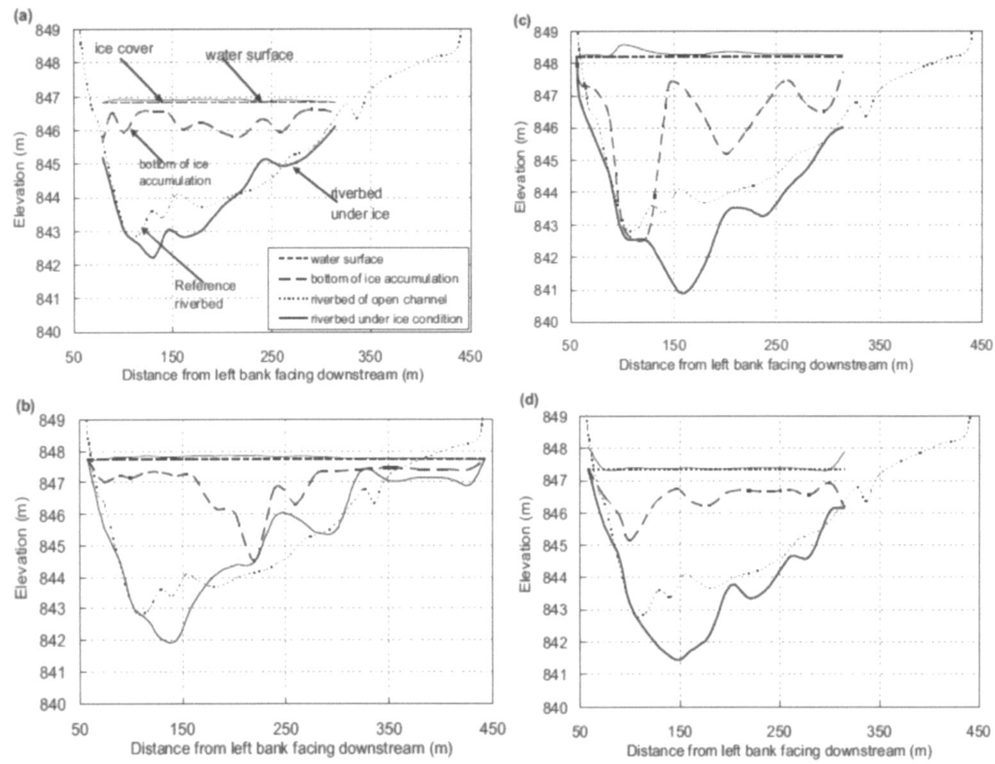

Fig. 8. Riverbed scour under ice cover conditions during the 1982 to 1983 ice jam season (from Sui et al. 2006).

During stable jamming periods in 1982-83 a number of cross sections on the Hequ gauging station were examined (Figure 8). As illustrated in Figure 8, the ice accumulation in the water profile is in the form of a hanging dam. The hanging dam formation and subsequent growth is responsible for the deformation of the river bed. As ice accumulation increases so does the amount of riverbed scour. In Figure 8a, the river had just undergone freeze up in January, a small amount of ice had accumulated, and the riverbed had been scoured by approximately a depth of 0.5 metres. Over the course of the next month, frazil ice continued to accumulate at the Hequ gauging station and a large ice dam was formed (Figure 8b, 8c). By the middle of February (Fig. 8c) the riverbed had been scoured by 2 meters. By around mid

March, temperature began to warm and frazil ice accumulation decreased. Ice accumulation at the Hequ gauging station decreased and sediment deposition occurred (Figure 8d).
Generally, as shown by the cross sections at the Hequ gauging station, the increase in ice accumulation in the form of a hanging dam, decreases the river cross section. The reduction in flow area will increase the velocity of the river flow and causes increased scouring of the riverbed. By examining riverbed scour depth as it relates to ice accumulation, Sui et al. (2006) developed the following equation,

$$\frac{A_D}{A_{IE}} = 0.202\frac{A_I}{A_{IE}} + 0.095 \qquad (9)$$

where A_D is the area of riverved scour in metres squared, A_{IE} is the cross sectional area under the hanging dam, A_I is the cross sectional area of the ice accumulation.

4.2 Nechako River, Canada

The next section is an examination of riverbed deformation at a river confluence in Prince George, Canada. The Nechako River in central British Columbia (BC), Canada, experienced a severe ice jam event that caused flooding and damage during the winter of 2007/2008. Ice jams have been documented for the past 100 years at the confluence of the Nechako and Fraser Rivers. During 1979 and 2008, cross section surveys along the Nechako River were completed (Figure 9). The Nechako River morphology in the 2 kilometre reach, close to the confluence has significantly changed. Here the Nechako River is more unstable. As the Nechako River approaches the confluence, the channel also becomes wider and the flow velocity decreases. As a result of a decrease in flow velocity, sediment deposition in the form of point bars and mid channel islands have evolved near the confluence. It is evident by examining the cross sections that sediment transport has been responsible for the change in channel morphology near the confluence.

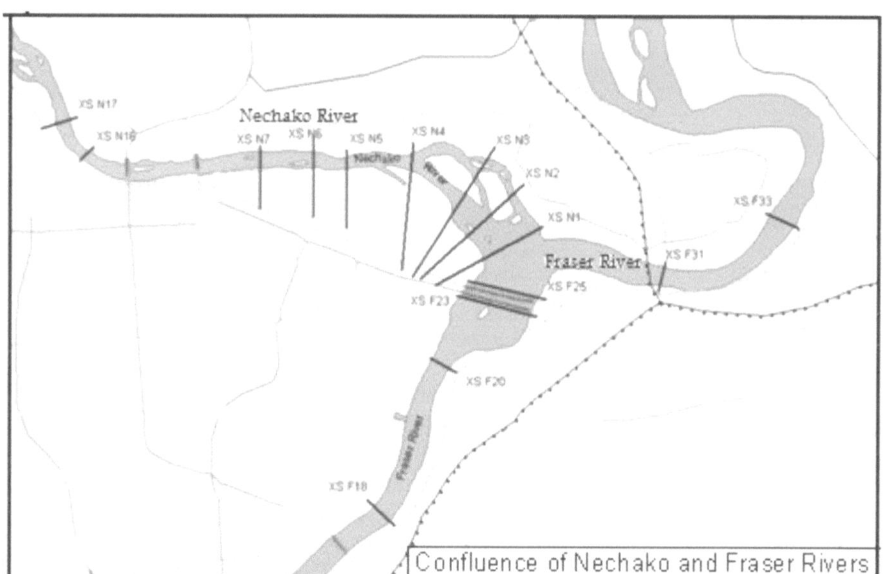

Fig. 9. Confluence of the Nechako River and Fraser River, Prince George, Canada.

The decrease in flow velocity and increase in sediment deposition at the confluence is reinforced by the Nechako River gradient (average bed elevation) as shown in Figure 10. The channel slope close to the river confluence decreases as the channel becomes flat. In addition, since the Nechako River is a secondary branch river of the Fraser River, flow current from the Fraser River may impede the flow out of the Nechako River resulting in backwater effects to the Nechako River. This phenomenon will clearly cause a decrease in flow velocity near river confluence in the Nechako River. When flow velocity or flow Froude Number is low enough (Sui et al, 2005), an ice cover near river confluence will be result.

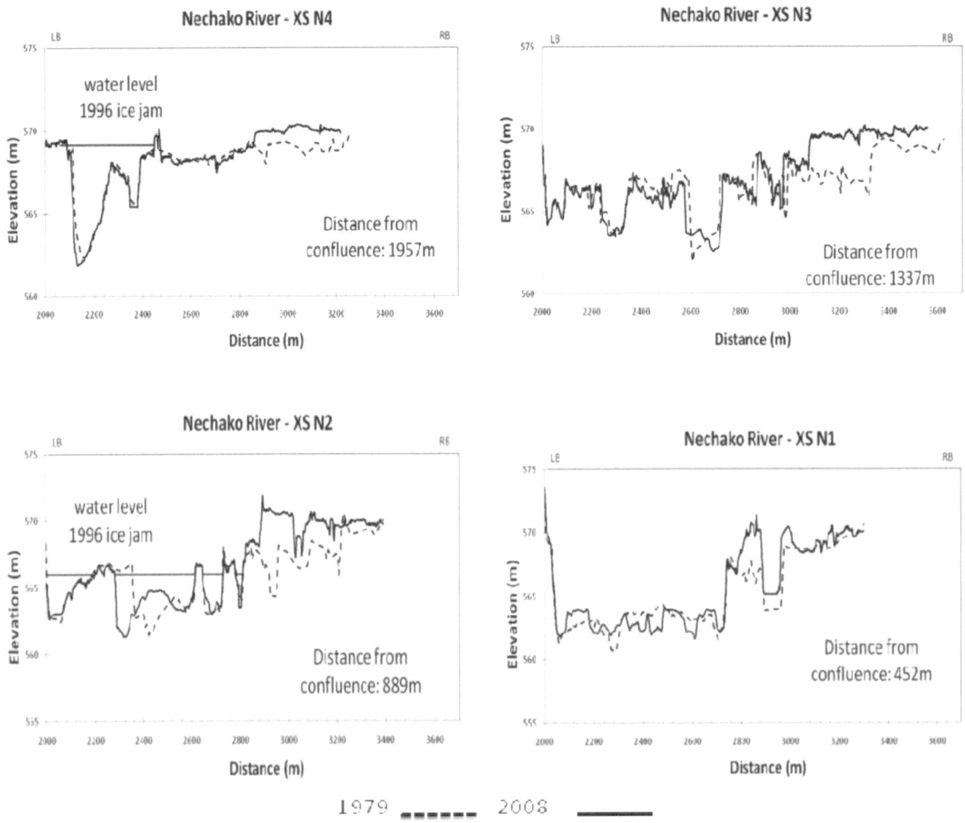

Fig. 10. Comparison of Nechako River cross sections near the river confluence for the 1996 and 2008 ice jams.

With the replenishment of frazil from the upstream river reach in midwinter, frazil ice will accumulate under ice cover initiated near the river confluence. Thus, the thickness of the ice jam will grow during winter period. This increased accumulation leads to a reduction of the flow area and a consequent rise in upstream water levels. With the increase in local flow velocity caused by the reduced cross section, frazil from the jam head is increasingly transported toward the toe, causing water levels there to increase as well. These mechanisms are maintained throughout the ice periods, causing the jam to grow and water level to increase.

5. Conclusion

This chapter examines characteristics of sediment transport and riverbed deformation under ice cover. Historically, ice jams have led to bridge collapse, flooding and damage to property. It is important to understand ice jam characteristics and subsequent sediment transport processes that occur simultaneously. Sediment transport under ice cover is different than that of open channel flow. For open channel flow the maximum flow velocity will occur at the water surface. The maximum flow velocity under ice cover depends on the roughness coefficients of the ice cover and the bed material. The location of maximum flow velocity will be closer to the surface with the smallest resistance coefficient.

Due to the boundary conditions imposed by river ice, ice jams can significantly deform a river bed as compared to deformation observed under open flow conditions (Shen and Wang, 1995; Sui et al, 2000). The presence of ice cover reduces the river channel cross section which will in turn increase the river flow velocity. Larger flow velocity will have greater kinetic energy and cause increased sediment transport and riverbed scour. Generally, as an ice jam continues to grow and develop, more scouring of the river bed occurs. Frazil ice development and the formation of hanging dams have a direct relationship with riverbed deformation; the larger the ice accumulation and hanging dam, the larger the depth of sediment scour. As temperatures warm and existing sediment entrapped in frazil ice is transported downstream until such time that jamming diminishes and the sediment is deposited. As presented in the examination of the Hequ River and Nechako River ice jam history, a riverbed can undergo significant changes to morphology as a result of sediment scour as it occurs under ice cover.

6. Acknowledgements

The authors would like to thank Dave Dyer, City of Prince George and staff at Northwest Hydraulic Consultants for providing ice jam data and information.

7. References

Andres. D. (1997). Freeze-up ice jams on the Nechako River at Prince George: analyses of the 1996 event. Report.

Andres. D & Doyle, P. (1984). Analysis of breakup and ice jams on the Athabasca River at Fort McMurray, Alberta. *Canadian Journal of Civil Engineering*, Vol. 11, 444-458.

Beltaos, S. & Burrell, B. (1998). Transport of metals on sediment during the spring breakup of river ice. Proceedings of the 14th IAHR International Symposium on Ice, Potsdam, N.Y. Vol. 2, 793–800.

Beltaos, S. (2002). Effects of climate on mid-winter ice jams. *Hydrological Processes*, Vol. 16, 789-804.

Beltaos, S. & Burrell, B. (2002). Extreme ice jam floods along the Saint John River, New Brunswick, Canada. The extremes of the extremes: extraordinary floods. Proceedings of symposium held at Reykjavik, Iceland, July 2000. IAHS publication No. 271, IAHS Press, Wallingford, 9-14.

Beltaos, A. & Burrell, B. (2005). Field measurements of ice-jam-release surges. *Canadian Journal of Civil Engineering*, Vol. 32, 699-711.

Ettema, R, & Muste, M. (2001). Laboratory Observations of Ice Jams in Channel Confluences. *Journal of Cold Region Engineering*, Vol. 15, No. 3, 34-58.

Ettema, R., Muste, M. & Kruger, A. (1999). Ice Jams in River Confluences. Report 99-6, US Army CRREL.

Fasken, G. (1963). Guide For Selecting Roughness Coefficient 'n' values for Channels. Soil Conservation Service, USDA.

Hains, D. & Zabilansky, L. (2007). Scour under ice: Potential contributing factor in the Schoharie Creek Bridge Collapse. From Current Practices in Cold Regions Engineering, in Proceedings of the 13th International Conference on Cold Regions Engineering, July 23-26, 2006, Maine, USA.

Jasek, M. (1997). Ice jam flood mechanisms on the Porcupine River at Old Crow, Yukon Territory. Proceedings of the 9th Workshop on River Ice, pp. 351-370, Frederiction, New Brunswick, September 1997.

Jasek, M. (2003). Ice jam release surges, ice runs, and breaking fronts: field measurements, physical descriptions, and research needs. *Canadian Journal of Civil Engineering*, Vol. 30, 113-127.

Limerinos, J. (1970). Determination of the Manning coefficient from measured bed roughness in natural channels. Water Supply Paper 1898-B, U.S. Geological Survey, Washington, D.C.

Prowse, T. (1986). Ice jam characteristics, Liard-Mackenzie rivers confluence. *Canadian Journal of Civil Engineering*, Vol. 13, 653-665.

Prowse, T. (1993). Suspended sediment concentration during river ice breakup. *Canadian Journal of Civil Engineering*, Vol. 20, No. 5, 872–875.

Prowse, T. & Beltaos, S. (2002). Climatic control of river-ice hydrology: a review. *Hydrological Processes*, Vol. 16, 805-822.

Prowse, T. & Bonsal, B. (2004). Historical trends in river-ice break-up: a review. *Nordic Hydrology*, Vol. 35, 281-293.

Prowse, T. & Conly, F. (1998). Effects of climatic variability and flow regulation on ice-jam flooding of a northern delta. *Hydrologic Processes*, Vol. 12, 1589-1610.

Shen, H. & Wang, D. (1995). Under cover transport and accumulation of frazil granules. *Journal of Hydraulic Engineering, ASCE*, Vol. 121, No. 2, 184-195.

Sui, J., Hicks, F. & Menounos, B. (2006). Observations of riverbed scour under a developing hanging ice dam. *Canadian Journal of Civil Engineering*, Vol. 33, 214-218.

Sui, J., Karney B. & Fang, D. (2005). Variation in Water Level under Ice-Jammed Condition - Field Investigation and Experimental Study. *Nordic Hydrology*, Vol.36, 65-84.

Sui, J., Wang, D. & Karney, B. (2000). Sediment concentration and deformation of riverbed in a frazil jammed river reach. *Canadian Journal of Civil Engineering*, Vol. 27, No. 6, 1120-1129

Sui, J., Wang, J., He, Y. & Krol, F. (2010). Velocity profiles and incipient motion of frazil particles under ice cover. *International Journal of Sediment Research*, Vol. 25, No. 1, 39-51.

Wang, J., Sui, J., & Karney, B. (2008). Incipient motion of non-cohesive sediment under ice cover - an experimental study. Journal of Hydrodynamics, Vol. 20, No. 1, 117-124.

Wardhana, K. & Hadipriono, F. (2003). Analysis of recent bridge failures in the United States. *Journal of Performance of Constructed Facilities*, Vol. 17, No. 3, 144-150.

Wuebben, J. & Gagnon, J. (1995). Ice Jam Flooding on the Missouri River near Williston,. North Dakota. Special Report 95-19, US Army CRREL, USA.

Zabilansky, L. (1996). Ice force and scour instrumentation for the White River, Vermont. US Army CRREL Special Report, 59pp, Hanover, USA.

14

Sediment Transport Circulation Pattern through Mesotidal Channels System

Silvia Susana Ginsberg[1,2] and Salvador Aliotta[1,3]
[1]*Instituto Argentino de Oceanografía, CCT-BBca- CONICET, Bahía Blanca,*
[2]*Departamento de Ingeniería Civil, Universidad Tecnológica Nacional-FRBB, Bahía Blanca,*
[3]*Departamento de Geología, Universidad Nacional del Sur, Bahía Blanca, Argentina*

1. Introduction

There are three modes of sediment in motion, which are recognized as rolling and/or sliding; jumped and suspension particles motion. These particles motion include two essential sediment transport modes, which are bedload and suspended load. The first corresponds to the part of the total load which is moving very close to the bed. On this opportunity, the sediment is transported by rolling and/or sliding along the floor or jumping immediately above the bed. The suspended load comprises the particles that are carried away at suspension in the moving fluid. Bedload transport occurs in low velocity flow and/or large grain sizes, while suspended load take place at high velocity flow and/or small grain sizes.

The sediment transport theory to point out that sediment capacity flow and sediment availability in channels regulates the motion of sediments. The sediment transport capacity is the maximum load that can be transported by flow and the sediment availability is the material supply that the flow can put in motion.

In general, the sediment load that moves in a rolling or saltating mode is smaller compared with the suspended load transported in a natural flow. However, the bedload sediment is important for its contribution to the morphological change of the channel. Moreover, it is the most important factor in determining the stability of natural channel, the bed topography, and therefore, the cause of the hydraulic resistance and some properties of the flow. Accordingly, on this chapter we focus on the bed sediment transport pointing to the pathway migration rather than the sediment transport rate itself.

The sediments transport as bedload in estuarine environment controls the bottom morphology. Therefore, estimating the transport of non-cohesive sediments, mobilized by the channel floor, has relevance to the understanding of many pressing environmental problems including eutrophication, contaminant transport, sediment bed erosion, siltation and waste disposal, which impacts on seabed stability and at times on ecosystem distribution. Also, sediments that are moving on the bed generate different bedforms such as shoal, dunes, point bar, etc., which can be hazards to navigation and to affect the flow conditions and the flanks channel stability.

There are different approaches to analyze and estimate the rates and pathways of sediment transport as bedload. Among them there are recording and analysis through indirect observations of the morphological and sedimentological features, the bedform geometry and their asymmetry, and the internal sedimentary structures, carried out with acoustic instrument (side-scan sonar, echo-sounder, multibeam, seismic, etc.), and/or the use of empirical formulas to estimating the quantity and net direction of sediment displaced as a function of the near-bed flow velocity. However, in tidal environment, the alternating ebb and flood currents as well as the effect of turbulence, become more difficult to estimate the bedload transport using empirical formulas from laboratory studies. Therefore, we consider that a best estimate of the bedload transport rates and directions should be in combination used empirical formulas with the techniques of indirect observations.
Numerous authors (Bagnold, 1956, 1963; Yalin, 1963; Engelund & Hansen,1967; Ackers & White, 1973; Madsen & Grant, 1976; Gadd et al., 1978; Langhorne, 1981) have given to know a number formulas for estimating the quantity of sediment transported as bedload as a function of the current velocity. The earliest formulas proposed to estimate bedload transport were mainly based on the concept that the sediment transport rate can be related to the bottom shear stress (Meyer-Peter & Müller, 1948; Einstein, 1950) and these formulas were valid for steady, uni-directional flows. In coastal and marine environments, the process of sediment transport becomes more complex due to the presence of oscillatory flows, and the interaction between steady and oscillatory flows (Bijker, 1967; Watanabe, 1982; Van Rijn, 1993). So, cited by Zheng-Gang Ji (2008) the bedload sediment transport formulas have the general semiempirical relation of

$$\frac{q_B}{\rho_s d_s \sqrt{g d_s}} = \Phi(\tau_0, \tau_{*cr})$$

where Φ is a function of the Shields parameter (τ_0) and its critical value (τ_{*cr}), q_B bedload transport rate (mass-per-unit time-per unit-width normal to the flow direction), d_s the sediment particle diameter, ρ_s the sediment density. Therefore, to predict of sediment transport rates, an important parameter is the shear stress (τ), which is the tangential force per unit area that a moving fluid exerts on the sediment bed. When the bottom shear stress (τ_0) exceeds the critical shear stress (τ_{*cr}), the sediment particles that are resting on the bed only start moving. The shear stress is related to the flow velocity above the bed (u) through $\tau = \rho\, u^2$, thus the flow velocity must exceed a critical value before particles start to move.
Many attempts have been made to determine the bedload transport sediment through the use of theoretical formulas, empirically deduced. However, to despite of many formulas existing, not yet was reached an agreement to accept of these equations as the more satisfactory for the determination of bedload transport. Therefore, the choice of an appropriate equation for tidal environments is still an open question.

2. Bedload sediment transport at tidal channels

2.1 Methods of analysis

Under all those considerations and taking account there are few studies about the movement of bed sediment at interconnected tidal channels system, the chapter purpose is to depict the sediment transport behavior as bedload in this type of sub-environment and determine the tidal flow as controlling factor of such transport pathways. Understand the processes and mechanisms controlling the rate and directions of sandy grain-size transport are very important because the maintenance of estuaries is closely linked with their transport paths. Specially, if the estuary is integrated by an interconnecting tidal channels system. Similarly, the construction of structures, oil pipelines, and sewages outfalls is dependent on the nature and stability of the bottom sediments. Therefore, the accurate knowledge of transport paths as bedload sediment is essential if the environmental managers be able to asses the ultimate destination of material.

Because the exchange of floor sediments between the interconnected channels is not well know, being subject of present investigations, and as our intend is to advance in this type of research, in this chapter we show how may be the pathways of bedload sediment transport at interconnected tidal channels system. Surprisingly, few field evidences supports studies on inferred the bedload transport rates and directions at an interconnecting tidal channels system. The majority studies dealing with specific aspects on intertidal dynamics as tidal asymmetries and morphodynamics of tidal channels but based in laboratory experiments and theoretical models (Friedrichs, 1995; Schuttelaars & de Swart, 2000; Lanzoni & Seminara, 2002; Bolla Pittaluga & Seminara, 2003; Tambroni et al., 2005). Under these considerations and based on studies carried out many years ago at Bahía Blanca estuary about sediment dynamics in interconnecting tidal channels, we evaluated the pathways and sediment transport rates as bedload at these sub-environments. A detailed set of in-situ field measurements include continuous monitoring of bottom currents during at least one tidal cycle (13 h) at different deployment sites; analysis and interpretation the morphological features as bedforms orientation (i.e. dunes, sand ribbons, point bar, shoals, etc) and analysis of sub-bottom strata characteristics that conform these features. All those data were collected during different campaigns, using high seismic-reflection profiles, side scan sonar records, bathymetric and current-meters data. In this context, we focusses our analysis mainly on the description of bedforms geometry, erosion-deposition process and, asymmetry and analysis of bottom currents in a tidal channels system of the Bahía Blanca estuary (Argentina).

2.2 General characteristics of channel systems

The regional coastal configuration of Bahía Blanca (Fig. 1) corresponds to a plain estuary formed by transgressive-regressive sedimentary processes during the latest postglacial events (Aliotta & Farinati, 1990). This mesomareal estuary is located to south of Buenos Aires Province (Argentina) and is characterizes because the freshwater input is very small compared to the tidal prism. There is extensive mudflats with one large sinuous main channel, where in some sectors the water depth is maintained by dredging (15 m) to ensure the passage of ships to ports and it is called Principal channel. The south side of the Principal channel is formed by an extensive intertidal area with an interconnected tidal channels system of sizes different, very sinuous or meandering and with general orientation NW-SE (Fig. 1). Because of the little contribution of runoff from terrestrial sources, the maintenance of channels depends on inflow and outflow tidal. The tidal regime of this estuary is a semidiurnal periodicity with mean range of around 2 m at the mouth and more than 4 m at its head, with tidal ranges 2.5 m for neap tides and of 3.42 m for spring tides. The tide wave propagates following a hypersynchronic mode, generating currents higher than 1.05 m s^{-1} (Perillo & Piccolo, 1991). This channel has tidal currents reversible with mean velocities, measurement recently by us with ADCP, of 1 m s^{-1} and 1.4 m s^{-1} for flood and ebb conditions, respectively. Maximum speeds of currents, along cross section to Principal channel, are occurring near the south coast.

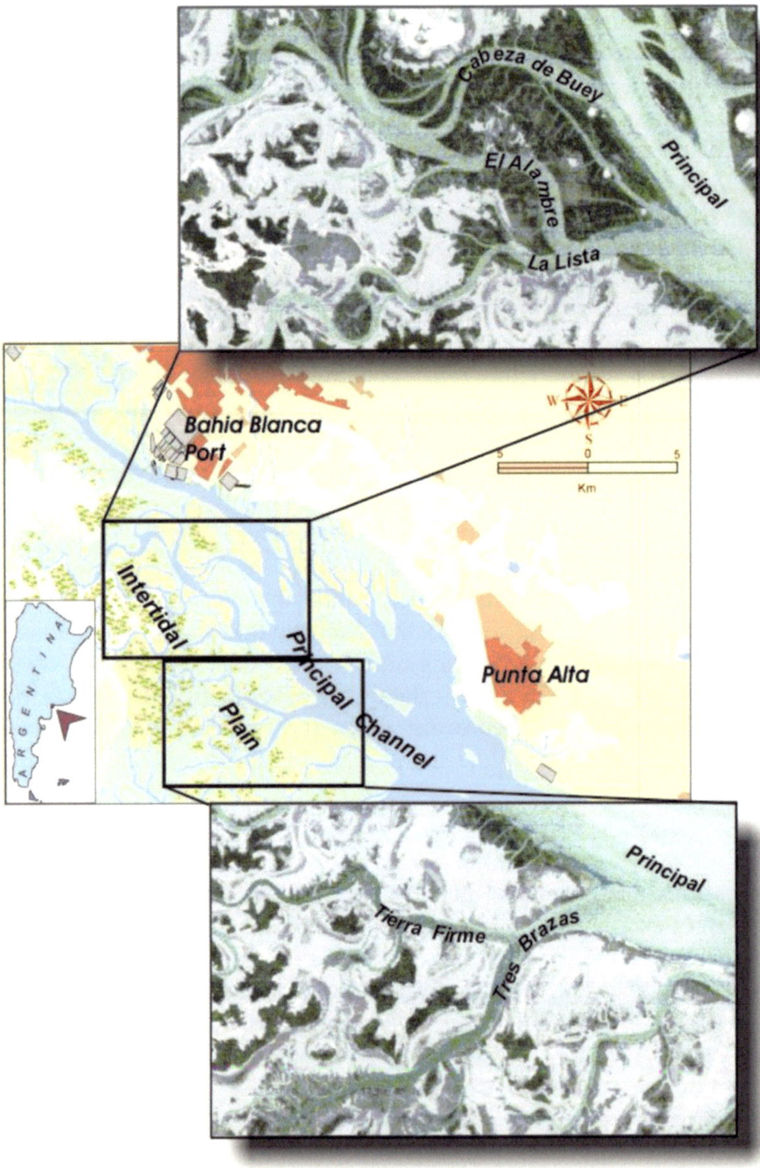

Fig. 1. Bahía Blanca Estuary. Intertidal plain with interconnected tidal channels system.

According to the different dimensions, tidal channels have been classified as large channels, creeks and gullies (Fig. 2). Generally, the smaller channels (gullies and creeks) flow into the large tidal channels and these latest flows into the Principal Channel. The width and depth ranges the former may from few tens of centimeters up to 1m, while the large channels could have reach more than 10m depth and width of 1km in their mouth (Ginsberg & Perillo, 2004). In plan view, the courses of large channels have shown a low to medium

sinuosity pattern while the smaller channels are predominantly strongly sinuous or meandering and their curvatures increases inward. The mouth of the large channels is highly dynamic area, with considerable movement of sediments and associated morphological changes, leading sometimes to an infilling as shoal.
In general, the topographic profile is different in relation to channel size (width depth ration varies) and the bottoms of the larger tidal channels are mostly sandy, enriched in shells and mud pebbles (Fig. 2). The bottom sediment of smaller channels and gullies of the intertidal zone are mainly muddy with a low percentage of sand fractions. The sub-bottom sedimentary material that characterizes sedimentologically the wide tidal plain of Bahía Blanca, was described by Aliotta et al. (2004) from a vertical stratigraphic profile made on the erosive flank of a large channel (Tierra Firme channel) (Fig. 3). It is composed at the bottom with gray-dark to brownish fine-sand and contains about 25% of mud. Upward of profile the percentage of silt + clay fraction increases more than 75%. A deposit - classified as clay silty sand - with horizontal parallel bedding made up of thin laminae of fine-to-very-fine sand (2 mm) and mud (4 mm) outcrops in the intertidal zone, which is characterized by escarpment levels. The sand/mud interlayered is eroded during the lateral displacement of channels. Studies in restricted sites indicate values of lateral displacement of up to 25 m yr^{-1} (Ginsberg & Perillo, 2004). Erosion can result in the formation of small terraces and cut bank rotational slumping of large blocks, where consolidated sediments after collapse can be accumulated on the bed along the bank (Fig. 4). Later all these sediments are transported as bedload. Depending on the strength of the currents, these deposits are carried away as bedload, normally following the channel currents dominance. Sometimes, large amounts of sediment are deposited as

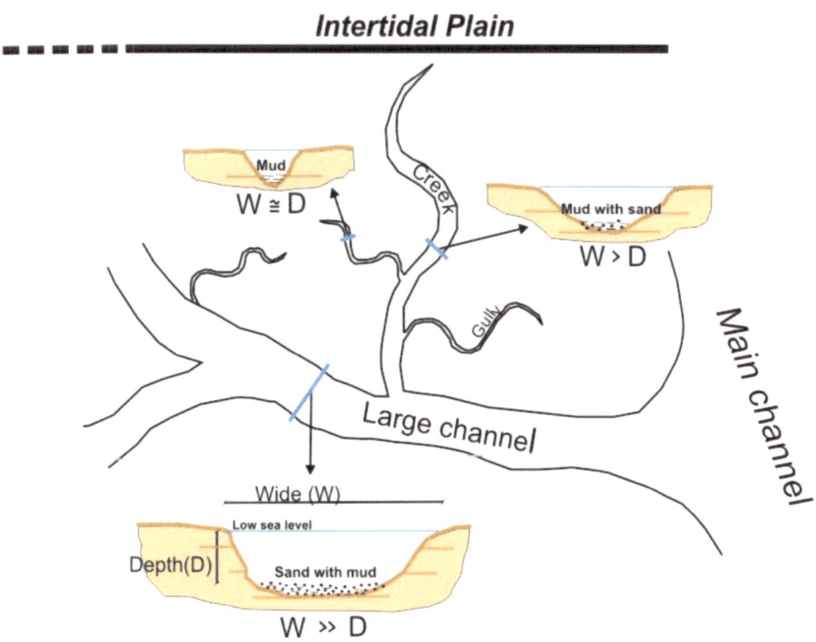

Fig. 2. Tidal courses classification. Classification of different channels based on their sizes range and bottom sediments

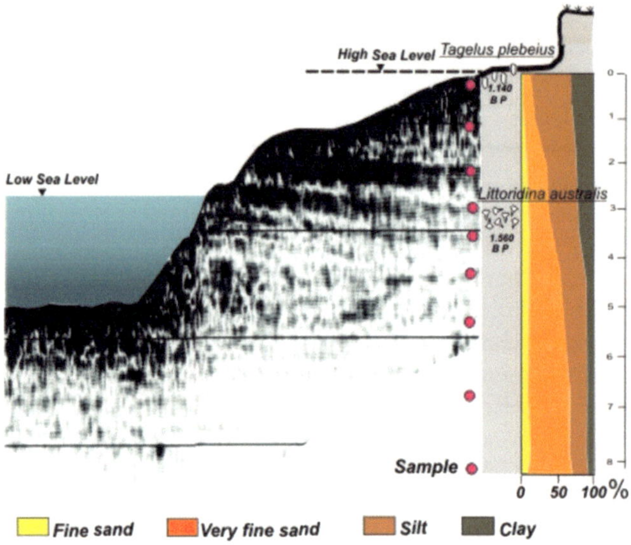

Fig. 3. Seismic-stratigraphic and sedimentological profile along erosive flank of a large channel.

shoals in mouth of tidal channels. The shoals lie partly below mean sea level, but most of the larger shoals have become intertidal. Usually, the floors of tidal channel consist of relatively coarse and fine sand, with mud balls. The shoals are sandy, sometimes with layers of sandy silt on top.

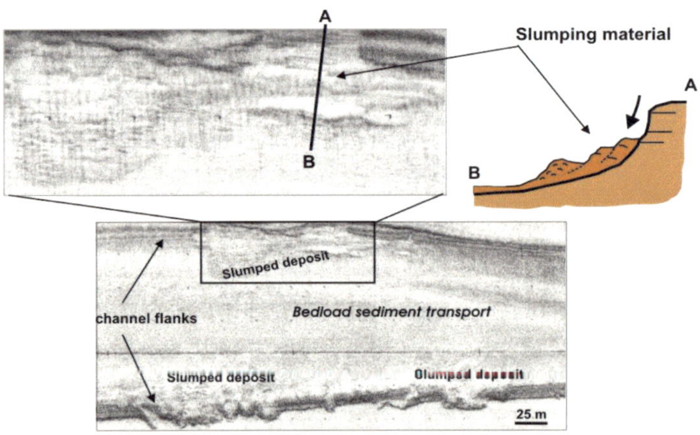

Fig. 4. Sediment slumped along channel flank. Side Scan Sonar record is showing this process.

3. Estimation of the bedload sediment transport and relationship with the morphodynamic

To improve our understanding of the processes and sediment transport as bedload in an interconnecting tidal channels system, on this chapter we focus in two systems at Bahía Blanca estuary (Fig. 1). One of them is composed of Cabeza de Buey, El Alambre, La Lista and Principal channels and the other system by Tierra Firme, Tres Brazas and Principal channels. At the first system, the Cabeza de Buey channel is connected to La Lista channel through El Alambre channel. The first two flow into the Principal channel. The second system corresponds to the Tierra Firme-Tres Brazas channels, where the first is a distributary of the second channel. The latter is flowing into the Principal channel.

Arguably, the most important factor controlling the sediment transport as bedload is the number and pattern of distributaries channels. The pattern in the transport of sediment through the distributaries channel generally follows the flow of water. However, variability in the sediment load carried by each branch, depend of the sediment availability at each channel. Recalling that sediment transport mainly depends on flow velocity, it is possible to relate the residual sediment transport to the peak values of velocity during the ebb and flood phases. In addition, is known that the magnitude and direction of the residual transport along a tidal cycle is mainly affected by tidal asymmetries which are favored by unequal duration and/or unequal magnitude of ebb and flood (Dronkers, 1986). In particular, flow asymmetry characterized by shorter flood duration and higher velocity flood current (i.e., flood dominance) induces a landward sediment transport (Lanzoni & Seminara, 2002). Conversely, shorter periods fall flow and maximum velocity ebb current (ebb dominance) cause a net outward sediment transport. To determine the flood/ebb dominance in the net transport through several sectors of channels has been made an approach depending on both the duration of each tidal phase and also its peak velocity. Tidal currents observations carried out in some tidal channels of Bahía Blanca estuary by Montesarchio & Lizasoain (1981); NEDECO-ARCONSULT (1983); Ginsberg & Perillo (1999, 2004); Vecchi (2009); Ginsberg & Aliotta, (2009) indicate that the behaviour of tidal currents during the flood phase is different from the ebb phase (Fig. 5). These measurements suggest the ebb peak is higher than the flood peak, but the duration of the ebb phase is shorter. For instance, evaluation of maximum velocity indicates that in general average maximum flood velocities have about 0.80 m s^{-1} and average maximum ebb velocities is about 1.20 m s^{-1}. Examination of currents duration indicates that flood time dominance is about 7 h while ebb period is 5 h. In particular, detailed analysis of the near-bed velocity in the Tres Brazas channel and its tributary (Tierra Firme channel) indicates the ebb dominance, with maximum flood velocities of about 0.60 m s^{-1} and maximum ebb velocities around 0.85 m s^{-1}. At El Alambre channel the maximum velocity is ebb-direction, which reaches 0.65 m s^{-1}, while flood is about 0.50 m s^{-1}, and at La Lista channel the observations indicate a dominance of the ebb current with average velocity of 0.50 m s^{-1} while flood current have maximum average values of 0.40 m s^{-1}. All these measurements indicate that the system is at present ebb-dominated because have a tidal asymmetry with a peak current velocities 30 % faster during ebb period and with relatively more longer duration of the flood currents. Therefore, under those considerations, all these measurements indicate that the strongest current is in ebb direction, suggesting a net sediment transport towards outer zone of channels.

Based on the velocity measurements above the bed, some of aforementioned specialists have predicted the net bedload sediment transport rate in some channels. Among them, Vecchi et al. (2008) and Vecchi (2009) calculation of the sediment transport found in the inner sector of La Lista and El Alambre channels values of the 0.07745 g cm^{-1} s^{-1} and 0.08773 g cm^{-1} s^{-1}, respectively, both in the ebb direction. At La Lista mouth, their estimates indicate that the sediment transport decreases to reach 0.02188 g cm^{-1} s^{-1}, also in ebb direction. Ginsberg & Perillo (1999), estimate the sediment transport at Tres Brazas and Tierra Firme channels, based on the theory where the bedload sediment transport is proportional to the stream power (Bagnold, 1966). The results of these studies showed the following values: 489 erg $cm^2\ s^{-1}$ for inner sector of Tres Brazas channel and 379 erg $cm^2\ s^{-1}$ for the other channel. All the studies discussed above demonstrate that in terms of bedload transport, the large channels are ebb-dominated and in consequence there is export of sediment from landward. These results showing that the velocity asymmetry at present-day channels, which is consistent with the residual sediment transport, controls the bed sediment movement towards Principal channel or outward area.

In order to have an over view complete of the sandy sediment mobility within different sectors of channels and knowing that bedform characteristics as well as the configurations of channel system are also determining in pathways sediment transport is given a complete review of different morphological features that several authors have found in each channel of both systems. The results indicate that bottom morphology consists of a wide variety of bedforms at scales ranging for centimeters to tens meters (Ginsberg & Perillo, 2004; Lizasoain, 2007; Ginsberg et al., 2009 a,b; Ginsberg & Aliotta, 2009). The location and distribution of bedform types reflects the patterns circulation of sediment transport. Accordingly, we considered that flood/ebb tidal dominance together with the channels morphology will give rise to illustrate sediment circulation as bedload at tidal interconnected channels. For instance, we can mention some junction zones of channels, which exert an important effect on the hydrodynamics, and therefore on sediment transport as bedload, increasing or decreasing the ability of the tidal flow to induce a scour holes or deposits, in these zones. Hence, the changing bottom features will affect the hydrodynamics and will cause significant changes in the transport pattern. In consequence, local changes as a meandering sector, point bars, shoals mouth and the junction channels zone could be responsible for the changes in transport within the channel system. As a result, we will use these features development in tidal channels as an important tool for evaluation of the bottom sediment transport.

The main characteristics in channels of Bahía Blanca are asymmetric dunes, whose distribution, pattern and morphology provide information on bottom currents on the assumption that they are in equilibrium under the current hydrodynamic conditions (Boothroyd, 1985; Ikehara & Kinoshita, 1994). In addition, sediment transport as bedload could be inferred from its asymmetry (Allen, 1968 a,b; Bokuniewicz et al., 1977; Swift & Freeland, 1978; Harris, 1988). Taking these concepts into account, the examination of the asymmetry of these dunes carried out by several authors in the systems considered here, have provided information on sediment transport as bedload as well as on sedimentary circulation pattern. From the different nomenclatures to available for the description of these features (Allen, 1968a; Boothroyd & Hubbard, 1975; Ashley, 1990; Berné et al., 1993) was chosen the classification scheme of Ashley (1990) which considered height (H) and wavelength (L) as the most important parameters. Two main bedforms scales were identified in the system considered namely medium dunes ($5\ m<L<10\ m$; $0.4\ m < H<0.75\ m$) and large dunes ($10m<L<100m$; $0.75\ m < H<5m$).

Medium dune is the most common bedform in various sites of different tidal channels of Bahía Blanca estuary (Aliotta et al., 2004; Lizasoain, 2007; Ginsberg et al., 2009b). In particular, the channel system of Tres Brazas-El Alambre (Ginsberg et al., 2009a) have been found these bedforms (L=8–10 m; H=0.4– 0.6 m) on a relatively thin sand belt, in the inner sector of Tres Brazas channel. Also, they found large dunes (L=12 m and H=0.8 m) in the mouth of Tres Brazas channel with their crestlines very sinuous and barchanoid structure with low lateral continuity. Their profile is asymmetric with the ebb current. Large dunes

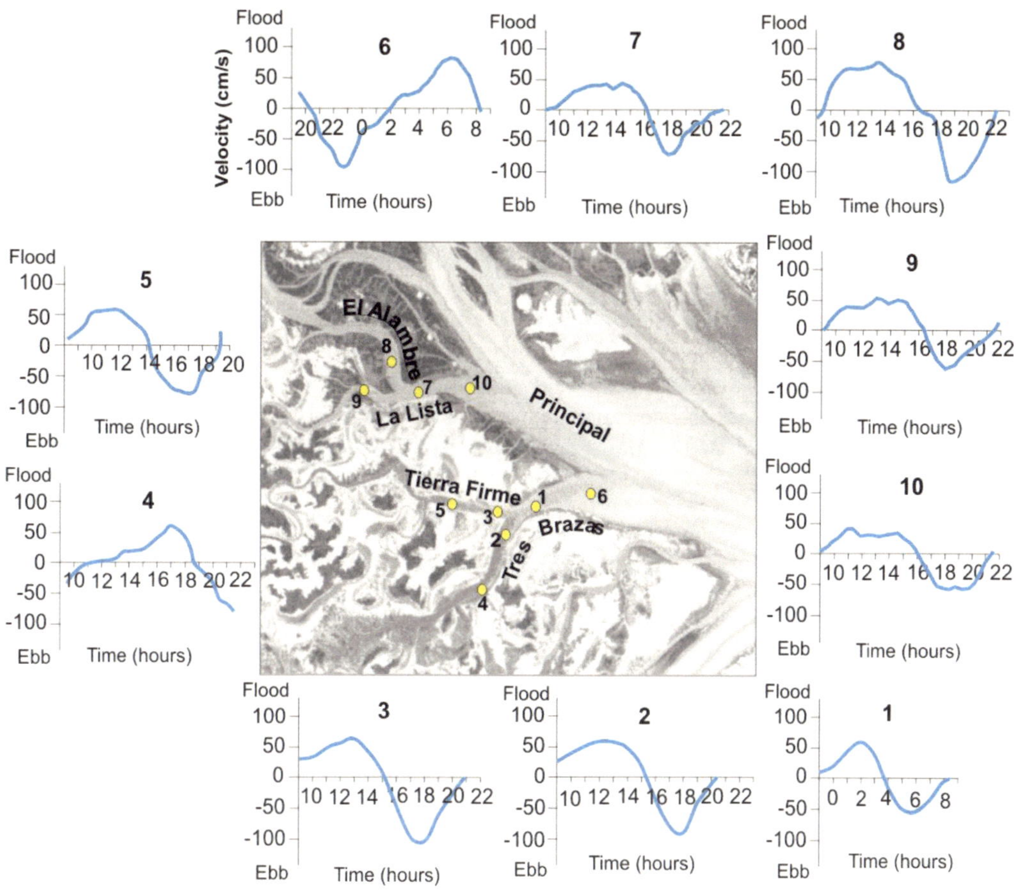

Fig. 5. Current measurements at tidal channels of Bahía Blanca estuary. The tidal asymmetry showing the difference of phases between flood and ebb.

are, in general, formed at a 5-6 m depth towards the inner part of Tres Brazas channel, and at higher depths they disappear, thus forming a sand bottom plane. These authors indicate that under different tidal conditions, the dunes always have displayed an asymmetric profile with the lee side towards the Principal channel. On the other hand, according to Ginsberg et al. (2009a) the morphological and seismic studies in the system Tres Brazas-Tierra Firme are also indicative of ebb dominance in both channels, although only in the inner part of Tres Brazas channel there is a significant availability of sandy sediment mobilized as bedload transport. This circulation pattern is evidenced by the formation of the dunes above mentioned. The sandy sediments coming from the inner zone of Tres Brazas channel are mobilized along their southern flank. The major part of the Tres Brazas channel sediments are transported along this channel towards the Principal channel and the other partly settles on mouth to form sand shoal. In contrast, the above mentioned authors have observed that the Tierra Firme channel have a low contribution of sandy sediment and no will result an important input of them.

Geomorphologically, tidal dominance on sediment transport as bedload produces also bed forms as sedimentary corps that can using to predict the pathways of sediment transport. So, based on point bar formation on meander bends, Aliotta et al. (2004) and Vecchi (2009) provide useful information to determine the net flood/ebb dominance in the bed-material transport direction through the channels. They have been described different point bars configurations and we assume that the difference on nature is in response to sediment availability that each channel has. Hence, Aliotta et al. (2004) note that the point bar formation at the inner side of Tierra Firme channel bends is characterized as bank-attached bar with lateral sedimentation. They observed through analyzing 3.5 kHz seismic data at Tierra Firme channel (Fig. 6), that the sedimentation on the point bar takes place laterally in the form of inclined beds. The downlap reflection terminations indicate progradation and migration of sedimentary body or point bar, which shows an oblique progradational pattern (Fig. 6, I) formed by sandy silt material supply. This type of seismic facies suggests (Mitchum et al., 1977) a relatively high hydrodynamic regime. These beds are curved and of short lateral extent. The layers curved show the sediment transport direction. The maximum ebb current velocity transports sediment as bedload outside of channel, and some of these materials tend to be deposited on point bar (Fig. 6, II).

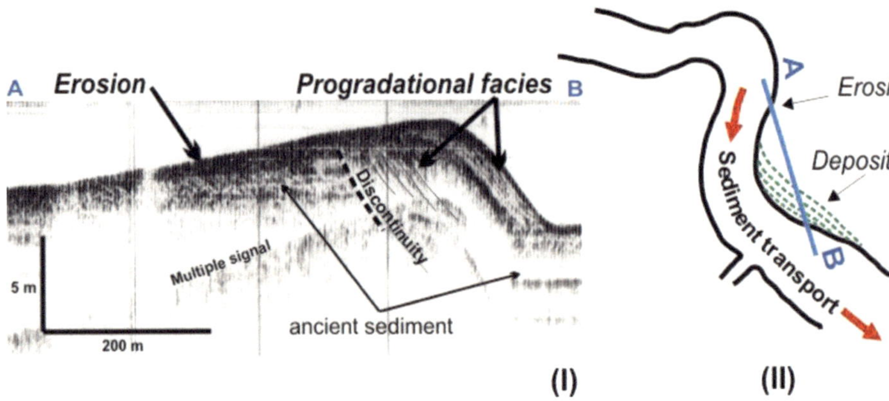

Fig. 6. Point bar in a tidal channel. (I) Seismic record showing the stratigraphic evidences of a point bar evolution. (1II) Scheme showing the dominant sediment transport direction.

Vecchi (2009), also observed the formation of a point bar in El Alambre channel (Fig. 7), but she recognized that the high amount sediment mobilized from inside the channel have a strong influence on natural development of this point bar. Is so those sediment mobilized by ebb dominance are accumulated on downstream portion of the point bar and cause a growth of this bar with development an ebb barb. Therefore, a short elongate tidal bar (spit bar) is formed as a bank attached to point bar (Fig. 7) with separation of the flow into mutually evasive flood-and ebb-dominated channels. Medium dunes with lee face oriented to ebb tide migrate on the surface of this bar.

As a resume, point bars are formed by aggradations during the channel migration by the ebb-domination, typifying meander bends in rivers (Fagherazzi et al., 2004). The current direction dominance obtained through this feature, indicates a current running parallel to the strike direction of the layers of longitudinal cross-bedding. The lateral sedimentation on the point bar demonstrates that bar sediment grow in the ebb-direction. The transformation of point bar at El Alambre channel as spit bar is the result of the increased sediment availability as bedload in this channel, relating to hydrodynamic regimes.

Other important area that modifies the sediment movement as bedload within the interconnected tidal channels system is the larger channels junction or bifurcation sector.

Several important changes in bed morphology are produced in the confluence/bifurcation at fluvial or tidal tributaries, being the most significant a deep scour hole formation as result from remobilization of sediment at the bottom. In this particular case, the spatial patterns of flow structure and bedload transport will depend on the morphological characteristics of the area. Various authors (Kjierfve et al., 1979; Ginsberg & Perillo, 1999; Aliotta et al., 2004; Ginsberg et al., 2009 a; Vecchi, 2009) have studied the dynamic conditions at the junction/bifurcation area of channels and they came to the conclusion that sediment movement in this zone follows a very complicated pattern. The flow bifurcation or junction creates steeper surfaces at the mouth of each confluent channel and a gentle sloping on the opposite downstream side. Although, erosion and deposition is take place at each face during the tidal cycle, there is always certain dominance of one of these processes based on the asymmetry of the tidal currents. As pointed out by Kjierfve et al. (1979) and Ginsberg & Perillo (1999) the nature of the tidal circulation associated to this zone and estimates of the sediment transport pathways are related to flood-dominance acting on the steep sides of tributaries mouth and the ebb-dominance acting over the gentler outer side. The latter authors observed that holes migrate toward the inner sector of channels, in an opposite pattern to fluvial holes. Analyzing the bedload circulation, they have found an important

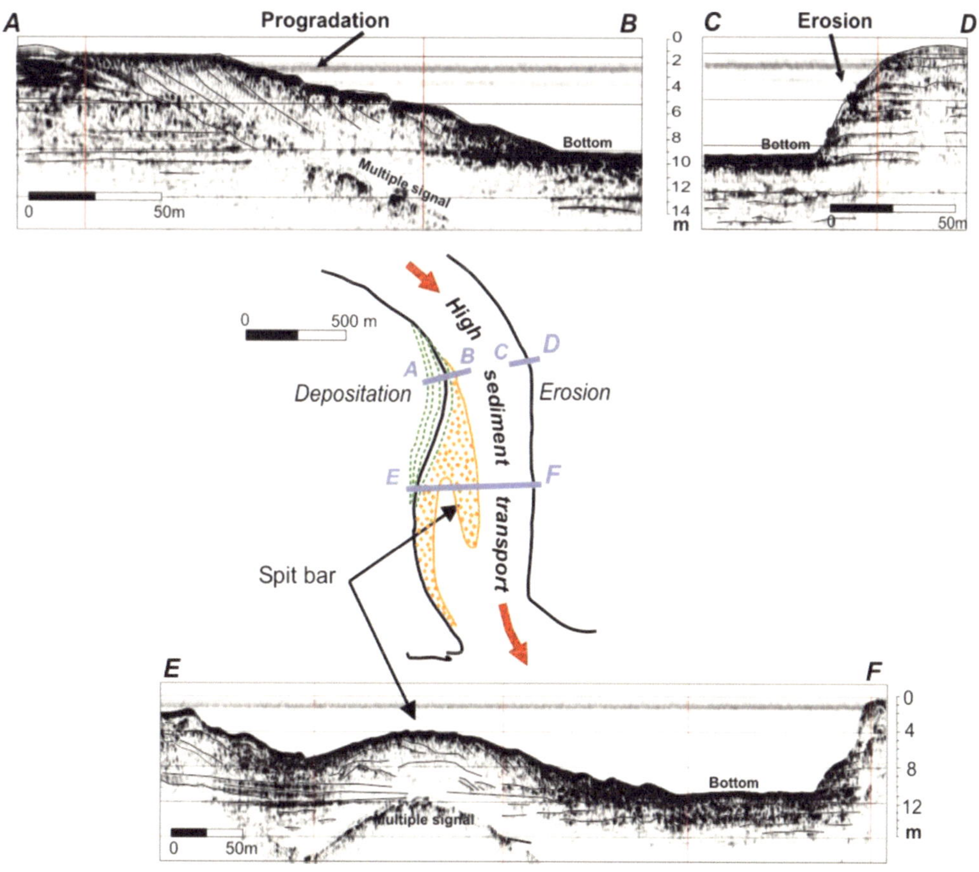

Fig. 7. Point bar in a tidal channel with high sediment transport rate. Seismic records that show erosion-deposition and sediment transport.

asymmetry in the transport direction toward the ebb current. On the basis of the currents measurements and near-bottom sediment flux, they have been recognized that flood-current eroding act the material within the hole and on the tributary channels, while most of the sediment is later transported in the ebb direction. Then, some material is deposited on the gentler outer flank of hole while another large fraction is actually exported towards the outer system.

Ginsberg et al. (2009 a), through seismic study in Tres Brazas-Tierra Firme channels, also described and interpreted the depositional and erosive sequences that characterize the hole developed at junction zone of both channels. This investigation has documented in the sub-bottom the existence of seismic sequences which found at confluence or bifurcation zone where the hole has the steeper slopes (Fig. 8A). One of these sequences corresponds to old sedimentary strata with parallel–subparallel bedding and high lateral continuity and the other consisting of cross-bedding and a complex sigmoid-oblique seismic progradational pattern. These facies occur in the inner area on the mouth of each tributary and takes place by erosive processes. The first facies, located in the lower part, is interpreted as flood flat deposits and the other, located in the upper, as a lateral migration of ancient stream beds, probably close to datum level. The materials showing alternating fine sand and silt layering. All these sediments are overlying by marine transgressive materials, corresponding to sand with containing different proportions of fragments of shells and pebbles. Contrary to this zone, in opposite direction, on the slope gentler flank is observed a progradational sequences mainly related to depositional processes. This sequence is characterized by oblique clinoforms which evidence a progressive lateral development of slightly sloping depositional surfaces (Fig. 8, B) with their downlap termination on an erosive discontinuity that forms in the ancient fluviodeltaic and marine transgressive materials. According to Ginsberg et al. (2009 a) this prograding clinoform pattern is similar to that produced by the accretion and migration of banks in tidal environments (Marsset et al., 1999; Tessier et al., 1999). These authors point out that the above-mentioned discontinuity is covered by 4-5 m of fine sandy sediment, and continuing laterally in the sub-bottom of the less steep slope of the hole.

Ginsberg et al. (2009 b) considered that turbulence flows are generated, whether when two flows combine during ebb tide or through flow division during flood tide, causing scour at the mouth of each tributaries, and extending into central deeper part and steeper flanks of the hole, resulting an increased in its depth and width. They also discuss the principal sediment transport pathway as bedload at the two channels and on junction zone. Bottom sediment circulation at channels confluence zone shows that sediment moving within distinct corridors, but ever there is an ebb-dominated sediment transport in both channels and the confluence zone. They concluded that largest material transport pathways occur essentially surrounding the scour hole (Fig. 8, C) and no bed sediment transport within of the confluence. They have also been revealed that sediment circulation is modifies on the face of low angle slope since the material is moves in a counter-clockwise direction (looking inward), causing introduce of sediments from upstream and thus an accretion effect. As mentioned above, this observation it is supported by the seismic configuration which evidences a sandy sedimentation process with prograding clinoforms on this face. The presence of downlap terminations on the ancient materials indicates that the deposition of these sediments occurs under flood current conditions (Fig. 8). These authors also point out that the distribution of sediment as bedload around this confluence or bifurcation depends on the availability of sediment that each tributary has and their capacity to mobilize them.

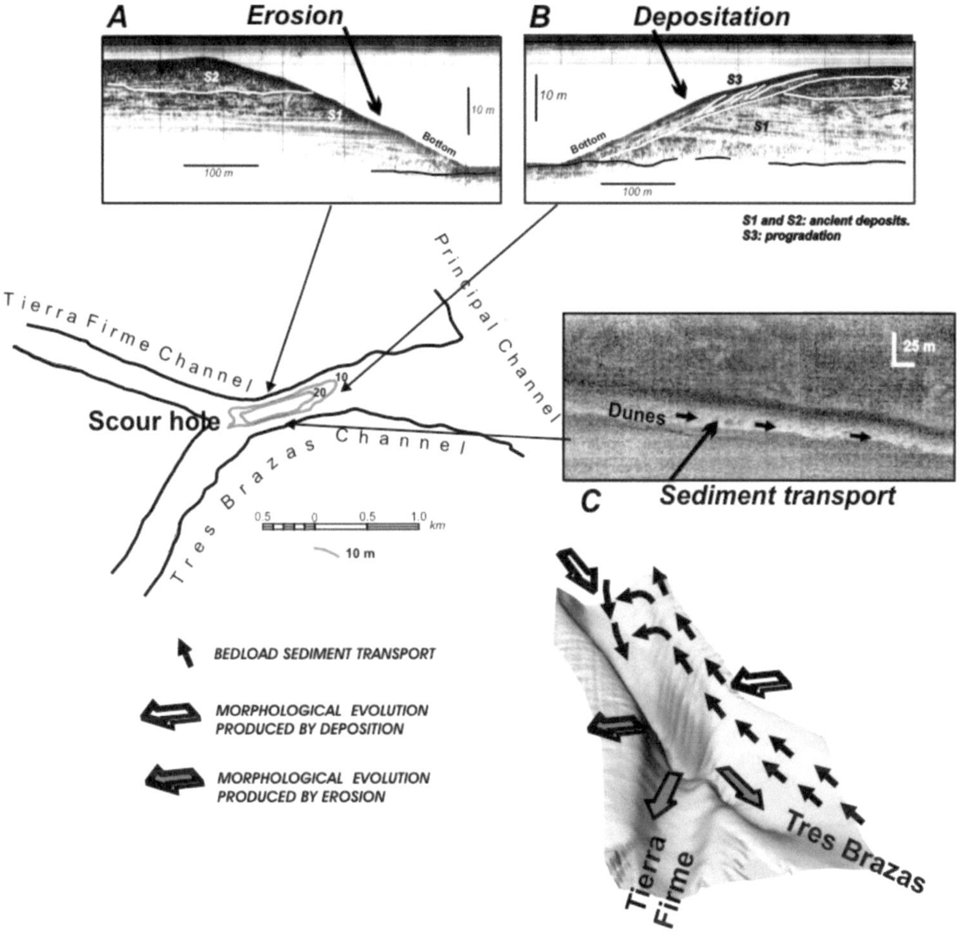

Fig. 8. Bedload sediment transport at confluence zone of Tres Brazas-Tierra Firme channels. A and B: seismic records, C: side scan sonar record.

Vecchi (2009), studied the confluence of La Lista-El Alambre channels, and found an analogous sediment transport pattern as described by Ginsberg and colleagues. She has been estimated the movement path of sediment as bedload at surrounding the confluence zone of these channels through analysis and interpretation of bedforms and seismostratigraphic sequences (Fig. 9). However, she have been observed an important sediment accretion along the steeper slope area of the hole, suggesting that a high sediment supply derived from the inner sectors of La Lista and El Alambre channels (Fig. 9, A), by ebb- dominance, is responsible for redistribution of sediment in these places. Her investigation has documented the existence of high-continuity parallel and sub-parallel reflectors that characterize the sub-bottom sedimentary sequences, with onlapping fill configuration at the flanks of the hole. Erosive sequences are obvious towards the lower parts of the sedimentary pile, indicating that contemporary scour processes continue to rework the bottom sediments of hole and the deeper sector of El Alambre mouth. The bedforms reveal the change in the current direction, which is aligned closely to the change in the topography and explain the ebb dominance. Therefore, the principal sediment transport pathway is to bypass through the thalweg towards outside channel, although amount sediment considerable is also transported and deposited over the flanks and around the periphery of the hole (Fig. 9, B). The results highlight the significant control of higher ebb velocity from the El Alambre channel over the bed morphology of the hole, since it is curved abruptly for aligned with the ebb dominance.

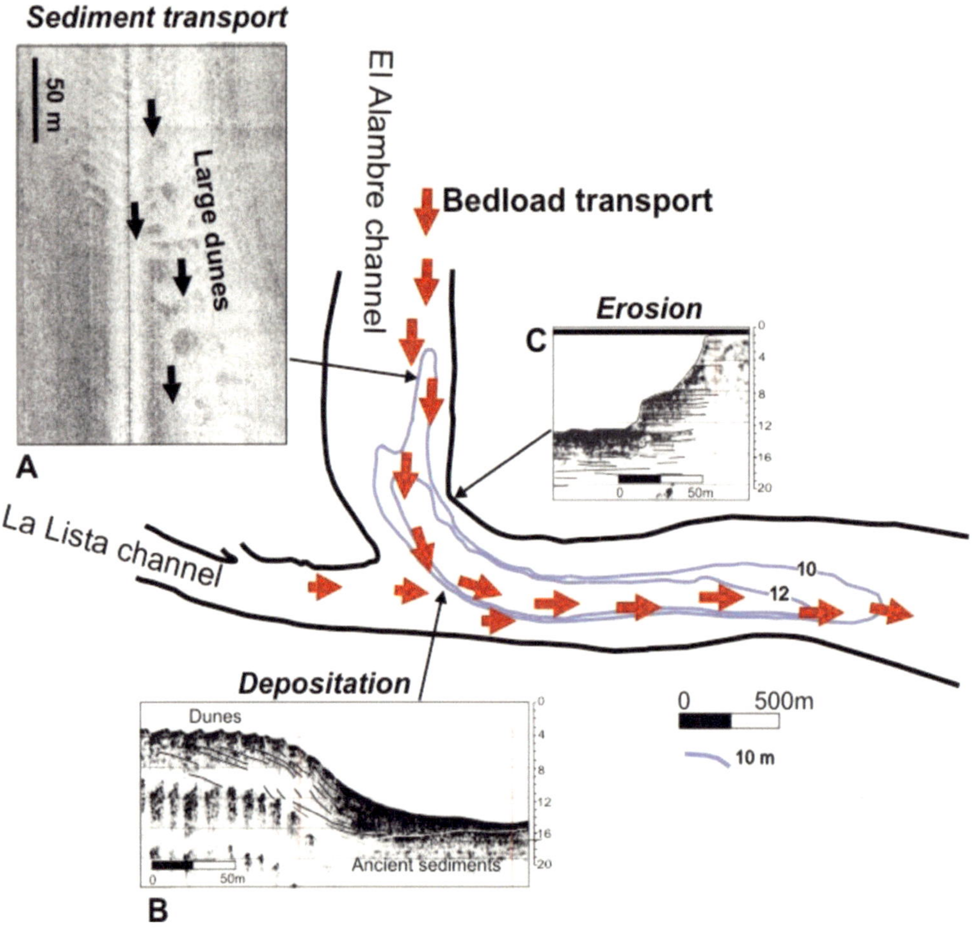

Fig. 9. Bedload sediment transport at confluence zone of El Alambre-La Lista channels. A: side scan sonar record with dunes, B and C: seismic records of the hole slopes

On the other hand, scour hole less deep than the confluence scour may be occur by the interaction of fluid flow with a natural obstacle that protrudes above the surrounding seabed as can be a rock. Caston (1979) indicated that these scour marks adjacent to these outcrops have been used to elucidate present bottom-flow regimes in the absence of other data. This process is characterized by the removal of bottom material from around the exposed features induced a sediment transport due to currents (Whitehouse, 1998; Vijaya Kumar et al., 2003; Myrhaug & Rue, 2005; Dey & Barbhuiya, 2006; Callagway et al. ,2009). According to Ram Babu et al. (2002), the erosion is produced when drag and lift forces of the flow are higher than the gravitational, frictional and cohesive forces that support the join sediment particles. Natural structures, such as rock basement exposed, constitute topographic highs, where significant scour occur around these highs suggesting that flows are diverted by the highs and erosion removes sediment preferentially close to the elevations. Like so, Ginsberg & Aliotta (2009) have been observed a high topographic in the Cabeza de Buey channel associated with a topographic control of the underlying basement at the bed of this channel (Fig. 10). The hydrodynamics at Cabeza de Buey channel (Fig. 1) is characterized by no or very low bed-material transport within this channel. The suspended sediment concentrations are generally high; the particles are fine, cohesive, and prone to flocculate. Therefore, the net bedload transport becomes less important and promotes suspended load transport. In consequence, as no bedforms were present along the channel, scour hole linked to this topographic irregularity will be taken into account to provide information about dominant bottom flow. In this sense, through characterization of seismic

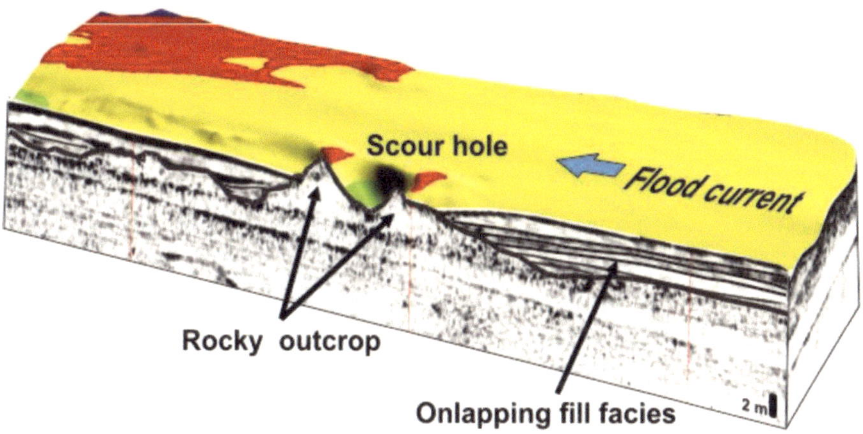

Fig. 10. 3D view of a rocky outcrop in Cabeza de Buey channel. The sub-bottom is a seismic record and the seabed was determined with multibeem survey.

facies and tidal currents data around this obstacle, these authors, have inferred the bottom sediment transport pathways. The deepest part of the scour occurs immediately upstream of the exposed bedrock, inward the channel. During ebb period within the hollow, the strongest currents form eddies which occurring at the deeper side. In contrast, the sediment deposition occurs on the flank of the exposed bedrock located towards the outflow channel, being characterized by the sedimentary sequences with onlapping fill configurations (Fig. 10). This provides an insight of the sediment transport around the exposed rock resulting in sediment removal by ebb dominance towards inner side of the channel and deposition of sediment occurring towards outer side of channel by flood dominance. Therefore, the hollow orientation around the exposed rock suggests that the ebb current affects removing sediment, while the flood current mobilizes a small amount material of bottom towards inside the channel. Ginsberg & Aliotta (2009), based on the analysis of flow, sedimentary structures and sedimentation-erosion patterns around of this high topographic suggest that, although there are less mobilized bedload material, the sediments would be transported inward by flood dominance.

Another bedform which can use to predict the pathways of sediment transport and are produced by bedload movement is the shoals-mouth. These sedimentary corps at the channels may be considered as terminus features of sand transport in the system as consequence of the ebb current dominance. A result of the sediment convergence towards single channel is the development at mouth of the channels an elongate shoal, indicating that a greater percentage of total sediment load is mobilized throughout this course by the ebb tidal, contributing to its formation.

In general, sand shoal present in large channel mouths are aligned parallel to the general direction of the strong tidal currents. They have an elongated shape and are submerged during high tide and exposed partially at low tide. In plan view, they are curvilinear and have asymmetrical cross-sections. Large and medium 3D dunes are present all over the shoals. Particularly, at tidal channel system of Bahía Blanca the shoal-mouth (Fig. 11) have this type of dunes with heights lower than 1m and wavelengths between 6 and 18 m. The morphological evolution of this elongated

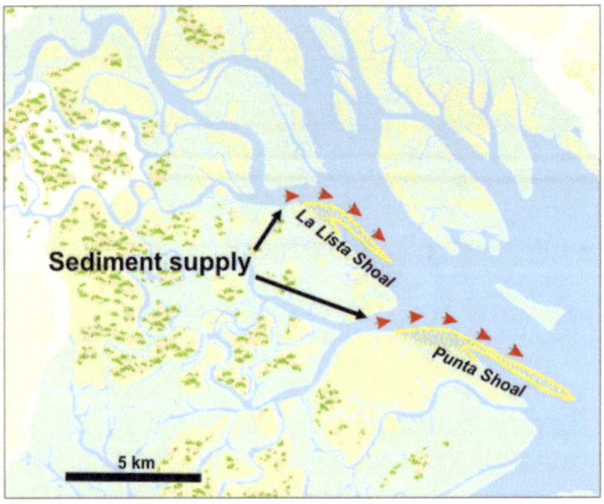

Fig. 11. Sand shoal mouth at large tidal channels. They are result of higher sediment supply.

sedimentary body is related to a residual sediment transport model with trajectories opposite to its two flanks. This process leads to a gradual increase in the height and longitudinal growth of the shoal. In the case of an interconnected tidal channels system, the convergence of sediment towards an only one channel exerts an important effect on the hydrodynamics since plays a significant role in forming sand deposits at the channel mouth due to an increasing in availability of sediment.

4. Conclusion

The sediment movement pattern as bedload in an interconnected channels system is a consequence of tidal currents combined with the system configuration, which result a net sediment transport through single channel. From this study seems more plausible that an ebb increase velocity combined with a short duration, will result in ebb dominance causing sediment loss from each channel. However, sediments are internally redistributed to be later mobilizing as bedload through only one channel. In Figure 12 can see that at Bahía Blanca estuary, the La Lista channel who makes up Cabeza de Buey-El Alambre-La Lista- Principal system, and Tres Brazas channel, which constitutes Tierra Firme-Tres Brazas-Principal channel system, are the channels where the sediment converges, before to be disperse into Principal channel. Additionally, not only the net transport as bedload will be controlled by ebb dominant at channel but it also will depend on whether the sediment availability at each channel is significantly high. Specially, in absence of data, field observations as bedform, point bar and other topographic and seismic-stratigraphic characteristics development in tidal channels are useful diagnostic features to providing an interpretation and predicting sediment mobilization throughout the interconnected tidal channel system.

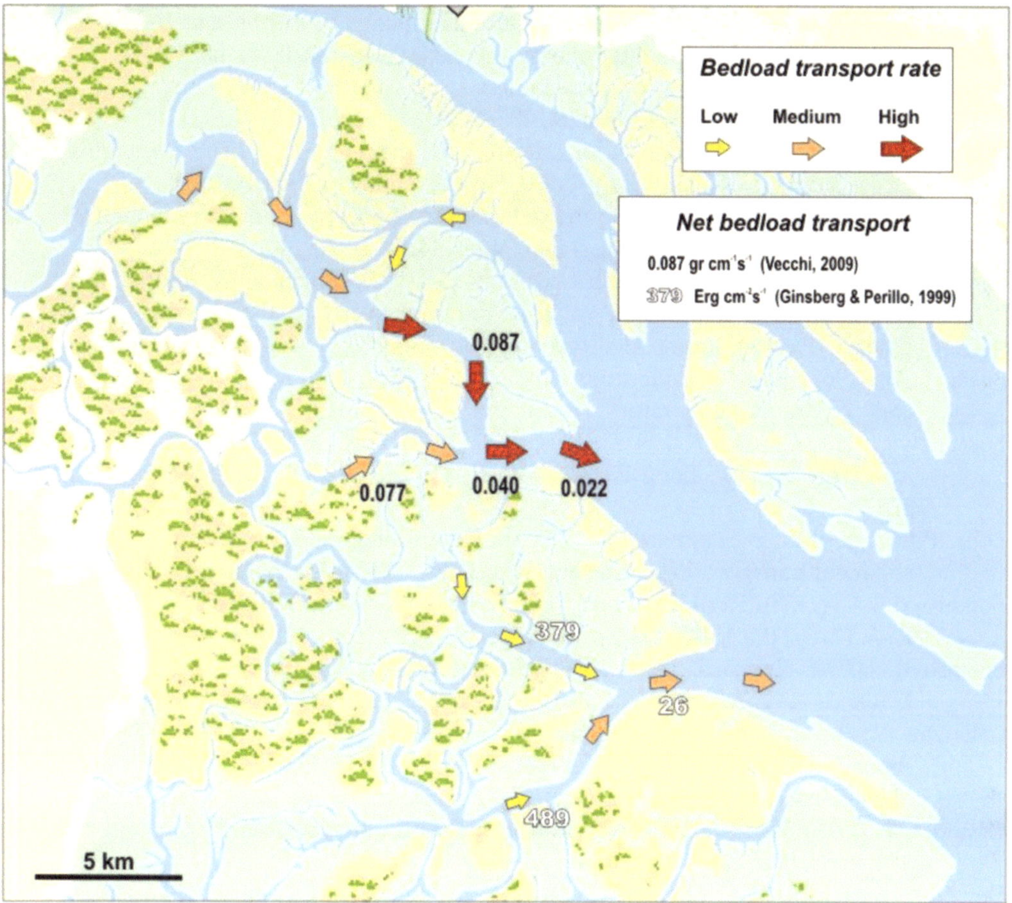

Fig. 12. Circulation pattern of sediment transport as bedload obtained of morphological, seismic and currents integration data. The arrows indicate the estimation of sediment transport rate according to availability sediment. The numbers indicate the flow capability for sediment transport.

5. References

Ackers, P. & White, W. R. (1973). Sediment transport: new approach and analysis. Jour. Hydraul. Div., Proc. ASCE, Vol. 99, HY 11, 2041-2060.

Aliotta, S. & Farinati, E. (1990). Stratigraphy of Holocene sand-shell ridges in the Bahía Blanca Estuary, Argentina. Marine Geology. 94, 353-360. ISSN: 0025-3227.

Aliotta, S.; Lizasoain, G. & Ginsberg, S.S. (2004). Dinámica sedimentaria y evolución morfológica de un profundo canal del estuario de Bahía Blanca. Rev. Asoc. Geol. Argent., 59(1), 14-28. ISSN 1851-8249

Allen, J.R.L. (1968a). Current Ripples: Their relation to patterns of water and sediment motion. North-Holland Publishing Company, Amsterdam.

Allen, J.R.L. (1968b). The nature and origin of bed form hierarchies. Sedimentology, 10, 16 - 182. ISSN 0037-0746

Ashley, G.M. (1990). Classification of large-scale subaqueous bedforms: a new look at an old problem. Journal Sedimentary Petrology, 60 (1), 160-172. ISSN: 0022-4472.

Bagnold, R.A. (1956). The flow of cohesionless grains in fluids. Proc. Royal Soc. Philos. Trans., London. Vol. 249, Series A, 235-297.

Bagnold, R. A. (1963). Mechanics of Marine Sedimentation. In: The Sea: Ideas and Observations, M.N. Hill (Ed.), Vol. 3, 507-28, Wiley Interscience, NY.

Bagnold, R. A. (1966). An approach to the sediment transport problem from general physics. U.S. Geological Survey Professional Paper, 442-1, 1-37.

Berné, S., Castaing, P., Le Drezen, E. & Lericolais, G. (1993). Morphology, internal structure, and reversal of asymmetry of large subtidal dunes in the entrance to Gironde Estuary (France). Journal Sedimentary Petrology, 63(5), 780-793. ISSN: 0022-4472.

Bijker, E. W. (1967). Some considerations about scales for coastal models with movable bed. Delft Hydraulics Laboratory, No. 50, pp. 142, Delft, The Netherlands.

Bokuniewicz, H.I.; Gordon, R.B. & Kasteus, K.A. (1977). Form and migration of sand waves in a large estuary, Long Island Sound. Marine Geology, 24, 185-199. ISSN: 0025-3227.

Bolla Pittaluga, M. & Seminara, G. (2003). Depth-integrated modeling of suspended sediment transport. Water Resour. Res., 39(5), 1137, 1-11. ISSN: 0043-1397

Boothroyd, J. C. (1985). Tidal Inlets and Tidal Deltas.In: Coastal Sedimentary Environments, R. A. Davis, (Ed.), 445-532, Springer-Verlag, New York, NY.

Boothroyd, J.C. & Hubbard, D.K. (1975). Genesis of bedforms in mesotidal estuaries. In Estuarine Research, Cronin, L. E. (Eds.), 217-234, London: Academic Press. New York.

Callaway A.; Smyth J.; Brown C.J.; Quinn R.; Service M. & Long D. (2009). The impact of scour processes on a smothered reef system in the Irish Sea. Estuarine, Coastal and Shelf Science, 84, 409–418. ISSN: 0272-7714.

Caston, G.F. (1979). Wreck marks: indicators of net sand transport. Marine Geology. 33, 193–204. ISSN: 0025-3227.

Dey, S. & Barbhuiya, A.K. (2006). Velocity and turbulence in a scour hole at a vertical wall abutment. Flow Measurement and Instrumentation, 17, 13–21. ISSN: 0955-5986.

Dronkers, J. (1986). Tidal asymmetry and estuarine morphology. Neth. J. Sea Res., 20,117–31. ISSN: 0077-7579.

Einstein, H.A. (1950). The bedload function for sediment transport in open channel flows. Technical Bulletin No. 1026, Soil Conservation Service, US Department of Agriculture, Washington, DC.

Engelund, F. & Hansen, E. (1967). A Monograph on Sediment Transport in Alluvial Stream. Teknisk Vorlag, Copenhagen V, Denmark.

Fagherazzi, S.; Marani, M. & Blum, L.K. (2004). Ecogeomorphology of Tidal Marshes. American Geophysical Union, Coastal and Estuarine Monograph Series, Washington.

Friedrichs, C. T. (1995). Stability shear stress and equilibrium cross-sectional geometry of sheltered tidal channels. J. Coastal Res., 11, 1062–1074. ISSN: 0749-0208.

Gadd, P. E.; Lavelle, J. W. & Swift, D. J. P. (1978). Estimates of sand transport on the New York shelf using near-bottom current-meter observations. Journal of Sedimentary Petrology, 48, 239-252. ISSN: 0022-4472.

Ginsberg, S. S. & Aliotta, S. (2009). Comportamiento de la corriente en un canal de marea controlado por un fondo rocoso. VII Jornadas Nacionales de Ciencias del Mar. Resúmenes CD: 119. Bahía Blanca, Argentina. ISBN 978-987-25479-0-5.

Ginsberg, S.S.; Aliotta, S. & Lizasoain, G. (2009a). Morphodynamics and seismostratigraphy of a deep hole at tidal channel confluence. Geomorphology, 104, 253- 261. ISSN: 0169-555X.

Ginsberg, S. S.; Aliotta, S. & Lizasoain, G. (2009b). Sistema interconectado de canales de marea: estudio de la circulación residual de sedimento como carga de fondo con métodos acústicos. Latin American Journal of Aquatic Research, 37(2), 231-245. ISSN 0718-560X

Ginsberg, S.S. & Perillo, G.M. (1999). Deep-scour holes at tidal channel junctions, Bahía Blanca estuary, Argentina. Marine Geology. 160, 171-182. ISSN: 0025-3227.

Ginsberg, S.S. & Perillo, G.M.E. (2004). Characteristics of Tidal Channels in a Mesotidal Estuary of Argentina. Journal of Coastal Research, 20 (2), 489-497. ISSN: 0749-0208.

Harris, P.T. (1988). Sediments, bedforms and bedload transport pathways on the continental shelf adjacent to Torres Strait, Australia - Papua New Guinea. Continental Shelf Research, 8(8), 979-1003. ISSN: 0278-4343.

Ikehara, K. & Kinoshita, Y. (1994). Distribution and origin of subaqueous dunes on the shelf of Japan. Marine Geology, 120, 75-87. ISSN: 0025-3227.

Kjierfve, B.; Shao, C.-C. & Stapor, F.W. (1979). Formation of deep scour holes at the junction of tidal creeks: a hypothesis. Marine Geology, 33, M9–M14. ISSN: 0025-3227.

Langhorne, D. N. (1981). An evaluation of Bagnold's dimensionless coefficient of proportionality using measurements of sand wave movement. Marine Geology, 43. 49-64. ISSN: 0025-3227.

Lanzoni, S. & Seminara, G. 2002. Long-term evolution and morphodynamic equilibrium of tidal channels. J. Geophys. Res., 107 (C1), 3001,1-13. ISSN 0148-0227.

Lizasoain, G.O. (2007). Estudio geomorfológico submarino en un ambiente estuarial mediante sistemas acústicos. Thesis, Universidad Nacional del Sur (inedited), 245 p., Bahía Blanca, Argentina.

Madsen, 0. S. & Grant. W. D. (1976). Sediment transport in the coastal environment. Massachusetts Institute of Technology, Ralph M. Parsons Laboratory of Water Resources and Hydrodynamics. Rep. 209, 105 p.

Marsset, T.; Tessier, B.; Reynaud, J.Y.; De Batist, M. & Plagnol, C. (1999). The Celtic Sea Banks: an example of sand body analysis from very high-resolution seismic data. Marine Geology, 158, 89- 109. ISSN: 0025-3227.

Meyer-Peter, R. & Müller, R. (1948). Formulas for bed-load transport. Proc. of the 2nd Meeting International Association of Hydraulic Research, pp. 39– 64, Stockholm, Sweden,.

Mitchum, J.R.; Vail, R.M. & Sangree, P.R. (1977). Seismic Stratigraphy and Global Changes of Sea Level, Part 6: Stratigraphic Interpretation of Seismic Reflection Patterns in Depositional Sequences. In: Seismic Stratigraphy applications to hydrocarbon exploration. Payton Ch. E. (Ed.). (26), 117-134, American Association of Petroleum Geologits, Memoir, Tulsa.

Montesarchio, L.A. & Lizasoain, W.O. (1981). Dinámica sedimentaria en la denominada ría de Bahía Blanca. InstitutoArgentino de Oceanografía. Contribución Cienífica, No 58, 202p. Bahía Blanca, Argentina.

Myrhaug, D. & Rue, H. (2005). Scour around group of slender vertical piles in random waves. Applied Ocean Research, 27, 56–63. ISSN: 0141-1187.

NEDECO-ARCONSULT (1983). Estudio de dragado al canal de acceso al puerto de Bahía Blanca. Technical Report. 3 Volumes, Buenos Aires, Argentina.

Perillo, G.M.E. & Piccolo, M.C. (1991). Tidal response in the Bahía Blanca Estuary. Journal of Coastal Research, 7, 437-449. ISSN: 0749-0208.

Ram Babu, M.; Narasimha Rao, S. & Sundar, V. (2002). A simplified instrumentation for measuring scour in silty clay around a vertical pile. Applied Ocean Research, 24, 355–360. ISSN: 0141-1187.

Schuttelaars, H.M. & de Swart, H.E. (2000). Multiple morphodynamic equilibria in tidal embayments. J. Geophys. Res., 105:24105–18. ISSN 0148-0227.

Swift, D.J.P. & Freeland, G.L. (1978). Sand waves and Current lineations on the inner shelf, Middle Atlantic Bight of North America. Mar. Geol., 48, 1257-1266.

Tambroni, N.; Bolla Pittaluga, M.& Seminara, G. (2005). Laboratory observations of the morphodynamic evolution of tidal channels and tidal inlets. J. Geophys. Res., 110, 1-23. ISSN 0148-0227.

Tessier, B.; Corbau, C.; Chamley, H. & Auffret, J.P. (1999). Internal Structure of Shoreface Banks Revealed by High-Resolution Seismic Reflection in a Macrotidal Environment (Dunkerque Area, Northern France). Journal of Coastal Research, 15(3), 593-606. ISSN: 0749-0208.

Van Rijn, L. C. (1993).Principles of sediment transport in rivers, estuaries and coastal seas. Aqua Publications (Eds.), Amsterdam, Netherlands.

Vecchi, L.G. (2009). Comportamiento morfodinámico e hidrosedimentológico de un sistema de canales mesomareales delestuario de Bahía Blanca. Thesis, Universidad Nacional del Sur (inedited), 155 p., Bahía Blanca, Argentina.

Vecchi, L.; Ginsberg, S.S. & Aliotta, S. (2008). Banco arenoso en un ambiente estuarial: Rasgos de fondo y dinámica sedimentaria. Revista de la Asociación Geológica Argentina, 63(3), 395-406. ISSN 1851-8249.

Vijaya Kumar, A.; Neelamani, S. & Narasimha Rao, S. (2003). Wave pressures and uplift forces on and scour around submarine pipelines in clayey soil. Ocean Engineering, 30, 271–295. ISSN: 0029-8018.

Watanabe, A. (1982). Numerical models of nearshore currents and beach deformation. Coastal Eng. Japan, JSCE, Vol. 25, 147-161. ISSN: 0578-5634.

Whitehouse, R. (1998). Scour at Marine Structures. Thomas Telford Publications, ISBN 0 7277 2655 2, London, UK.

Yalin, M. S. (1963). An expression for bedload transportation. J. Hydraul. Div., Proc. ASCE (HY3), Vol. 89, 221-250.

Zhen-Gang J. (2008). Sediment transport, In: Hydrodynamics and Water Quality: Modeling Rivers, Lakes, and Estuaries. John Wiley & Sons, Inc. (Ed.), 113-200, ISBN: 978-0-470-13543-3

15

Continuous Monitoring of Suspended Sediment Load in Semi-arid Environments

Tiziana Bisantino, Francesco Gentile and Giuliana Trisorio Liuzzi
University of Bari "Aldo Moro" (Dep. PROGESA)
Italy

1. Introduction

Sediment transport in water courses is an indicator of soil eroded from agricultural land, and the intensity of the phenomenon provides a measure of land degradation and the associated reduction in the global soil resource. Suspended sediment load is a useful indicator for assessing the effects of landuse changes and engineering practices in watercourses.
The investigation of the trend in the sediment loads has different constraints in terms of available data. Sediment load data are lacking for rivers in many areas of the world, particularly in developing countries where changing sediment yields might be expected (Walling & Fang, 2003).
Continuous river monitoring is essential to effectively measure suspended sediment loads during storm events and to accurately describe the sediment transport dynamic. The paper focuses on the application of technologies for continuous monitoring of suspended sediment concentration in rivers in semi-arid environments.
In the first part of the paper, methods for measuring suspended sediment concentration are reviewed. Technologies to continuously monitor suspended sediment overcome traditional methods requiring routine collection and analysis of water samples. Among the available instruments, based on optical principles, pressure difference and acoustic backscatter principles, the optical technology largely spread as turbidity is considered a good "surrogate" for suspended sediment (Gippel, 1995; Lewis, 1996; Lenzi & Marchi, 2000; Seeger et al., 2004;) and particularly suitable for high suspended concentrations (López-Tarazón et al., 2009; Gentile et al., 2010). It provides reliable data when the point measurements can be correlated to the river's mean cross section concentration value, the effects of biological fouling can be minimized, and the concentrations remain below the sensor's upper measurement limit (Gray & Gartner, 2009).
In the second part of the paper the continuous monitoring of the suspended sediment concentration in a semi-arid watershed is used to analyze the sediment transport dynamics. In semi-arid areas the seasonality of the hydrological processes and the strong interannual variation in precipitation rates enhance the role of infrequent flood events (Soler et al., 2007). As a consequence suspended sediment concentrations in rivers are generally high, as they compensate for the infrequency of runoff in producing high annual unit sediment yields (Walling & Kleo, 1979; Alexandrov at al., 2007; Achite & Ouillon, 2007). In these areas suspended sediment transport provides problems for water-resource management where channels are impounded as high rates of sedimentation occur in reservoirs.

High sediment yields in semi-arid environments can be explained in terms of the interaction between erosive energy and vegetation density even if climatic seasonality, relief, basin lithology and the extent of human activity combine to influence the global pattern of erosion processes. The estimation of the sediment transport during the events is necessary for the calculation of long-term sediment yields from basins, as one single event may represent the transport of several 'normal' years (Wolman & Gerson, 1978).

A measuring station in the Carapelle torrent (Northern Puglia - Southern Italy) was established in 2007 to continuously monitor suspended sediment concentration. The station was set up at Ordona bridge and delimits a catchment area of 506.2 km^2. It is equipped with a dual function infrared sensor (turbidity/suspended solids), a remote data acquisition system and an ultrasound stage meter for water level monitoring. The sensor was laboratory tested and a field calibration stage was carried out (Gentile et al., 2010).

High temporal resolution data, recorded over a 3-year period (2007-2009), were then analysed to investigate the relationships between suspended sediment concentration (SSC) and discharge (Q) in flood flows. During intense flood events the concentrations of sediment are high, more than 20 g/l. The concentration of suspended sediment varies hysteretically with water discharge and there are group-types of response. The SSC-Q curves revealed that in the Carapelle stream clockwise, counterclockwise, mixed loops and also no hysteresis are possible and can be related to the event intensity and to the sediment availability. The unsteadiness of the flow mainly influence the sediment transport capacity of the rising and falling limb of the hydrograph. Sediment load generally closely depends on flood volume while the maximum suspended sediment concentration is correlated with peak discharge.

2. Measurement of suspended sediments in water courses

2.1 Suspended sediment sampling

Different technologies are available to estimate suspended sediment concentrations in water courses (Garcia, 2008). Sediment measurement can be carried out sampling the water-sediment mixture to determine the mean suspended sediment concentration and the particle size distribution. The different sampling equipments can be summarized in: single-stage, point-integrating, depth-integrating and pumping type samplers.

If the suspended sediment is uniformly distributed within a stream cross section, the suspended sediment concentration can be simply determined by means of a *single-stage* sampler. It consists of a container (isokinetic or not) that is deployed at a specific flow depth and instantaneously collects the samples. This method allows a reasonably estimation of the sediment concentration at the sampled point in small streams and under slow flow conditions. Rooseboom and Annandale (1981) showed that sampling in bottles generally gives concentrations about 25% lower than results obtained from more sophisticated techniques.

Generally suspended sediment concentration in natural streams varies in time and space, from the water surface to the streambed and laterally across the stream. Concentration increases from a minimum at the water surface to a maximum at or near the streambed. In order to take into account this variability *point-integrating* and *depth integrated* samplers can be used.

Point-integrating samplers take the sample over an extended period of time to determine the average concentration at a point. A point integrating sampler has a nozzle that points directly into the streamflow. An electrically valve mechanism is used to start and stop the sampling process. To eliminate a sudden inrush after opening of the intake nozzle, the air pressure in the bottle is balanced with the hydrostatic pressure before opening of the valve. The time sampling can vary from 1 to 30 minutes.

The *depth integrated* sampler is lowered at a uniform rate from the water surface to the streambed, instantly reversed, and then raised again to the water surface. The sampler continues to take its sample throughout the time of submergence. At least one sample should be taken at each vertical selected in the cross-section of the stream. A clean bottle is used for each sample. This type of samplers collects velocity-weighted samples.

Important storm flows are infrequent and difficult to predict, and when they occur, trained personnel may not be available to collect the required information. As a consequence automated data collection is essential to effectively capture such events. *Automatic samplers* pump a small sample into a series of bottles, at predetermined times and intervals, triggered by predetermined flow (Pavanelli & Bigi, 2005) or turbidity conditions (Lewis, 1996).

2.2 Acoustic backscatter systems

The acoustic backscatter (ABS) instruments transmit high frequency ultrasound beams towards the measurement volume. Suspended sediments scatter a portion of this sound back to the transducer. The strength of the backscattered signal allows the calculation of sediment concentration. Suspended sediments significantly scatter underwater sound at megahertz frequencies (0.5 -5 MHz) and the sediment concentration and the grain size control the backscattered intensity (Thorne & Buckingham, 2004).

The acoustic backscattering method is based on the sonar equation (Medwin & Clay, 1997). Since the acoustic backscatter system is an indirect method of measurement, an inversion algorithm is required for determining sediment concentration with measured backscattered signal strength. The acoustic backscatter equations provide the basis for the development of such an algorithm (Thorne & Meral, 2007).

The backscatter amplitude depends on the concentration, particle size, and acoustic frequency. In recent years, the development of multi-frequency acoustic backscatter systems overcame the uncertainties associated with measurements from single frequency instruments. Several experiments have been made to use the backscattered signal of the Acoustic Doppler Current Profiler (ADCP) for determining the suspended sediment concentrations. The multi-frequencies method using commercial ADCP frequencies is able to investigate concentration and grain size field inside the silt-sand range, with typical acoustic shadow zones (Guerrero & Lamberti, 2008). Similarly, the Acoustic Doppler Velocity Meter (ADV) was used to determine the suspended sediment concentrations.

The acoustic backscatter system requires accurate in-situ calibration and is not suitable for the upper concentration range (>10 g/l). This technique is sensitive to the presence of flocculated materials, air bubbles, phytoplankton and seems to provide less precise results, in terms of concentration, than the optical instruments, when large quantities of fine material are transported (Fugate & Friedrichs, 2002).

2.3 Optical turbidity

Methods for measuring suspended sediments via optical turbidity derive from the pioneering work of Whipple and Jackson around the year 1900 that lead to a candlebased turbidity standard. In 1926, Kingsbury and Clark discovered formazin and improved the consistency in standards formulation (Sadar, 1999). Turbidity measurement standards changed in the 1970's when the nephelometric turbidimeter, or nephelometer, was developed which determines turbidity by the light scattered at an angle of 90° from the incident beam.

In recent years the use of the optical technology for river monitoring largely spread as it is considered a good “surrogate” for suspended sediment (Gippel, 1995; Lewis, 1996). These instruments are based on the theory that an incident beam in a mixture is subject to absorption, scattering and transmission caused by the presence of suspended particles. The light source can vary in wavelength and type. The scattered beam can be measured by one or more electronic photodetectors positioned at $\alpha=90°$ (nephelometric scattering), $\alpha<90°$ (forward scattering) or at $\alpha>90°$ (backscattering) to the direction of the incident beam.

These instruments are sensitive to the grain size distribution of the material. This feature has little relevance in small watersheds because they are generally characterised by homogenous pluviometric regimes and erosion times, but can be significant in large watersheds (Lenzi and Marchi, 2000). Variations in grain size distribution, associated with floods, can lead to nonlinear turbidity/suspended sediment concentration relationships. In practice, the degree of curvature is usually very small and linear models perform quite well in estimating loads (Lewis, 2003).

The interferences associated with optical measuring are the colour of the particles and the watery medium, which absorb light in some bands of the visible spectrum, altering the characteristics of the transmissed light. Further interference is caused by the ratio between the size and shape of the particles and the wavelength of the incident beam: particles of a comparable size to that of the incident beam wavelength transmit light symmetrically, whereas larger particles transmit an irregular light in all directions. The scattered light, the density of the particles and the content of dissolved particles such as pollutants and organic matter can lead to errors of measurement since they interfere with the diffusion of light between the particles and the photodetector (Gippel, 1989; Foster et al., 1992; Sadar, 1998). These interferences are minimized by the ratio detection system that uses a combined system of photodetectors and a specific algorithm that calculates the intensity of the scattered light (Sadar, 1999). Another advantage of this system is that it can measure in a broad range of concentrations, therefore it is suitable for suspended sediment monitoring in different environments.

2.4 Laser diffraction

A new generation of systems for measuring the size distribution and concentration of suspended sediments is based on small-forward angle scattering technology, formerly called laser diffraction. At small angles, the scattered light is composed primarily of light diffracted by a particle. Laser scattering instruments direct a laser beam through the water sample, and the suspended particles scatter, absorb and reflect the beam. The scattered laser beam is received by a ring detector that allows measurement of the scattering angle. Particle size and sediment concentration can be calculated using this angle.

These systems replace old technology optical backscatter sensors or single frequency acoustic sensors which suffer from calibration changes with particle size and composition.

The laser scattering and transmissiometry devices measure scattering of light at multiple angles. Inversion of this data leads to sediment concentration and size distribution in 32 size classes. The size classes span a 200:1 range of particles (1.25 to 250; 2.5 to 500; or 7.5 to 1500 microns). In addition, these systems also measure beam attenuation, which is used to de-attenuate measured multi-angle scattering. Thus, these instruments have the

transmissometer function built in. The detailed size distribution of the laser instrument exceeds in information content that offered by the multi-frequency acoustic. These instruments measure sediment concentration in a range of 10 – 3.000 mg/l and are not suitable for high suspended sediment transport.
This technology can be influenced by few compact aggregates that may scatter the light beam as individual entities, altering the actual size distribution of the material (Pedocchi and Garcia, 2006).

3. Suspended sediment transport in semiarid watersheds

Semi-arid climate describes regions receiving low annual rainfall, having nutrient-poor soils and short-grass or shrublands vegetation. Climate is characterized by extreme variability in precipitation and is subject to droughts and infrequent rainfall periods and subsequent flooding. In these areas actual evapotranspiration represents an important feature and it is recognised as the main hydrologic loss (50-60 % of mean annual rainfall).
In the semi-arid climatic zone hydrological processes are largely variable both in time and space due to the high variability of rainfall regime, in addition to the influence of topography and the spatial distribution of geology, soil and land-use. Suspended sediment load is generally high and it reaches maximum values at the beginning of the flood season and after dry periods.
In semi-arid environments (if gully erosion is not active), sheet erosion is the main source of sediments. In the western United States sediment load of rivers is derived mainly from semiarid watersheds (Wilcox & Wood, 1989).
Vegetation influences soil erosion and agriculture likely represents the dominant cause of catchment disturbance and accelerated erosion in most areas of the world (Walling & Fang, 2003).
About 40% of Spanish territory is seriously affected by erosion. One of these areas, the central Ebro valley in northeast Spain, provides a case-study of climate and human impact on soil erosion. The occurrence of convective storms coupled with the high erodibility of the soils gives rise to very fragile agrosystems in which soil essentially constitutes a non-renewable resource (Navas et al., 1997).
The River Isábena in the Ebro valley drains areas of highly erodible sediments (badlands) that occupy a small portion of the basin but are the main source of sediment. This leads to intense suspended sediment transport during most flood events. Thirty-four flood were monitored using a turbidity probe and different hydrological and sedimentary responses of the cachment to similar rainfall were observed. This variability can be attributed to the sediment availability and to the antecedent soil moisture conditions (López-Tarazón et al., 2010).
In the Eshtemoa basin (Israel) vegetation is sparse, small areas are cultivated for growing winter wheat and the steeper hillslopes are used for grazing. The Yatir forest covers a portion of the basin. An automatic sampling programme was esthablished in the period 1994-1998 and 9 flood events were monitored in the winter seasons. The range of suspended sediment concentration in the Eshtemoa basin covers similar orders of magnitude of other semi-arid and arid-zone drainage basins (Alexandrov et al., 2003). Distinct seasonal differences are identified and related to the prevalence of cellular, convectively enhanced storms in autumn and spring and frontal storms in winter (Alexandrov et al., 2009).
In northwest Algeria Megnounif et al. (2007) investigated the variability and seasonality of suspended sediment transport in the Wadi Sabdou collecting water samples during 19 flood

events (1988-1993). They distinguished two periods of active erosion and high sediment yield. The first in autumn due to the sediment accumulated during the dry summer and the second in spring, at the end of the wet season, when the collapsing of the river banks contributes to increase the sediment transport. During the wet period of winter and spring the relationships between sediment and discharge produces hysteresis mainly in the form of mixed loop.

An attempt to explain the relationship between sediment concentration and discharge was carried out by Benkhaled & Remini (2003) monitoring 13 flood events in the Wadi Wahrane (Algeria). The high intensity and the variability of concentration of sediments is due to seasonal effect and hysteresis.

A summary description of the basins monitored in semi-arid environments is reported in table 1.

River	Basin area (km^2)	N° events	Q_p (m^3/s)	SSC_{max} (g/l)	References
Eshtemoa (Israel)	119	9	0.24-84.4	15.5-186.0	Alexandrov et al. (2003)
Isábena (Spain)	445	34	1.52-78.7	0.1-90.0	López-Tarazón et al. (2010)
Sabdou (Algeria)	256	3*	29.2-106.0	22.5-115.0	Megnounif et al. (2007)
Wahrane (Algeria)	270	13	0.9-200.0	63.0-250.0**	Benkhaled & Remini (2003)
*For three events of the total 19 monitored SSC_{max} and Q_p are available. **The event Q_p=1 m^3/s, SSC_{max}~500 g/l was excluded from the dataset					

Table 1. Basins monitored in semi-arid environments.

4. Monitoring suspended sediment in Southern Italy

The experimental station measuring suspended sediment concentration (fig. 1; tab. 2) is located in the Carapelle torrent (Ordona-Castelluccio dei Sauri bridge). The torrent originates from flyschoid formations of the Daunia Mountains and develops into the alluvial fan of the Tavoliere plain. The plain and the low hilly areas are mainly used for cereal cultivation and olive growing, whereas the higher slopes are occupied by woods and pasture. The climate is typically Mediterranean, with rainfalls ranging from 450 to 800 mm/year and average temperatures ranging from 10 to 16 °C.

The station is equipped with a remote data transmission ultrasound stage meter and a stage recorder. In addition, an infrared optic probe (Gentile et al., 2010) was implemented for the measurement of the suspended sediment concentration (fig. 2). The instrument is housed in a shelter tube through a pulley, a float and a counterweight group. The housing device, anchored to a bridge pier, protects the instrument from the impact of any flowing coarse material and prevents any potential measuring errors caused by incident radiant energy straying into the infrared field. Solar infrared radiation decreases of approximately 63% below 50 mm of clean water, and this information is important particularly in regard to surveys carried out near surface level. The instrument is controlled through a data acquisition system that is power supplied by solar panels.

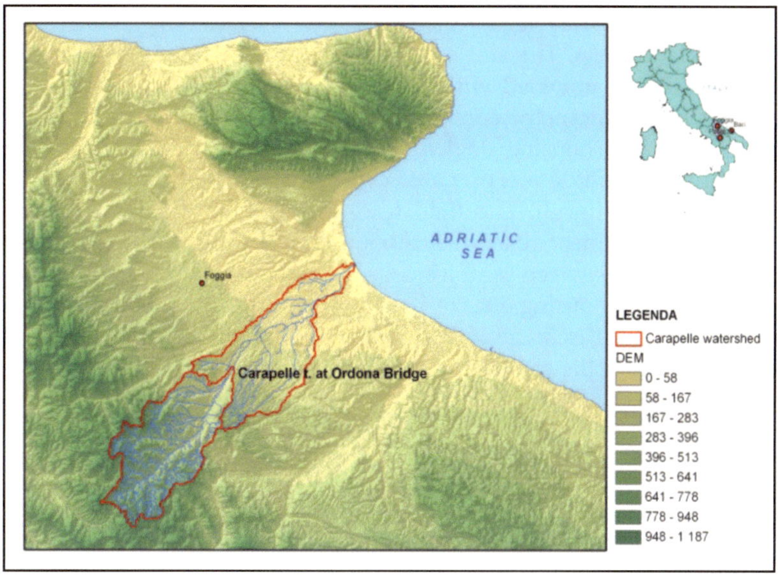

Fig. 1. The Carapelle watershed in Southern Italy

Watershed area	km^2	506.2
Maximum altitude	m a.s.l	1075.0
Average altitude	m a.s.l.	466.0
Minimum altitude	m a.s.l	120.0
Main channel length	km	52.2
Main channel slope	%	1.8
Mean watershed slope	%	8.2

Table 2. Main characteristics of the Carapelle watershed at Ordona bridge.

Fig. 2. The suspended sediment measuring station and the optical device (instrument).

The instrument was laboratory tested using mixtures of different grain size distributions and sediment concentrations. The aim was to determine the existing relationship between the optical data and that obtained with standard methods and to assess the instrument response to different granulometric contents of the mixture. Afterwards, the instrument was field-tested to verify the housing device and the calibration curve of the optical probe (Gentile et al., 2010). This phase was executed collecting 90 samples during the flood seasons in the period 2007-2009 (fig. 3). The field testing confirmed the results achieved in the laboratory, evidencing a linear relationship between the optical and the gravimetric data (fig. 4). The comparison between samples collected inside and outside the tube allowed to verify that the instrument housing does not interfere with the measuring process.
The concentration and the discharge of 27 flood events, monitored at half-hourly scale, were then considered to analyze the sediment transport dynamic (tab. 3).

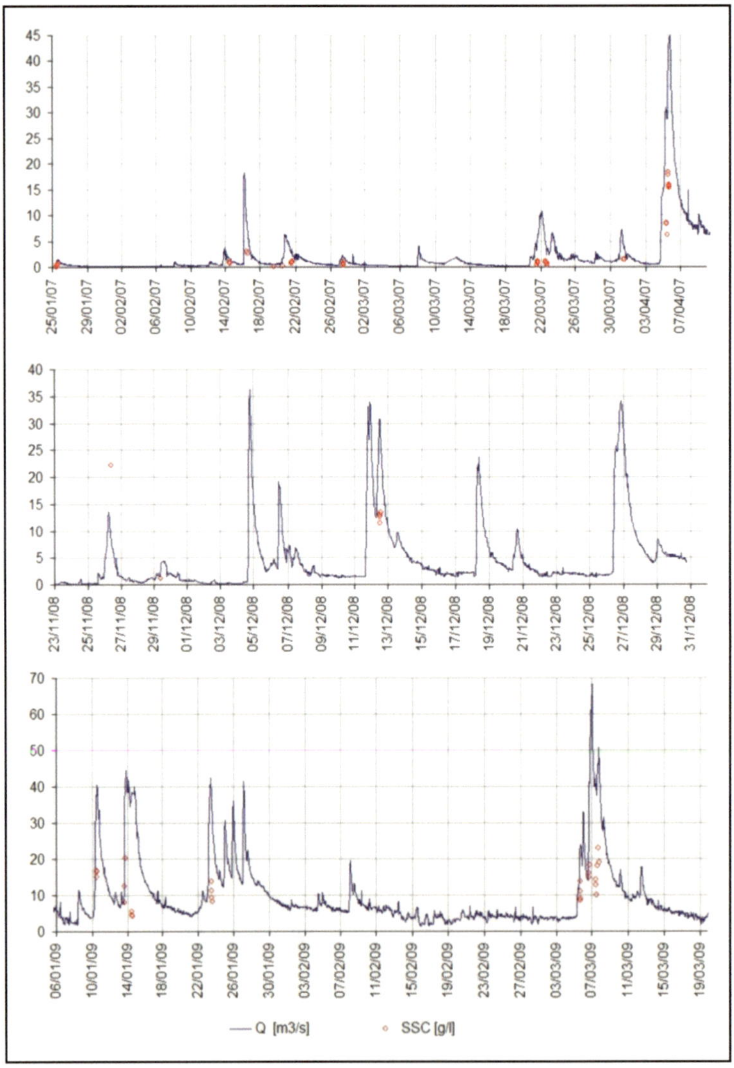

Fig. 3. Sampling of suspended materials during the flood seasons 2007-2009.

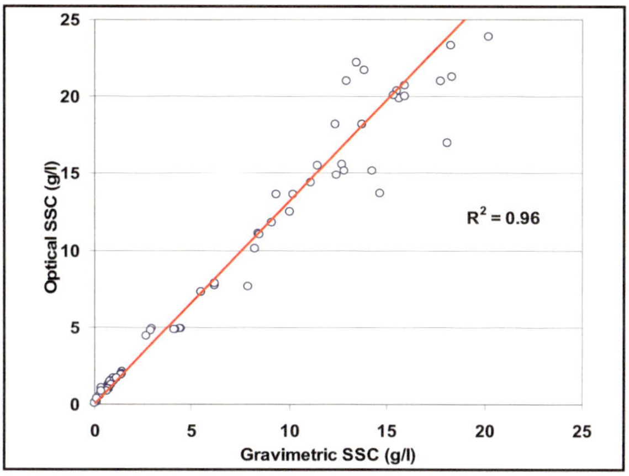

Fig. 4. Field calibration of the optical probe.

Event	Q_p (m^3/s)	Flood volume (m^3)	SSC_{max} (g/l)	Sediment load (t)
27-Feb-07	2.3	167481	2.1	202.7
7-Mar-07	4.1	188457	2.3	232.4
21-Mar-07	10.7	974542	2.5	2184.1
24-Mar-07	6.7	484555	1.4	462.2
28-Mar-07	14.4	600895	3.7	1097.0
31-Mar-07	7.3	470471	2.1	593.0
5-Apr-07	44.9	5551463	20.7	36124.0
9-Apr-07	10.2	54442	9.5	1279.5
23-Jan-08	1.6	187033	2.9	358.0
6-Mar-08	26.7	4076370	18.5	18231.6
6-Dec-08	19.1	987512	32.8	11336.1
18-Dec-08	23.8	1335547	26.6	13064.8
20-Dec-08	6.2	672826	4.9	1997.0
26-Dec-08	34.2	2679841	17.4	25503.3
8-Jan-09	11.6	917056	19.2	5155.6
10-Jan-09	40.6	4133587	22.4	23694.0
14-Jan-09	44.6	6561790	41.2	59652.7
23-Jan-09	42.6	3870310	14.8	21959.2
8-Feb-09	19.4	1418963	28.8	10137.1
5-Mar-09	33.0	1489752	43.0	28208.6
7-Mar-09	68.4	3947932	34.7	87947.0
10-Mar-09	17.1	730348	7.8	2131.3
12-Mar-09	17.7	756934	5.78	2726.8
20-Mar-09	69.2	4014736	20.45	43746.0
22-Mar-09	33.0	2779452	7.77	9879.0
24-Mar-09	21.7	3298477	7.77	10867.6
2-Apr-09	22.9	1630998	27.75	14973.4

Table 3. Hydrologic characteristics of the flood events monitored in the Carapelle torrent.

5. Results

During the monitoring seasons 2007-2009 twenty-seven events, having continuous data and absence of anomalies in the time series, have been selected. Most of the relationships between sediment concentration and discharge (fig. 5), registered at half-hourly scale, revealed the existence of hysteresis (clockwise, counterclockwise, mixed-shaped). This means that the relationship between suspended sediment concentration and discharge is quite different for the rising limb of the hydrograph than for the falling one. The tendency for sediment concentration to have different values at identical stream discharges is the primary drawback to the application of a rating curve during a storm flow. In some cases there can be the absence of hysteresis as shown in figure 5.

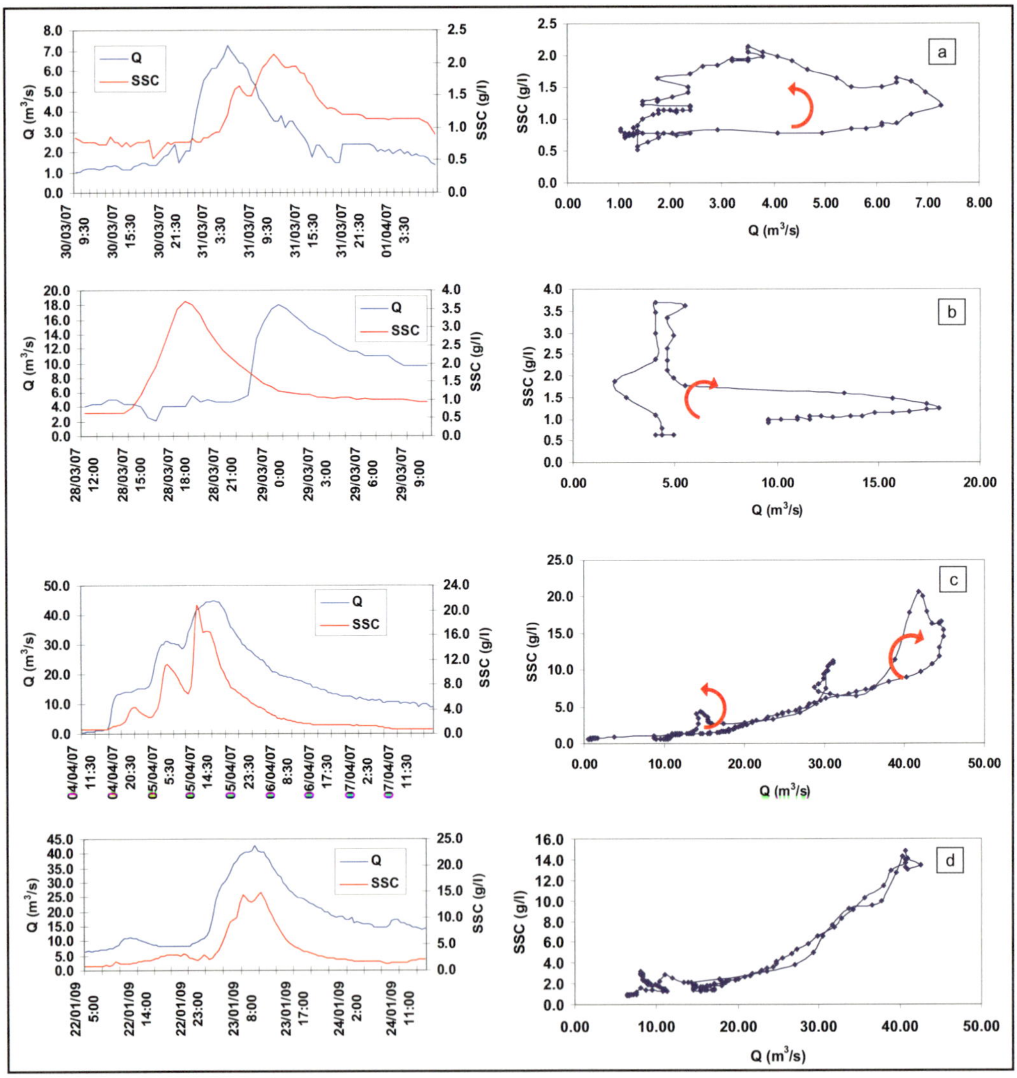

Fig. 5. Types of relationships between sediment concentration and discharge during flood events in the Carapelle torrent.

Counterclockwise loops can be observed during all the flood season and prevail in the moderate events. The sediment concentration increases in the falling limb of the hydrograph due to the distance between the sediment sources (mainly hillslopes) and the measuring station (Brasington and Richards, 2000).
Clockwise loops occur during the most intense events and the sediment concentration start to decrease in the rising limb of the hydrograph. This type of response indicates that sediments delayed in the riverbed contribute to sediment transport (Williams, 1989) but their availability decreases during the event. This type of loop frequently occurs at the end of the rainy season as it is influenced by the depletion of sediments produced by the previous floods (Campbell, 1985).
Mixed-shaped loops occur in multi-peaked floods, for example when an intense flood immediately follows a moderate flood. This kind of loop can be generated also during the steady states of the flow (De Sutter et al., 2001).
When the absence of hysteresis is observed the sediment transport capacity is maximum during the flood.
The unsteadiness of the flow influences sediment transport as the turbulence intensity can change during a flood. The streamwise and vertical components of the turbulence are generally larger in the rising limb of the hydrograph than in the falling one. The increased intensity in the rising limb can be caused by the flood wave overtaking the base flow. The turbulence intensity decreases immediately in the falling limb, having the minimum value in the middle of this limb, and then increases to the value for a steady flow condition. High turbulence intensities maximizes the transport capacity and the sediment concentration is highest when the duration of the rising limb of the hydrograph is short (De Sutter et at., 2001). When the decrease of the turbulence occurs the sediment transport capacity reduces in considerable way. This phenomenon has been observed in some floods of the Carapelle (fig. 6): when the flow becomes stable and the turbulence reduces an abrupt decrease of the values of the concentration occurs.

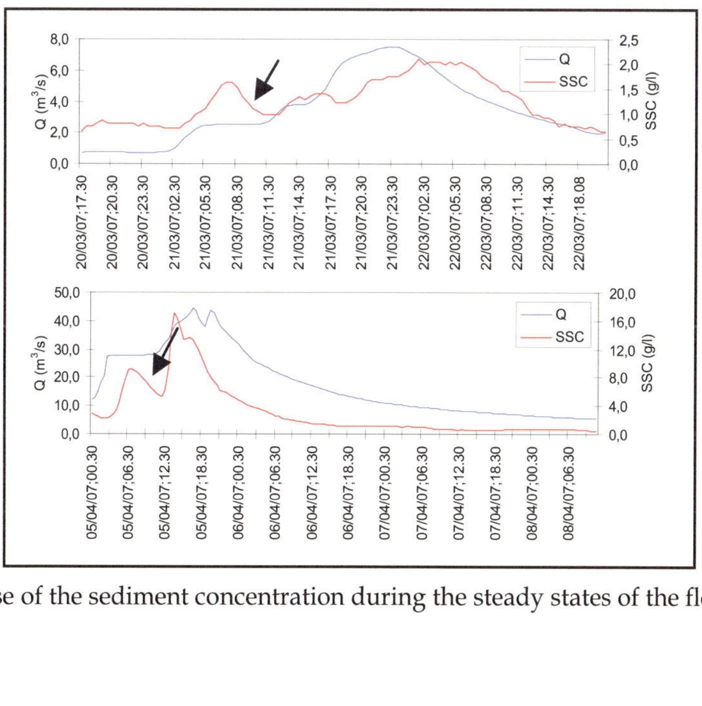

Fig. 6. Decrease of the sediment concentration during the steady states of the flow.

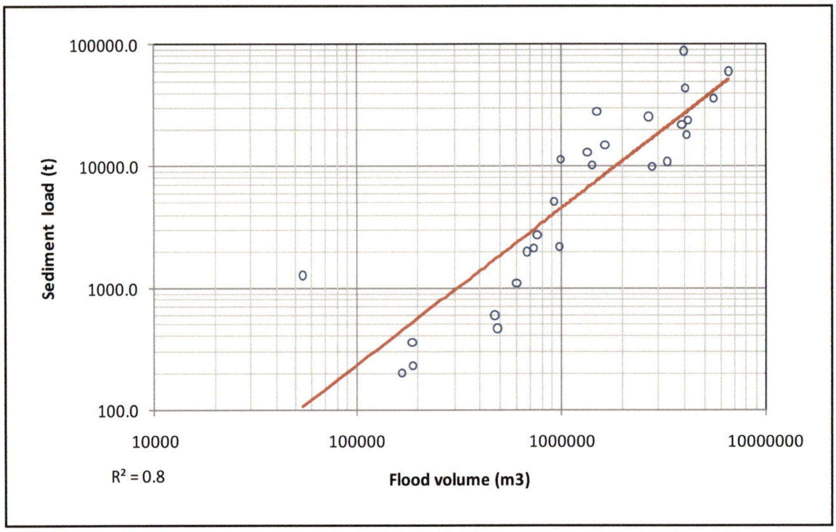

Fig. 7. Sediment load *vs* flood volume

Due to the hysteresis effect the regression scatter of the rating curve SSC = aQ^b between suspended sediment concentration SSC and discharge Q (a and b are empirically derived regression coefficients) is often great for the same river. As a consequence the dependency of the sediment load from the flood volume and of the SSC_{max} from Q_p was assessed. The best-fit regression relationship (fig. 7) between the sediment load [t] and the flood volume [m^3] is the power-law relationship ($r^2 = 0.8$):

$$y=10^{-4}\,x^{1.3} \qquad (1)$$

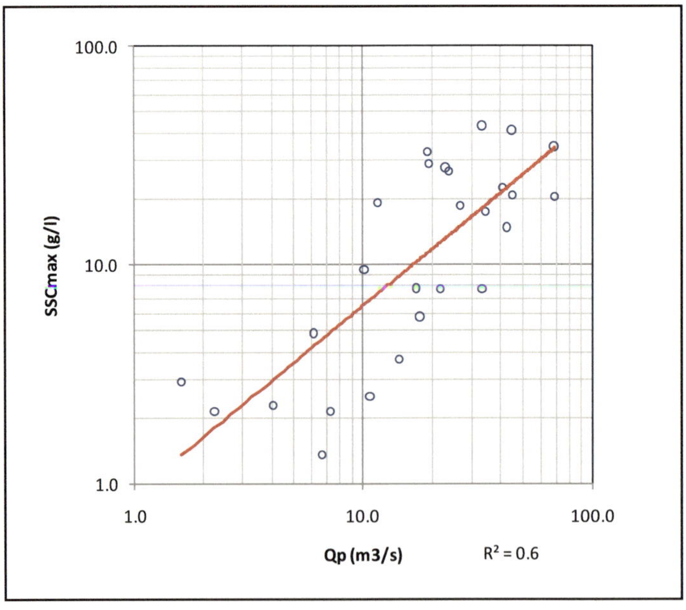

Fig. 8. Maximum suspended sediment concentration *vs* peak discharge.

The SSC_{max} reached during the events has a good correlation with the peak discharge. The r^2 =0.6 means that the variation of SSC_{max} are explained up to 60% by those of Q_p and by 40% by other factors such as the sediment supply, rainfall intensities and land use variations. The best-fit regression relationship (fig. 8) between SSC_{max} and Q_p is the power-law relationship:

$$y=0.89\, x^{0.86} \tag{1}$$

The time delay T between SSC_{max} and Q_p registered during the events varies with the flood intensity (Gentile et al., 2010). Plotting the time delay T versus peak discharge Qp a decreasing trend was observed (fig. 9). Positive values of T indicate that SSC_{max} follows the peak discharge and the hysteresis is counter-clockwise; negative values of T indicate that SSC_{max} occurs before the peak discharge and the hysteresis is clockwise. A time delay equals to zero defines the threshold at which discharge exerts a strong influence on the sediment load and determines the advance of SSC peak value.

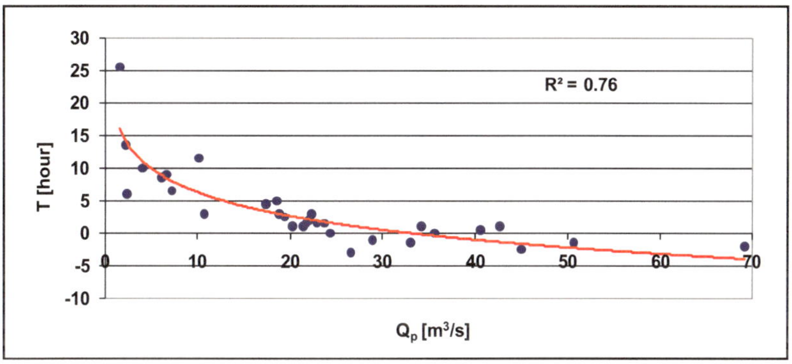

Fig. 9. Time delay T between SSC_{max} and peak discharge *vs* Q_p.

On the basis of the previous results, the variability of the SSC_{max} values among the events was evaluated dividing the floods in moderate events (Q < 30 m^3/s) and intense events (Q > 30 m^3/s). Some considerations on the hydrological response of the watershed are reported.

Three moderate events of the season 2007 (fig. 10), with increasing peak discharges, show similar maximum SSC values. At the same time the moderate flood of 5 Mar 2009 (fig. 11) has higher concentrations than the following having comparable intensities.

On the other hand the intense event of the season 2009 (fig. 12) has markedly higher sediment concentrations than the following floods that show increasing peak discharges.

The reasons of such behaviour can be attributed to the presence in the first flood also of material coming from the river bed. The sediment concentrations mobilized during the following floods mainly derive from hillslopes and a best correlation between SSC_{max} and Q_p can be found.

This can be referred to the sediment depletion from hillslopes that influenced the last events. The relationship between the maximum SSC and Q_p at event scale was evaluated considering the experimental values of the Carapelle torrent and those reported by Alexandrov et al. (2003), Benkhaled & Remini (2003), Megnounif et al. (2007) and López-Tarazón et al. (2010) in order to determine a possible trend of the suspended sediment transport in semi-arid zones. A general trend exists between the SSC_{max} and Q_p (fig. 13) but, as expected, a large scatter in sediment concentration at similar water discharge was observed (r^2=0.3).

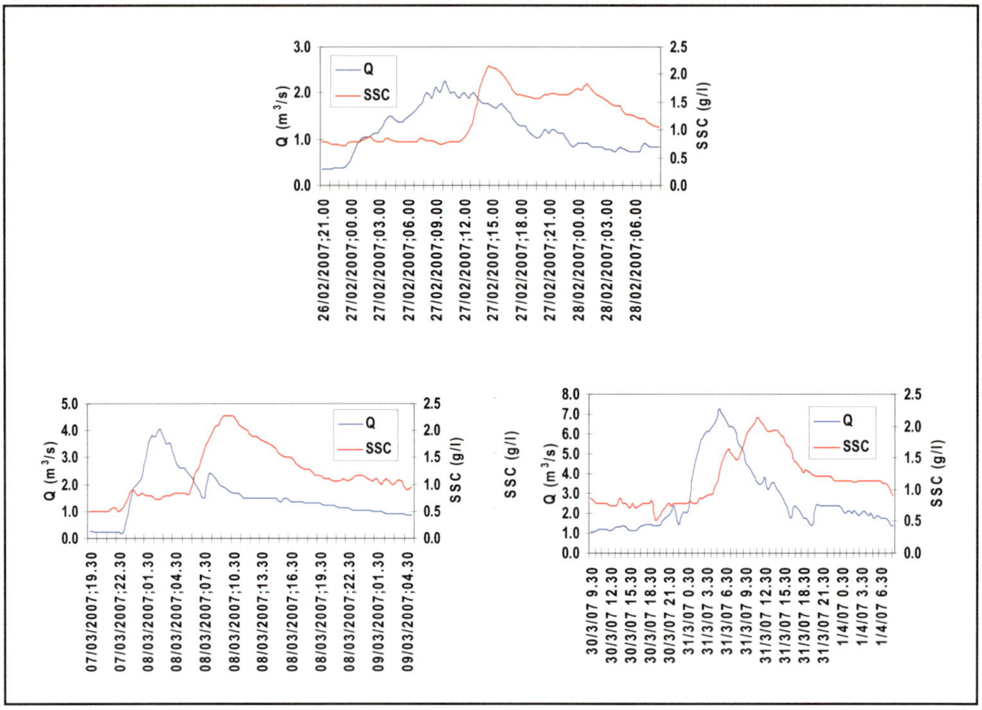

Fig. 10. Consecutive events of moderate intensity of the flood season 2007.

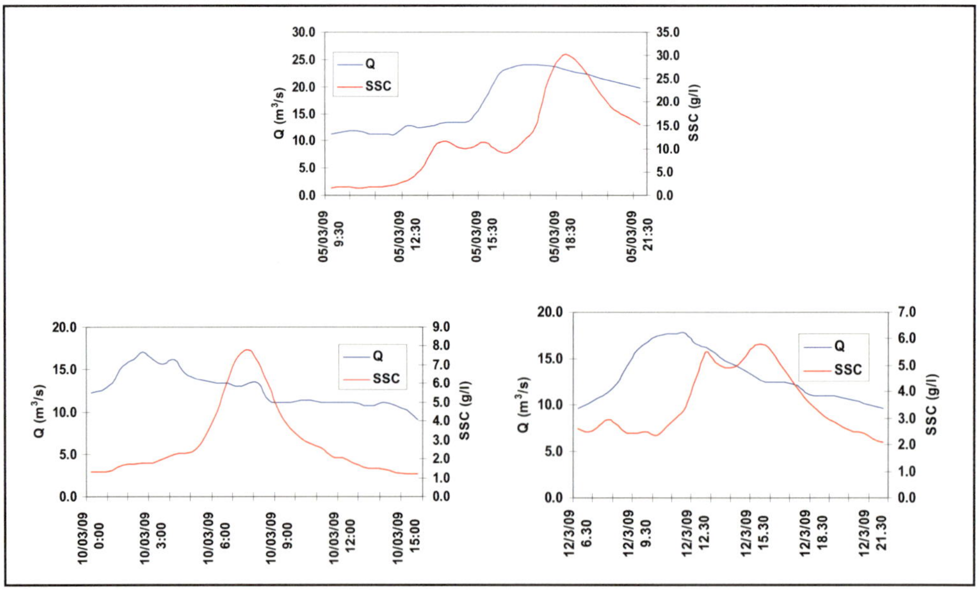

Fig. 11. Consecutive events of moderate intensity of the flood season 2009.

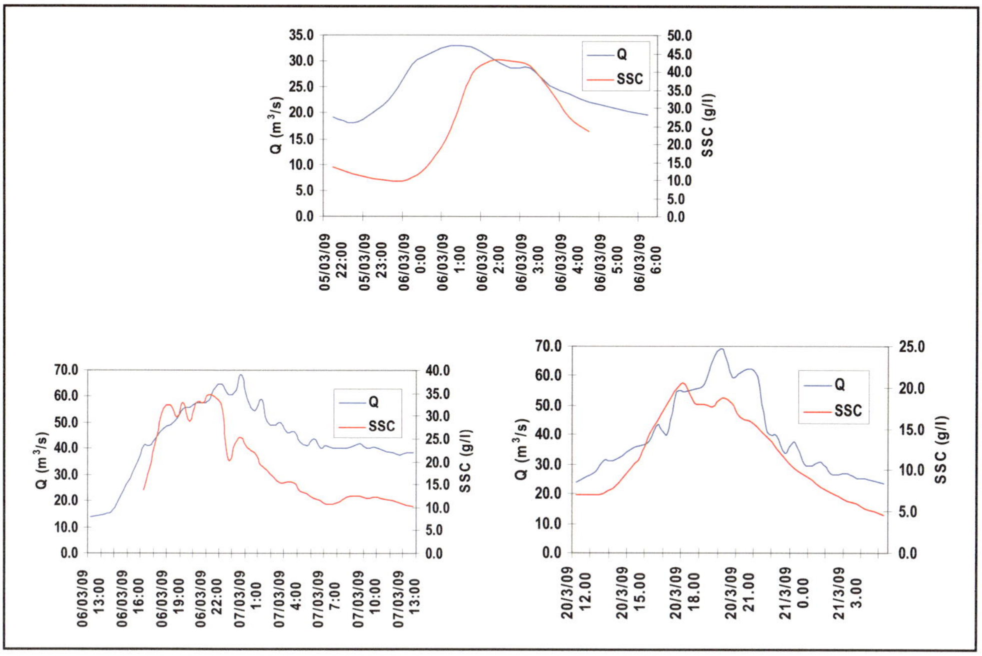

Fig. 12. Consecutive events of high intensity of the flood season 2009.

Rating parameters *a* and *b* of a river strongly depend on flow regime and do not vary in the same proportions (Achite & Ouillon, 2007). Generally the coefficient *b* varies in range 0.3–2.5, while *a* can vary by several orders of magnitude). In arid zone the coefficient *b* is smaller (usually $b < 1$) as compared to typical perennial rivers of the temperate regions where $b > 1$. Intermittent rivers in semi-arid environments seem to have *b* coefficients lower than 1, similar to ephemeral rivers in arid zones. In this work the coefficients of the relationship SSC_{max} - Q_p obtained for five experimental basins are quite consistent with those reported in literature and can be considered typical of the semi-arid environments.

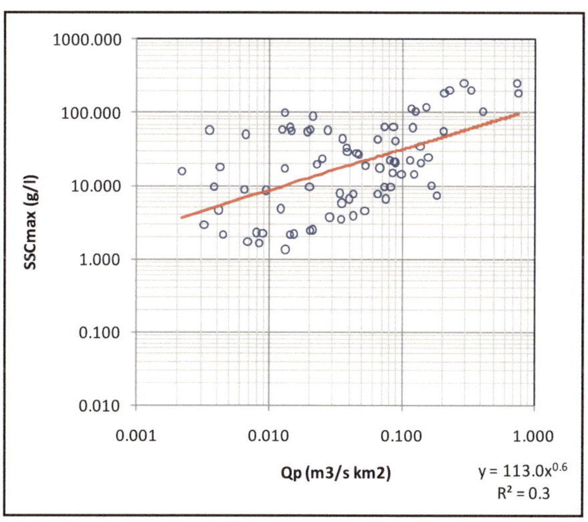

Fig. 13. Trend of the suspended sediment concentration in five semi-arid basins.

6. Conclusions

In recent years many different instruments have been used to determine sediment concentration in rivers. They have different operational modes and range of measurement.

In a semi-arid basin of Southern Italy an optical submerged probe is used for continuous monitoring. The data collected over the period 2007-2009 provided the basis for the analysis of the suspended sediment transport at the event scale.

As in other semi-arid basins, hysteresis between SSC and discharge during flood events has been observed. The main reasons of such non-linear hydrological response can be found in the different flood intensities and in the variations of turbulence that characterize the unsteadiness of the flow.

An excellent correlation exists between sediment load and flood volume and a good agreement has been found also between SSC_{max} and peak discharge. The analysis of the decreasing trend of the time delay between SSC_{max} and Q_p in relation to flood intensity allows to consider separately the behaviour of moderate and intense events.

A general relationship between the maximum SSC and Q_p at the event scale can be determined considering the experimental values collected in five semi-arid basins. The rating parameters a and b of the relationship estabilish hydrological conditions that characterize the semi-arid environments.

7. References

Achite, M. & Ouillon, S. (2007). Suspended sediment transport in a semiarid watershed, Wadi Abd, Algeria (1973–1995). *Journal of Hydrology*, 343, 187-202.

Alexandrov, Y.; Cohen, H.; Laronne, J.B. & Reid, I. (2009). Suspended sediment load, bed load, and dissolved load yields from a semiarid drainage basin: A 15-year study. *Water Res. Res.*, Vol. 45, W08408, 13 pp.

Alexandrov, Y.; Laronne, J.B. & Reid, I. (2003). Suspended sediment concentration and its variation with water discharge in a dryland ephemeral channel, northern Negev, Israel. *Journal of Arid Environments,* 53, 73–84.

Alexandrov, Y.; Laronne, J.B. & Reid, I. (2007). Intra-event and inter-seasonal behaviour of suspended sediment in flash floods of the semiarid northern Negev, Israel. *Geomorphology*, 85, 85–97.

Benkhaled, A. & Remini, B. (2003). Temporal variability of sediment concentration and hysteresis phenomena in the Wadi Wahrane basin, Algeria. *Hydrological Science Journal*, 48(2), 243-255.

Brasington, J. & Richards, K., (2000). Turbidity and suspended sediment dynamics in small catchments in the Nepal Middle Hills. *Hydrol. Process.*, 14, 2559–2574.

Campbell, G.S. (1985). *Soil Physics with BASIC.* In: Transport Models for Soil-Plant Systems Developments in Soil Science Vol. 14, Elsevier, Amsterdam, p. 150.

De Sutter, R.; Verhoeven, R. & Krein, A. (2001). Simulation of sediment transport during flood events: laboratory work and field experiment. *Hydrol. Sc.* 46(4): 599-610.

Foster, I. D. L.; Millington, R. & Grew, R. G., (1992). The impact of particle size controls on stream turbidity measurement; some implications for suspended sediment yield estimation. *Erosion and Sediment Transport Monitoring Programmes in River Basins,* 210, 51-62.

Fugate, D.C. & Friedrichs, C.T. (2002). Determining concentration and fall velocity of estuarine particle population using ADV, OBS and LISST. *Continental Shelf Research 22,* 1867-1886.

Garcia, M.H. (2008). *Sedimentation Engineering: Processes, Measurements, Modeling, and Practice,* ASCE Manuals and Reports on Engineering Practice No. 110 - Society of Civil Engineering, ISBN 13: 978-0-7844-0814-8, US.

Gentile, F. ; Bisantino, T. ; Corbino, R. ; Milillo, F. ; Romano, G. & Trisorio Liuzzi G. (2010). Monitoring and analysis of suspended sediment transport dynamics in the Carapelle torrent (southern Italy). *Catena,* 80, 1-8.

Gippel, C. J. (1995). Potential of turbidity monitoring for measuring the transport of suspended solids in streams. *Hydrological Processes,* 9, 83-97.

Gray, J.R. & Gartner, J.W. (2009). Technological advances in suspended-sediment surrogate monitoring. *Water Res. Res.,* 45, W00D29, 20 pp.

Guerrero, M. & Lamberti, A. (2008). Field measurements of concentration and mean grain size using 2 ADCP working at different frequencies. *Proc. of the workshop Particle in Europe.* ISMAR-CNR Bologna, Italy, 13-14 October 2008.

Lenzi, M. A. & Marchi, L. (2000). Suspended sediment load during floods in a small stream of the Dolomites northeastern Italy. *Catena,* 39, 267–282.

Lewis, J. (1996). Turbidity-controlled suspended sediment sampling for runoff-event load estimation. *Water Res. Res.,* 32(7), 2299-2310.

Lewis, J. (2003). Turbidity-controlled sampling for suspended sediment load estimation. In: Bogen, J. Tharan Fergus and Des Walling (eds.), *Erosion and Sediment Transport Measurement in Rivers: Technological and Methodological Advances* (Proc. Oslo Workshop, 19-20 June 2002). IAHS Publ. 283, 13-20.

López-Tarazón, J.A. ; Batalla, R.J. ; Vericat, D. & Balasch, J.C. (2010). Rainfall, runoff and sediment transport relations in a mesoscale mountainous catchment: The River Isábena (Ebro basin). *Catena,* 82, 23-34.

Medwin, H. & Clay, C.S. (1997). *Fundamentals of Acoustical Oceanography.* Boston: Academic Press; 712 pp.

Megnounif, A.; Terfous, A. ; Ghenaim, A. & Poulet, J.B. (2007). Key processes influencing erosion and sediment transport in a semi-arid Mediterranean area: the Upper Tafna catchment, Algeria. *Hydrological Sciences Journal,* 52, 6, 1271- 1284.

Navas, A.; Garcia-Ruiz, J. M.; Machin, J.; Lasanta, T.; Valero, B.; Walling, D. E. & Quine. T.A. (1997). *Soil erosion on dry farming land in two changing environments of the central Ebro Valley, Spain.* Human Impact on Erosion and Sedimentation (Proc. of Rabat Symposium S6, April 1997). IAHS Publ. no. 245.

Pavanelli, D. & Bigi, A.A. (2005). New Indirect Method to estimate Suspended Sediment Concentration in a River Monitoring Programme. *Biosystems Engineering,* 92 (4), 513–520.

Pedocchi, F. & Garcia, M. H. (2006). Evaluation of the LISST-ST instrument for suspended particle size distribution and settling velocity measurements. *Continental Shelf Research,* 26, 943-958.

Rijn, L.C. van. (2007). *Manual sediment transport measurements in rivers, estuaries and coastal seas,* Aquapublications, The Netherlands. 500 p.

Sadar, M. J. (1998). *Turbidity science.* Technical Information Series, Hach Company, 26 pp. http://www.hach.com.

Sadar, M. J. (1999). *Turbidimeter Instrument Comparison: Low-level Sample Measurements.* Technical Information Series, Hach Company, 56 pp. http://www.hach.com/

Seeger, M.; Errea, M.P.; Beguería, S.J.; Arnáez, Martí-Bono C. & García-Ruiz, J.M. (2004). Catchment soil moisture and rainfall characteristics as determinant factors for discharge/suspended sediment hysteretic loops in a small headwater catchment in the Spanish Pyrenees. *Journal of Hydrology,* 288, 299–311.

Soler, M.; Regüés, D.; Latron, J. & Gallart, F. (2007). Frequency-magnitude relationships for precipitation, stream flow and sediment load events in a small Mediterranean basin (Vallcebre basin, Eastern Pyrenees). *Catena,* 71, 164-171.

Thorne, P.D. & Buckingham, M.J. (2004). Measurements of scattering by suspensions of irregularly shaped sand particles and comparison with a single parameter modified sphere model. *The Journal of Acoustical Society of America,* 116 (5), 2876–2889.

Thorne, P.D. & Meral, R., (2007). Formulations for the scattering properties of suspended sandy sediments for use in the application of acoustics to sediment transport processes. *Continental Shelf Research,* 28, 309-317.

Walling, D.E. & Fang, D. (2003). Recent trends in the suspended sediment loads of the world's rivers. *Global and Planetary Change,* 39, 111- 126.

Walling, D. E. & Kleo, A. H. A. (1979). Sediment yields of rivers in areas of low precipitation: a global view. *IAHS-AISH Publ.,* 128, 479-493.

Wilcox, B. P. & Wood, M.K. (1989). Factors influencing interrill erosion from semiarid slopes in New Mexico. *Journal of Range Management,* 42(1), 66-70.

Williams, G.P. (1989). Sediment concentration versus water discharge during single hydrologic events in rivers. *Journal of Hydrology,* 111, 89–106.

Wolman, M. G. & Gerson R. (1978). Relative scales of time and effectiveness of climate in watershed geomorphology. *Earth Surf. Processes,* 3, 189-208.

16

Modification of Mackinawite with L-Cysteine: Synthesis, Characterization, and Implications to Mercury Immobilization in Sediment

Marcia R. M. Chaves[1], Kalliat T. Valsaraj[2], Ronald D. DeLaune[3], Robert P. Gambrell[3] and Pedro M. Buchler[1]
[1]*Department of Chemical Engineering, University of Sao Paulo; Sao Paulo,*
[2]*Department of Chemical Engineering,*
Louisiana State University, Baton Rouge, Louisiana,
[3]*Department of Oceanography and Coastal Sciences,*
Louisiana State University, Baton Rouge, Louisiana,
[1]*Brazil*
[2,3]*USA*

1. Introduction

Environmental remediation of mercury contaminated sediment includes the inhibition or minimization of mercury methylation. Mackinawite (FeS) is an excellent material that inhibits mercury methylation (Liu et al, 2009). Effective remediation of sediment contaminated with mercury is essential in minimizing the contamination of fish and shellfish and, consequently, the human exposure to methyl mercury. In-situ capping (ISC) is one of the remediation methods that have been found to reduce the mercury transport from sediment to the overlying water. It consists of placing a proper layer of isolating material between the layers of contaminated sediment and overlying water (Palermo, 1998; Liu et al., 2007). This method is attractive due to the reduction in mobility and availability of the contaminant, and requires fewer infrastructures associated with handling, dewatering, treatment and disposal. Mackinawite was reported as an effective isolating material for use as a sediment cap (Liu et al, 2009).

Although pure mackinawite can immobilize mercury, it has a low stability when exposed to oxidized conditions (Burton et al, 2009; Liu et al, 2008; Wolthers et al, 2005). Under field conditions, the capping material has to be in contact with the oxigenated water column during the capping building. This process can oxidize the mackinawite, making this material unable to reacting with the contaminant. Therefore, is important to make the FeS more resistant to oxidation. To obtain stable mackinawite, we proposed to modify its surface using L-cysteine as an organic ligand, considering ferredoxin as a model. Ferredoxins are proteins that have an inorganic active core constituent of Fe_2S_2 or Fe_4S_4 bonded to L-cysteine groups. This core participates in electron-transfer, contributing to important biological functions such as respiration and photosynthesis (Fontecave and Ollagnier-de-Choudens,

2008). Piuille et al (1995) reported that the enzyme pyruvate-ferrodoxin oxyreductase was stable in O_2 (atmospheric air) and the L-cysteine groups seemed to provide the oxidation stability.

Some investigators have tested iron sulfide oxidation at environmental pH ranges (Burton et al, 2009; Bourdoiseau et al, 2008). Most of the literature is related to the oxidation of reduced iron sulfide compounds such as pyrite, troilite, and pyrrhotite under acidic conditions, for solving acid mine drainage issues (Moses et al, 2003). Mackinawite is considered a precursor to other iron-sulfides (FeS_2, $Fe_{1-x}S$, $Fe_{1+x}S$, Fe_3S_4), and should follow the same pattern of oxidation, occurring via intermediate polysulfide species, resulting in FeOOH and elemental S^0 or iron sulfate. Mackinawite oxidation can occur in biotic or abiotic conditions, but is not microbially mediated (Burton et al, 2009). In attempting to improve the oxidation resistance of pyrite, Zhang et al (2003) reported the suppression of pyrite oxidation by about 90%, through the polymerization of diacetylene groups in the lipid tails, producing a cross-linked layer on the mineral surface. Along the same lines, Belzile et al (1997) used a series of coating agents (acetyl acetone, humic acids, amonium lignosulfonates, oxalic acid and sodium silicate) to increase oxidation inhibition above 40%.

Our current study on anoxic sediment contaminated with mercury is to investigate the capability of mackinawite modified with L-cysteine to sequestrate mercury. It involves the effects of L-cysteine on mackinawite structure, oxidation stability, mercury uptake capacity, and inhibition of mercury methylation capacity. In this chapter, we present the synthesis and characterization of mackinawite modified with L-cysteine, the effect of this modification on the structure and oxidation stability of mackinawite, and its implication to mercury immobilization in contaminated sediment using capping technology.

2. Synthesis of modified mackinawite

Synthetic mackinawite was prepared from deoxygenated solutions of $(NH_4)_2FeSO_4 \cdot 6\ H_2O$ (Mohr's salt; Sigma-Aldrich), and $Na_2S\ 9\ H_2O$ (Acros Chemicals), using the method described by Liu et al (2008). The experiments were conducted under N_2, using a glove box, at room temperature (25°C), and all solutions were previously deoxygenated by purging with N_2 (ultra-pure) for 30 minutes.

The unmodified mackinawite (in this work is designed as FeS) was prepared by addition of 100 ml Mohr's salt (0.4 mol/l) to 100 mL of $Na_2S.9\ H_2O$ (0.4 mol/l) in a three-necked flask, which was magnetically stirred for 15 minutes. The suspension was filtered using a 0.45μm membrane filter in a vacuum filtration system. The solid was placed in a HDPE centrifuge tube, N_2 was added to filled the empty space, the tube was tightly closed, placed into a vacuum-sealed bag, and stored in a freezer. After being frozen, the solid was removed from the freezer and returned to the reaction flask with 200 ml of water (HPLC grade) previously deoxygenated, stirred for 15 minutes and filtered. This procedure was repeated twice, and the solid was dried under a N_2 flow, and stored in 1.5 ml vials, preserved in a vacuum-sealed bag, until be used.

The modified mackinawite (in this work is designed as FeS-Cys) was prepared following the same procedure, except that 2.0 g of L-cysteine was dissolved in the Mohr's salt solution before reaction with the sodium sulfide solution. Although the reaction between Mohr's salt and sodium sulfide was effective, the gravimetric yield was 75%, due to solid losses during the filtration and drying process.

3. Mackinawite characterization

The characterization was performed by determination of mackinawite composition; crystallinity; specific surface area and porosity; morphology; oxidation resistance; and mercury uptake capacity. These characteristics were obtained using the techniques, instruments, and procedures as described following.

The chemical composition of the solids was determined by a Perkin Elmer 2400 CHNS Elemental Analyzer, and evaluated by FTIR (Nicolet 6700 FTIR), EDS (Jeol 840A), and XPS (Kratos AXIS 165 XPS/Electron Spectroscopy, with X-ray source of Al Kα (1253,6 eV), hybrid mode, 15kV of accelerating voltage,10 mA of current).

The crystallinity of the solids was obtained using a Bruker/Siemens D5000 automated powder X-ray diffractometer with CuKα. The data were collected in the interval from 2 to 70 degrees 2θ at a rate of 0.02 degrees per 2 seconds. The results were interpreted using the JADE software V6.1.

The SEM images and EDS spectra were obtained from samples coated with gold, using a Scanning Electron Microprobe (JEOL 840A) with 15 kV of accelerating voltage. TEM images were obtained from samples deposited on 200# Cu - Lacey carbon-coated discs, through the High Resolution Transmission Electron Microscope operated at 200 kV of accelerating voltage.

The specific surface area was measured following the multipoint N_2-BET adsorption method, using a QuantaChrome 500 instrument. The sample was transferred to the bulb tube of the instrument under a N_2 flow inside a glove box, sealed with Parafilm®, and transferred rapidly to the work station where it was degassed at room temperature overnight.

The oxidation experiments were carried out in a 250 mL three-necked flask, open to the atmosphere. The flask was placed in a water bath on a stirrer plate. In the neck of the flask were placed pH, E_H and dissolved oxygen (DO) sensors. Water (250 mL HPLC grade) was added to the flask; the pH was controlled with additions of 0.1 M NaOH and/or HCl (trace metal grade) solutions. Mackinawite (0.5 g) was added to the water and stirred during the experiment.

The Ross Ultra combination pH sensor (ORION 8156BNUWP), with external automatic temperature compensation (ATC) probe, was calibrated before each measurement using commercial 4 and 7 pH buffer solutions, connected to pH/mV/^{o}C meter (Oakton pH 510 series). The Eh sensor, a platinum sure-flow combination Redox-ORP type (Orion 89678BNWP), was calibrated using ORP standard solution (Orion 967901) at the same temperature of the experiment, connected to a Cole Parmer pH/mV/^{o}C meter, giving direct Eh measurement. The dissolved oxygen was determined using a Mettler Toledo InLab® 605 sensor, with internal ATC probe, connected to Mettler Toledo SevenGo™ Pro SG6 DO meter, which was calibrated following the manual instructions, using oxygen-saturated water.

The experiments were conducted under both short and long time scales. Short time-scale (5 hours) experiments were evaluated at pH (4 to 8) and temperature (25°C to 45°C). Long time-scale experiments were performed at pH 6 and 25°C for 24 hours. Immediately at the end of each experiment, the solid was filtered using a 0.45μm Pall GN Metricel® mixed cellulose membrane filter, in a vacuum filtration system, dried under N_2 flow, and stored in 1.5 mL vials with N_2 to avoid posterior solids oxidation, sealed with parafilm®, and preserved in a vacuum-sealed bag.

The solids were evaluated using FTIR, obtained through powder direct analysis (Nicolet 6700 FTIR) collected by accumulation of 36 scans, 4 cm^{-1} resolution, and using air as background. The crystallinity of the solids was obtained using a Bruker/Siemens D5000, as described early.

3.1 Chemical composition

EDS Analysis

The chemical composition was evaluated through energy dispersive spectroscopy (EDS) obtained on a scanning electron microscope (SEM). Figure 1 shows the spectra of samples of mackinawite, unmodified (FeS) and modified with L-cysteine (FeS-Cys). The presence of oxygen was observed in the unmodified mackinawite, but not in the sample modified with cysteine. This indicates a possible partial oxidation of the sample of unmodified mackinawite, and also some resistance to further oxidation of modified mackinawite.

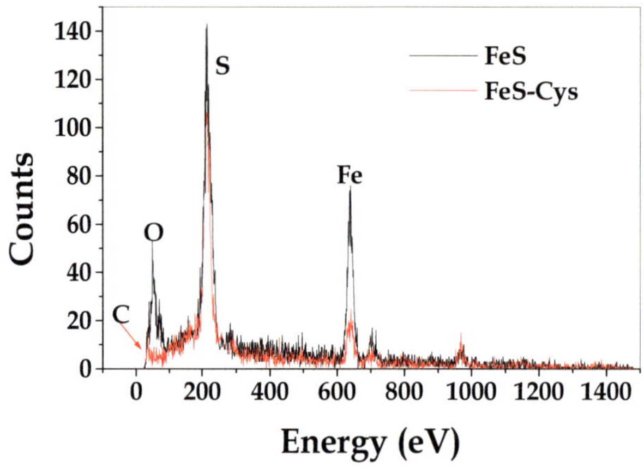

Fig. 1. Spectra of EDS of the mackinawites studied.

FTIR Analysis

The chemical composition was also evaluated through FTIR of the samples. The iron sulfides do not present clear absorption bands at lower wavenumber, in the infrared spectrum. Although it is not possible to distinguish the types of vibration between the atoms of iron and sulfur present in mackinawite, was possible to judge which bands are due to the presence of FeS, and which bands are characteristic of L-cysteine. In support of this evaluation, spectrum was obtained of Mohr's salt solution containing L-cysteine dissolved under the same conditions used in the preparation of modified mackinawite. The interaction between iron and L-cysteine may occur through interaction of iron-oxygen, iron-sulfur or iron-nitrogen, depending on the pH of the system.

The pH below 5 of the iron-cysteine solution facilitated the protonation of the cysteine. In this case the iron preferentially binds to the sulfur displaying a spectrum similar to mackinawite. The amount of cysteine was very lower than the iron; thus, the complex resulted was formed by attachment of one cysteine per atom of iron. From figure 2, it can be observed that the solution of iron and L-cysteine (Fe-Cys) showed bands characteristic of asymmetric stretching of carboxylate $(C\overset{...}{-}O)_2$ at 1638 cm^{-1} typical of amino acid, a weak band at 1451 cm^{-1} characteristic of ammonium ion NH_4^+, and another at 1096 cm^{-1} indicating the presence of sulfate (SO_4^{2-}) constituents of Mohr's salt. A broad band, centered in the region of 3300 cm^{-1}, comprises the vibration spectrum of the water in the solution superimposed on the band in 3340 cm^{-1}, relative to the N-H stretching in protonated amine

(NH_3^{+}) of L-cysteine. The absence of the features of the connection with H-S at 2551 cm^{-1}, shows that iron is linked to the sulfur. We observed that mackinawite modified with cysteine showed an infrared spectrum almost identical to the original mackinawite, differing only by the presence of a broad band of low intensity, centered at 1583 cm^{-1}, assigned to the presence of carboxylate ($(C\cdots O)_2$). The bands of low intensity in the region of 2200 cm^{-1} to 1900 cm^{-1} are attributed to the vibration of the Fe-S bond.

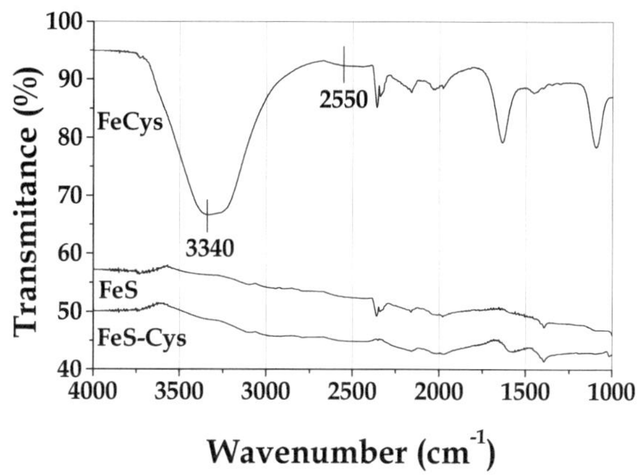

Fig. 2. FTIR of the iron-cysteine complex (FeCys), unmodified (FeS) and modified (FeS-Cys) mackinawite.

XPS Analysis

The X-ray photoelectron spectroscopy (XPS) is a characterization technique used to evaluate the surface of the sample. For mackinawite is particularly very useful because it can identify the presence of oxidation products on the surface of the sample, and the oxidation state of elements involved.

The survey spectra of modified and unmodified mackinawite indicated the presence of the elements C, O, Fe and S. The carbon is a ubiquitous contaminant, being found in all samples. Nevertheless, its presence results in a well-defined spectrum for modified mackinawite (Fig. 3a).

The spectrum consisted of two broad bands, which were decomposed using the software Origin 8.1 for Gaussian peaks. The identified peaks are characteristic of cysteine at 284.5 eV referring to the C-C bond, at 285.5 eV assigned to C-OH and C-N bonds. The peak at 288.4 eV and characteristic of the presence of C=O, found in both the cysteine as the adsorbed CO_2 from the air.

The XPS spectra of Fe $2p_{3/2}$ of the unmodified mackinawite (Fig. 3b) and mackinawite modified with cisteine (Fig. 3c) showed a broad band, result of combination of severals peaks, characteristic of the weathered surface of mackinawite. This corrosion was atributted to the exposure of the samples to ambient atmosphere during the sample preparation prior to insertion in the evacuated chamber of the XPS equipment.

The spectrum of unmodified mackinawite showed peaks at 707.9, 709.8 and 711.6, characteristic of Fe (II)-S, Fe (III)-S and Fe (III)-O bonds. For modified mackinawite, the peaks were identified at 707.7, 709.0 and 711.2, characteristic of Fe (II)-S, Fe (II)-O and Fe (III)-O. The presence of these peaks reveals a surface composed by mackinawite and oxy (hydroxy) of iron, such as goethite or lepidocrocite FeOOH.

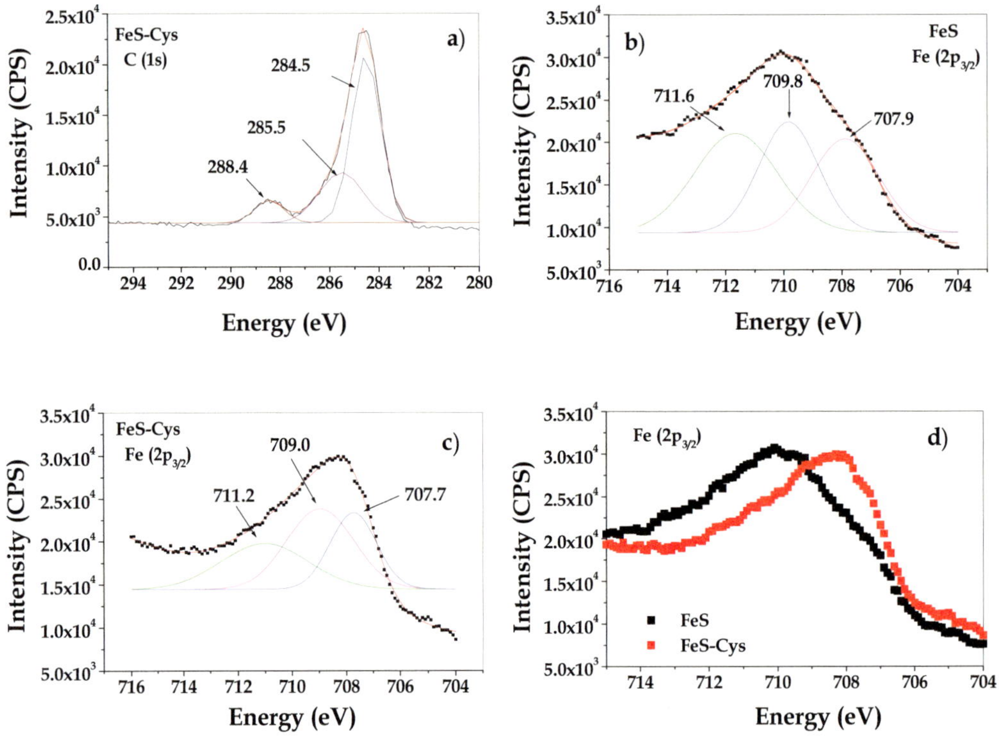

Fig. 3. XPS spectra of C (1s) a) FeS-Cys and Fe ($2p_{3/2}$): b) FeS; c) FeS-Cys; d) merge of the Fe ($2p_{3/2)}$ curves. Errors for energy ± 0.3 eV.

By merging the curves into one graph, it can be observed that the spectrum presented by unmodified mackinawite is a result of greater contribution of oxidation products of its surface; modified mackinawite presented a relatively less weathered surface (Fig. 3d). This demonstrates the efficiency of cysteine to increase the oxidation resistance of mackinawite.
Similar to iron, the bands resulting from the sulfur 2p spectrum was decomposed, obtaining the peaks related to mackinawite and products of oxidation (Fig. 4a and 4b).
The peaks of unmodifed mackinawite were identified at 160.1 eV and 161.4 eV characteristics of (S^{2-}), 162.4 eV characteristic of (S_2^{2-}), 163.6 eV characteristic of (S_n^{2-}), 164.1 characteristic of (S^0), and a broad band centered at 169.0 eV attributed the presence of (SO_4^{2}). It is important to note the presence of polysulfides (S_n^{2-}) and elemental sulfur (S^0), typical for the oxidation of the surface mackinawite. The presence of sulfate is attributed more to the inefficient removal by mackinawite washing during the synthesis process, than the oxidation process. These results are in agreement with those obtained by other researchers (Nesbitt and Muir (1994); Behra et al (2001), Boursiquot et al, 2001; Mullet et al, 2001).
For modified mackinawite the identified peaks were centered at 160.3 and 161.2 eV characteristic of (S^{2-}), 161.8 eV characteristic of (S_2^{2-}), 163.0 eV characteristic of (S_n^{2-}), and 164.5 characteristic of (S^0).
The peak at 168.1 eV, attributed to sulfate (SO_4^{2-}) adsorbed, is well-defined with lower energy than that presented by unmodified mackinawite. This could be a result of the

adsorption of sulfate on cysteine, probably through hydrogen bonds between the hydroxyl and oxygen of the sulfate, with the amine and the carboxylate groups of the cysteine. Thus it made the sulfate removal more difficult, explaining the greater contribution in the spectrum. The results showed that the cysteine influences the oxidation process, since the modified mackinawite presented the sulfur 2p spectrum with greater contribution of the peaks characteristic of sulfide, and lower peak intensity of the polysulfide groups. For unmodified mackinawite, the formation of the spectrum of sulfur showed higher contribution of groups S_2^{2-}, polysulfides and elemental sulfur, ie, products of oxidation on its surface.

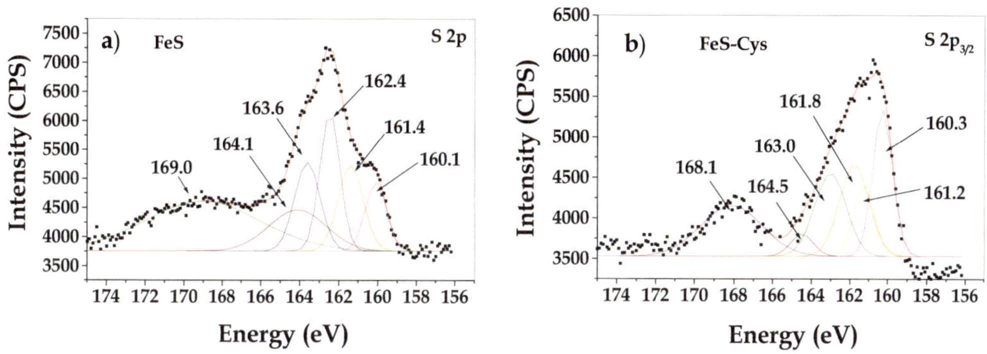

Fig. 4. XPS spectra of S 2p: a) FeS and b) FeS-Cys - Errors for energy ± 0.3 eV.

The XPS spectra of O 1s for both mackinawites are shown in the figures 5a and 5b. For unmodified mackinawite we identified the peaks at 529.4 eV attributed to species (O^{2-}), 531.6 eV attributed to the presence of hydroxyl (OH^-) present in iron hydroxide, the peaks at 533.0 eV attributed to adsorbed water molecule, and 534.2 eV attributed to the adsorbed (CO_2) from the air. For the modified mackinawite these peaks were identified at 529.3 eV, 530.7 eV, and 532.3 eV. The peak characteristic CO_2 adsorption was not present, and the intensity of the peak for adsorbed water was much lower than that reported in unmodified mackinawite, confirming the higher hydrophobic character of its surface, caused by the presence of cysteine.

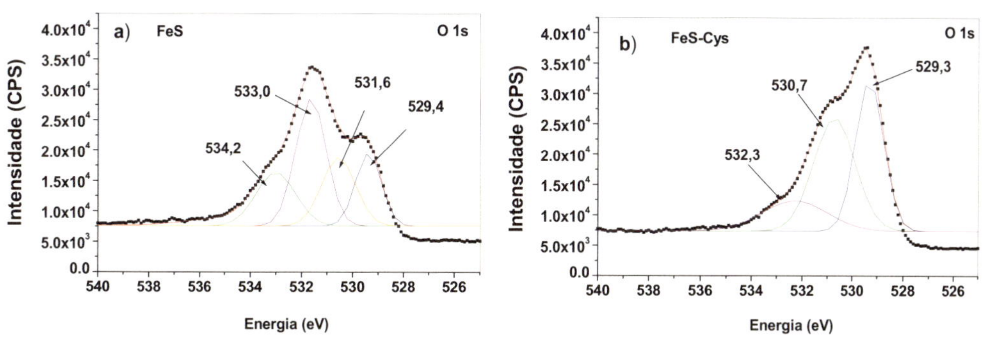

Fig. 5. XPS spectra of O 1s: a) FeS; b) FeS-Cys. Errors for energy ± 0.3 eV.

Elemental (CHNS) Analysis

Using elemental analysis (CHNS) and ICP-OES analysis to determine iron content, the chemical composition of unmodified mackinawite was determined as $FeS_{0.86}$. Sweeney and Kaplan (1973) also found that FeS has a composition of $FeS_{0.87-0.92}$. For modified mackinawite, the carbon content is due to the amount of L-cysteine in the sample. As L-cysteine contains a sulfur atom in its molecule, the quantity of sulfur was subtracted from the total, resulting in the amount of sulfur corresponding to mackinawite (Equation 1).

$$S_{mackinawite} = S_{total} - S_{cysteine} \tag{1}$$

Considering the mass of carbon in the sample being equal to 0.6 g corresponding to L-cysteine, and each molecule of L-cysteine has 3 carbons and a sulfur atom, the amount of cysteine and hence of sulfur was determined to be 16.7 mmol (or 2 g). This amount is in line with the amount of cysteine added during the preparation of modified mackinawite (2.00 g), meaning that all the cysteine was reacted. Thus, the chemical composition of mackinawite modified with L-cysteine was determined as $FeS_{0.735}.Cys_{0.0133}$. The contents of each element are presented in table 1.

Sample	C (% wt)	H (% wt)	N (% wt)	S (% wt)	Fe[a] (% wt)
FeS	nd	nd	nd	32.1	67.9
FeS-Cys	0.6	2.1	1.2	28.9	67.2

[a] Determined by ICP_OES.

Table 1. Chemical analysis of unmodified (FeS) and modified (FeS-Cys) mackinawite.

3.2 Crystalline structure

X-ray diffraction analysis by the powder method confirmed the nano-crystalline structure of mackinawite. Figure 6 presents the XRPD pattern for the samples. There were low intensity peaks, characteristic of compounds with low crystallinity and/or nano-size of particles. The addition of L-cysteine to mackinawite modified the X-ray diffraction patterns. The main change is the increased intensity of the peak attributed to the (001) spacing with a small lower angle displacement of the peak. The change in chemical composition and increased (001) spacing indicate that L-cysteine reacts with the iron, when mixed during the synthesis, remains within the structure of mackinawite, occupying the spaces between the layers. This results in the observed lattice expansion.

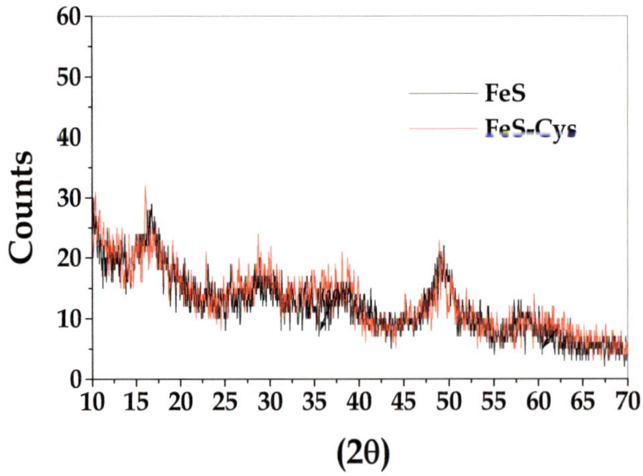

Fig. 6. XRPD pattern of FeS and FeS-Cys

Using the length and angle between the atoms, it was possible to calculate the length of the cysteine molecule, as approximately 0.432 nm (Ayyaar and Srinivasan, 1965). The lattice spacing for the mackinawite is 0.529 nm. This is large enough for the L-cysteine molecule to occupy and expand the lattice. This is aided by the electrostatic repulsion between the charges in the cysteine and on the surface of mackinawite.

The X-ray diffraction pattern of disordered mackinawite is not well-defined. However, it was possible to determine the diffraction of the characteristic peak of the plane (001) for unmodified and modified mackinawite - see Table 2.

	FeS[a]	FeS-Cys[a]	FeS-ref[b]	FeS[c]
hkl	d (Å)	d (Å)	d (Å)	D (Å)
001	5.2913	5.5187	5.196	5.0328

[a] Synthetic disordered mackinawite aged for 3 h (this study)
[b] Synthetic mackinawite original - aged for 3 days (JEONG et al, 2008)
[c] Crystallized mackinawite (LENNIE et al, 1995)

Table 2. XRPD parameter for synthetic mackinawite

3.3 Particle size and morphology

The particle size, and morphology of the solids was evaluated using a Nitrogen adsorption /desorption isotherm (BET), Scanning Electron Microscopy (SEM), and Transmission Electron Microscopy (TEM).

The nitrogen adsorption isotherms (BET) for mackinawite are shown in figure 7. The mackinawite exhibited typical adsorption curves (type III), but with a characteristic H3 type hysteresis. The isotherm of type III is unusual, and is exhibited by materials with low interaction with the adsorbate (N_2). The adsorption-desorption hysteresis is indicative of the type of pores on the surface of a solid. The type H3, where desorption occurs along the whole P/P_0 curve, results from clusters of particles of the lamellar type giving rise to the cleft-like

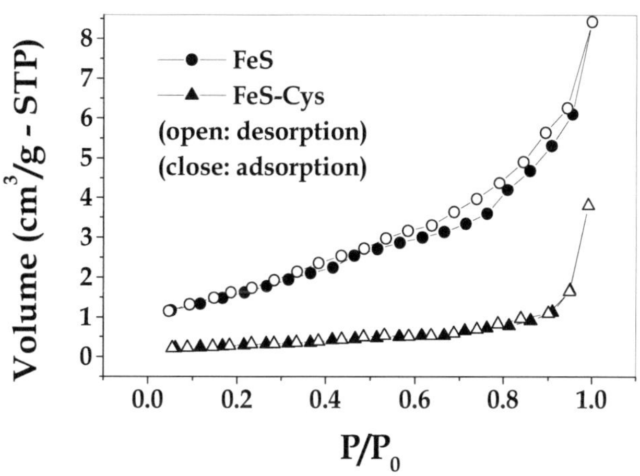

Fig. 7. N_2 adsorption/desorption isotherms of the unmodified mackinawite (FeS) and modified mackinawite (FeS-Cys). Average pore diameter and total pore volume: FeS 8.74 nm; 1.30×10^{-2} cm^3/g - FeS-Cys 23.09 nm; 0.59×10^{-2} cm^3/g.

pores (Sing, 1985). This feature confirms the layered structure of mackinawite prepared. This characteristic is important in metal ion adsorption from the liquid medium. The values of average diameter and total pore volume of the mackinawite are shown in figure 7.
Results show that L-cysteine increases the average pore diameter, but decreases the total pore volume. The specific surface area was determined by N_2 adsorption/desorption technique, using the BET equation; the values observed experimentally are shown in table 3. Mackinawite is a nanosized material, the specific surface area determined by BET is underestimated, due the nitrogen does not cover properly all surface area, result of porous blocked by cysteine. Thus, it is useful just to make a comparison, and determine the effect of cysteine on the mackinawite surface. The experimental result of the unmodified mackinawite is in agreement with values described in the literature, ranging from 4.7 to 80 m^2/g. The effect of L-cysteine on the microstructure reduces significantly the specific surface area, and thus could influence the adsorption capacity of the modified mackinawite.
The SEM images present the shape of the clusters of unmodified and modified mackinawite (Fig. 8). The original mackinawite had clusters of particles mean spherical in shape, while modified mackinawite showed more elongated larger clusters. The effect of L-cysteine on the morphology of mackinawite can be explained based on the attachment of L-cysteine during the synthesis, between the layers of mackinawite. When L-cysteine is mixed with Mohr's salt, the complex $[HO_2CCH(NH_2)CH_2S]Fe^+$ is formed in solution. The reaction of this solution with sodium sulfate solution, results in a solid containing a portion of the carbon

Material	Specific surface area (m^2/g)	Particle diameter (nm)	Reference
FeS	5.97	5.0[a]	This work
FeS-Cys	1.02	5.0[a]	This work
FeS	7.0	210[b]	Taylor et al. (1979)
FeS	36.5	40[b]	Rickard (1997)
FeS	80	18[b]	Widler (2002)

[a] determined by TEM analysis; [b] calculated using BET specific surface area.

Table 3. Specific surface area and particle diameter for synthetic mackinawite

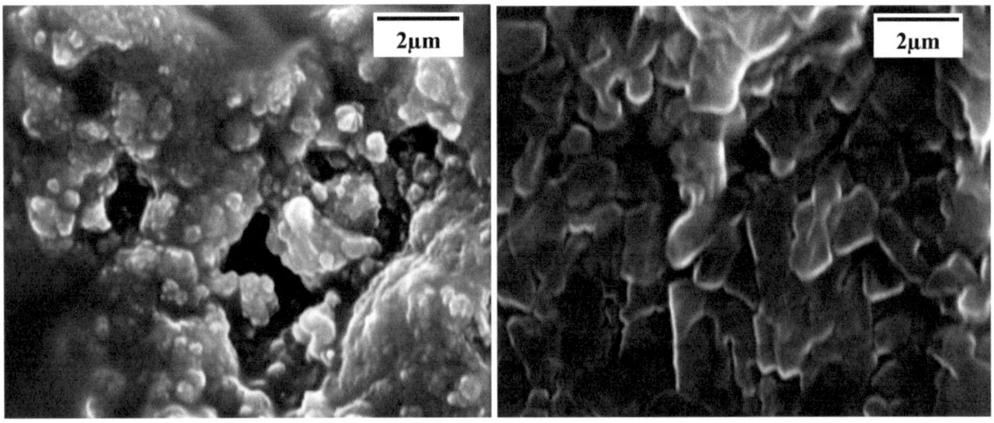

Fig. 8. Scanning Electron Microscopy: a) unmodified mackinawite; b) modified mackinawite. (10,000 x magnification).

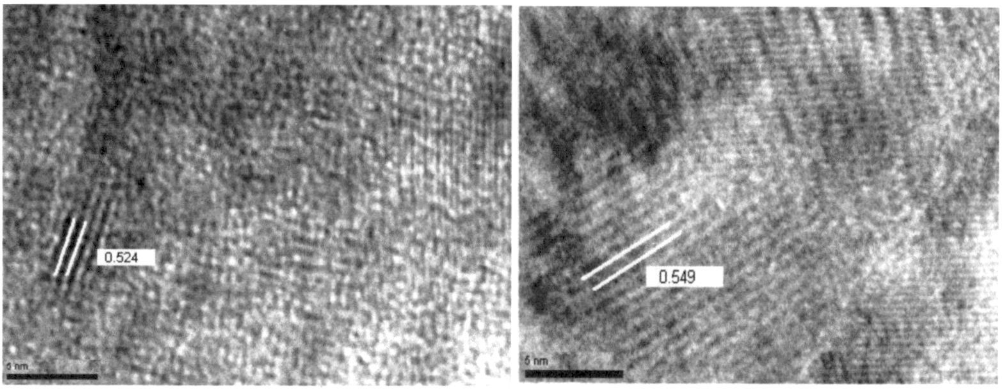

Fig. 9. High Resolution Transmission Electron Microscopy of a) unmodified mackinawite; b) modified mackinawite.

chain of L-cysteine, on the surface. The interaction of amine and carboxylate groups with atoms of iron in the adjacent layer, should limit the configuration to a lamellar shape. Thus, the presence of L-cysteine between the sheets of the FeS, forces the clusters to form plates.
The unmodified and modified mackinawite analyzed by TEM (Fig. 9a and 9b) presented particles in average diameter of 5 nm. The samples were poor crystallized, but some crystals allowed to identify the space between the layers (d_{001}) as being 5.24 Å and 5.49 Å for unmodified and modified mackinawite, respectively. These values are close to those observed using XRPD analysis, and support the information that the cysteine is located between the layers in the mackinawite structure, resulting in the lattice expansion.

3.4 Effect of L-cysteine on mackinawite oxidation

Mackinawite oxidation – Short time-scale experiments

The solubility of mackinawite (FeS) in water with pH lower than the pH of point of zero charge (pH_{pzc}=7.5), is a pH-dependent reaction (equation 2). Above pH_{pzc}, the solubility is a pH-independent reaction (equation 3), and the aqueous species are neutral and consist in Fe_nS_n or clusters with Fe:S ratio of 1(Rickard, 2006). The oxidative dissolution of FeS is a process that can be described by serveral reactions (equations 4-6), being the pH a important parameter.

$$FeS_{(S)} + H_2O + H^+ - HS^- + Fe^{2+} \quad (2)$$

$$nFeS_{(S)} - Fe_nS_{n\,(aq)} \quad (3)$$

$$2\,FeS_{(S)} + \tfrac{9}{2}\,O_2 + 3\,H_2O - 2FeO(OH) + 2SO_4^{2-} + 4\,H^+ \quad (4)$$

$$FeS_{(S)} + \tfrac{3}{2}\,O_2 + \tfrac{1}{2}\,H_2O - \tfrac{1}{8}\,S_8^0 + FeOOH \quad (5)$$

$$\tfrac{1}{8}\,S_8^0 + \tfrac{3}{2}\,O_2 + H_2O - SO_4^{2-} + 2H^+ \quad (6)$$

When mackinawite was added to water, a fast increase of pH to a maximum occurred within the first minute, which decreased slowly to no more than one unit below the maximum (Fig. 10). This behavior indicates the existence of two processes, viz., the fast protonation of the surface, followed by mackinawite surface oxidation. The protonation of mackinawite surface is favored due to the high content of sulfur atoms.

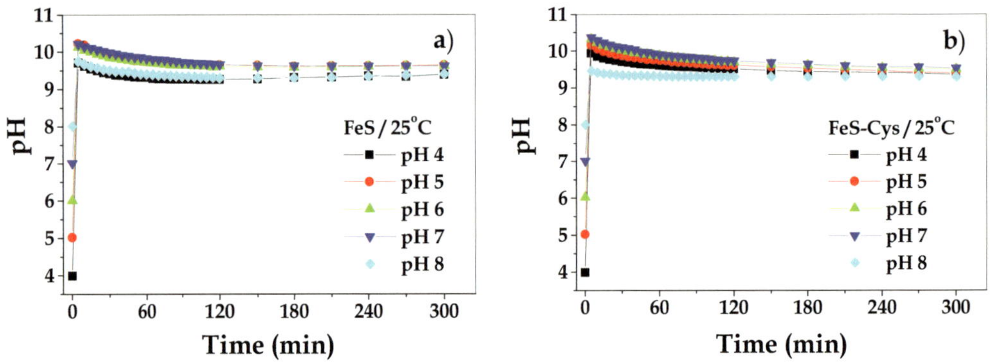

Fig. 10. pH solution during the mackinawite oxidation process (short time).

Mackinawite when added to water (2 g/l) produced a strongly reduced environment within the first minute of contact, resulting from surface protonation and sorption of dissolved oxygen. The redox potential increased gradually during the following 5 hours, attributed to gradual increase of amount of oxidized species in solution.

The redox characteristics of unmodified mackinawite were not sensibly affected by initial pH, whereas that of the modified mackinawite showed sensible to initial pH (Fig. 11). The Eh was more negative at initial pH values between 4 and 7 indicating progressive resistance to oxidation.

The modified mackinawite showed the same trend of Eh behavior presented by unmodified mackinawite in initial pH 4 to 7; however, the Eh was more sensible to initial pH than the unmodified mackinawite. It can be explained by the variable charge surface of cysteine, and its protonation/deprotonation with pH change.

The Eh of solutions containing FeS-Cys was lower than that of FeS under the same conditions throughout the experiments, indicating that cysteine increased the mackinawite oxidation resistance, as showed earlier by XPS analysis. This can be observed through the XRPD (Fig. 12) and FTIR (Fig 13) analyses of the solids obtained after 5 hours of experiment. The effect of initial pH is more visible to the peak at 14 (2θ), characteristic of lepidrocrocite. The intensity of this peak decrease wiht increasing pH range 4-6, and increase with decreasing the pH range 7-8. This behavior indicates that the oxidation process decreased until pH 6, increasing to higher pHs. This is in agreement with the mackinawite surface chemistry (Mullet et al, 2002).

There is a lower intensity of the peaks caracteristic of sulfur (S) and lepdocrocite (L), and the highest band intensity characteristic of mackinawite (M) in the XRPD pattern of the FeS-Cys in relation to the XRPD pattern of the FeS.

The FTIR spectra of FeS oxidized at different initial pHs showed bands at 668, 743, 795 and 885 cm^{-1}, and 3127 cm^{-1}, associated with the stretch modes and δOH νOH of lepidocrocite, respectively. A band of low intensity at 3730 cm^{-1} and due to the stretching of the OH free connection, similar to the silanol on the surface of silica. Elemental sulfur (S^0) has no stretch modes in the spectrum. The bands at 1200, 1120 and 996 cm^{-1} can be attributed to sulfate adsorbed on the surface of lepdocrocite. This is similar to sulfate adsorbed on goethite (Peak et al., 1999). The band expanded in 2000 and 2150 cm^{-1} indicate the stretching of the Fe-S bound in mackinawite.

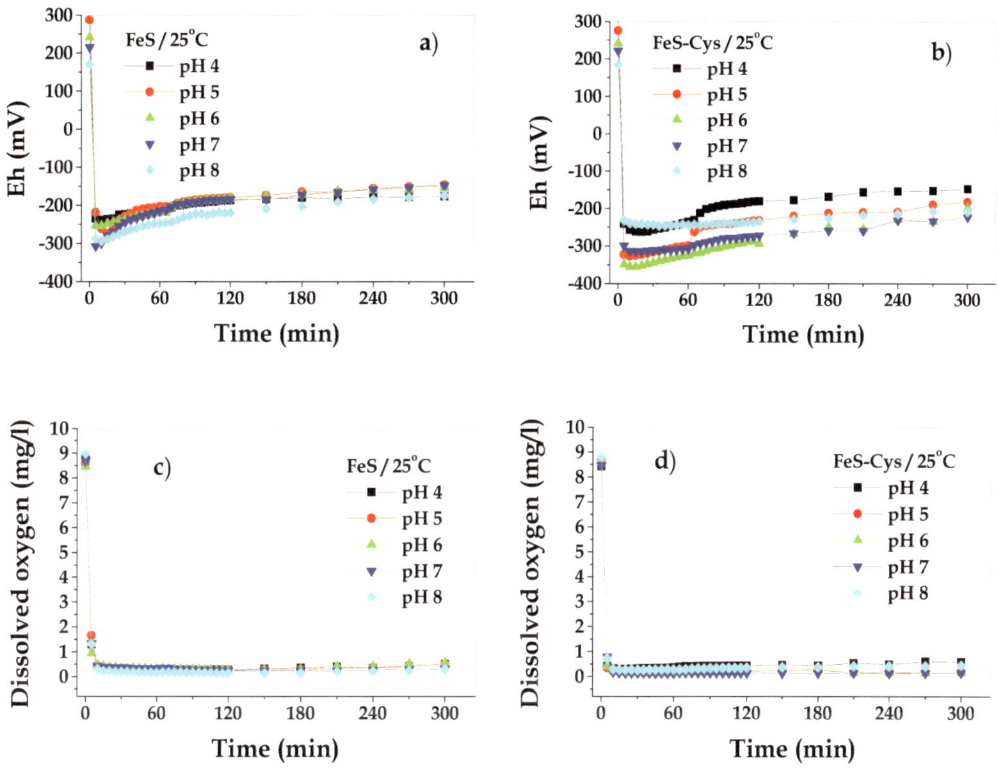

Fig. 11. Eh and DO of solution during the oxidation of mackinanwite (short time).

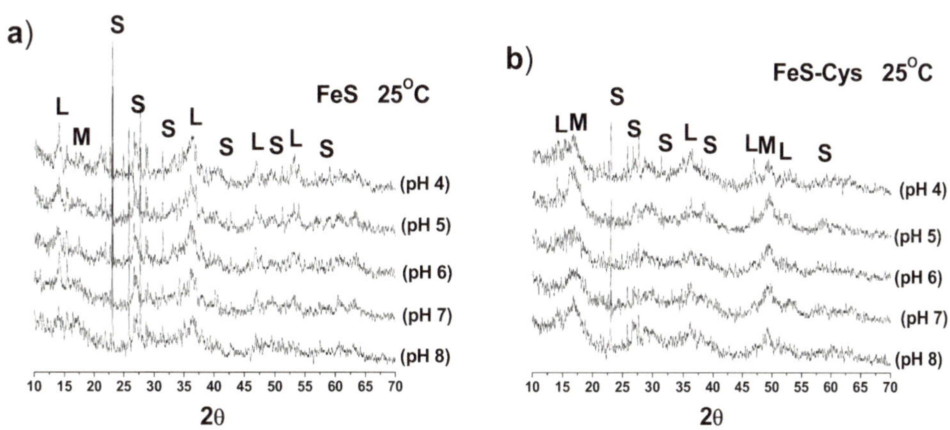

Fig. 12. XRPD pattern of the solids after oxidation experiments (short time). (L) lepidocrocite; (S) sulfur; (M) mackinawite.

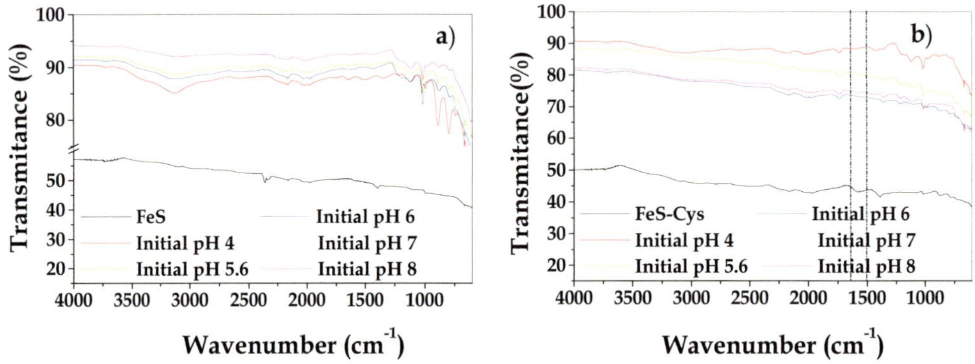

Fig. 13. FTIR spectra of a) unmodified and b) modified mackinawite after oxidation processes (short time)

The FeS-Cys showed the characteristic bands of the OH vibration of lepdocrocite (668, 739, 794, 883, and 3128 cm^{-1}), the low intensity band in the 3730 cm^{-1} is attributed to stretching of free OH, and bands associated to the presence of sulfate (1200, 1120 and 996 cm^{-1}).

Similar to the XRPD, the FTIR spectrum of the FeS shows visible difference, with greater intensity of the peaks associated with the presence of sulfate and lepdocrocite, in relation to FeS-Cys, indicating greater susceptibility to oxidation.

Effect of the temperature on the mackinawite oxidation

The effect of temperature on the mackinawite oxidation was also evaluated (Fig. 14). It was observed that an increase in temperature inhibits the oxidation process. The oxidation reaction is dependent on the dissolved oxygen in water, which decreases at higher temperatures. We observed a difference in the Eh of the solution of FeS and FeS-Cys at 25°C and 35°C, but in 45°C all the samples showed the same behavior, due the critical concentration of dissolved oxygen have limited the oxidation process.

By comparison, the FeS showed a higher degree of oxidation than the FeS-Cys at different temperatures, observed through the peak of characteristic X-ray diffraction of the oxidation products (lepidocrocite and sulfur) (Fig. 15).

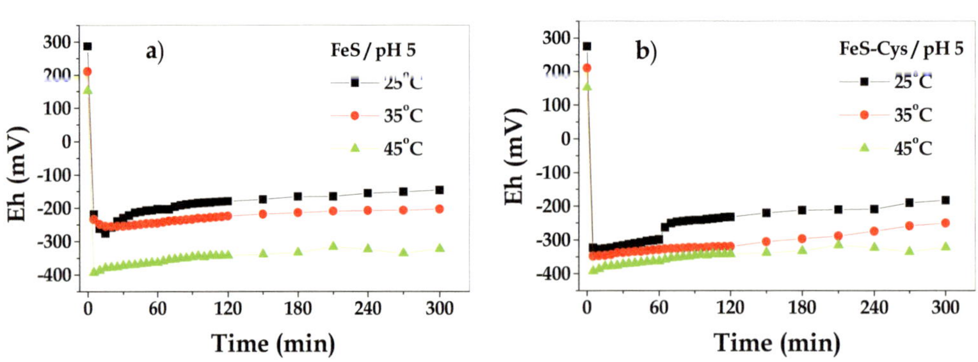

Fig. 14. Effect of temperature on the mackinawite oxidation process.

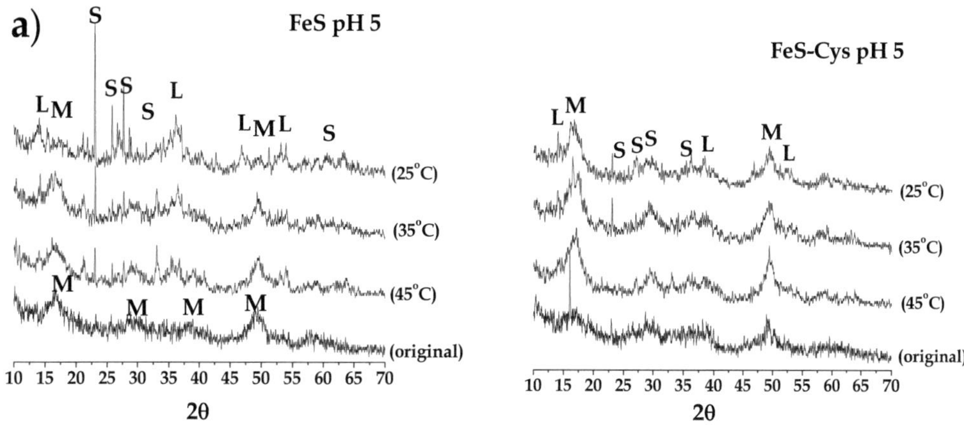

Fig. 15. XRPD pattern of the solids after oxidation experiments in diferent temperatures (short time).

Mackinawite oxidation – Long time-scale experiments

The unmodified mackinawite (FeS) has low resistance to oxidation (Liu et al, 2008, Wolters et al, 2005). In preliminary analysis (not shown), we observed that FeS oxidized in less than 24 hours. Thus, this period was selected to evaluate the oxidation process in experiments of long time scale.

The redox potential of the solutions followed the same pattern as that described for the short time experiments. After strong initial reduction, the Eh increased during the next 24 hours. The FeS-Cys showed lower Eh throughout the experiment, compared to the FeS, indicating increased stability towards oxidation (Fig. 16a). In the graph of Eh vs. time, it is possible to observe two distinct regions (a and b). The boundary between these regions was defined as a function of change of slope of the curve, denoting a change in the kinetics of the process. The point that occured this process was accompanied by color change of suspension. For FeS, after nine hours of agitation in the experimental conditions described, the suspension changed from black to red very quickly, attributed to the oxidation of iron (II) of the mackinawite to iron (III). The modified mackinawite showed the same pattern, except that the region was extended up to 14 hours.

The time required for the onset of advanced oxidation was considered as a parameter to evaluate the oxidation resistance, since the experiments were performed under the same conditions, varying only the type of solid. Thus, the increase in 5 hours observed for the modified mackinawite represented an increase of 55% in relation to unmodified mackinawite.

The results are in agreement with the model proposed by Chiriță et al (2008) for the oxidative dissolution of a mixture of 4M pyrrhotite and troilite. According to this model and the experimental results, the mechanism of oxidation of mackinawite is based on the process of protonation and surface oxidation (phase a), and the collapse of the surface with the fast oxidation of the bulk solid (phase b).

The mackinawite surface is protonated upon contact with the water (equation 7). Protonated groups react with adsorbed oxygen, making the surface-rich in sulfide groups (equation 8). Sulfide groups ($\equiv S_m^{2-}$) prevent the diffusion of oxygen to the bulk solid, resulting in lower oxidation rate, which was called inhibition oxidation phase (phase a). The decrease in pH, observed experimentally supports this information.

The iron released into the solution reacts with oxygen and is oxidized to iron oxide-hydroxide, which precipitates as lepdocrocite (equation 9).

$$\equiv FeSH^0 + H^+ \rightarrow \equiv FeSH_2^+ \qquad (7)$$

please, send me this eq with second proof form **in mathtype** (8)

$$Fe^{2+} + O_2 + H^+ \rightarrow FeO(OH) \qquad (9)$$

The inhibition and fast oxidation of mackinawite also are characterized by the concentration of dissolved oxygen (Fig. 16c). When mackinawite (FeS and FeS-Cys) was added to water, the dissolved oxygen concentration decreased rapidly, making the system anoxic within five minutes. The system remained below 0.5 mg/l during 9 and 14 hours, respectively, for FeS and FeS-Cys. Then, the DO increased until the saturation value of approximately 7 mg/l. The amount of oxygen consumed in this process was determined using the Fick's law of diffusion, the diffusion coefficient of oxygen in water (1.96x10^{-5} cm^2/s), the area of the reaction flask, and the gradient of concentration of oxygen, considering 1mm of the surface water (Table 4).
As shown in figure 16c the dissolved oxygen was continuously consumed during the inhibition phase. Chirițǎ and Descostes (2008) proposed that the impermeability of the oxidized surface layer is a function of the length and position of the polysulfide chains.

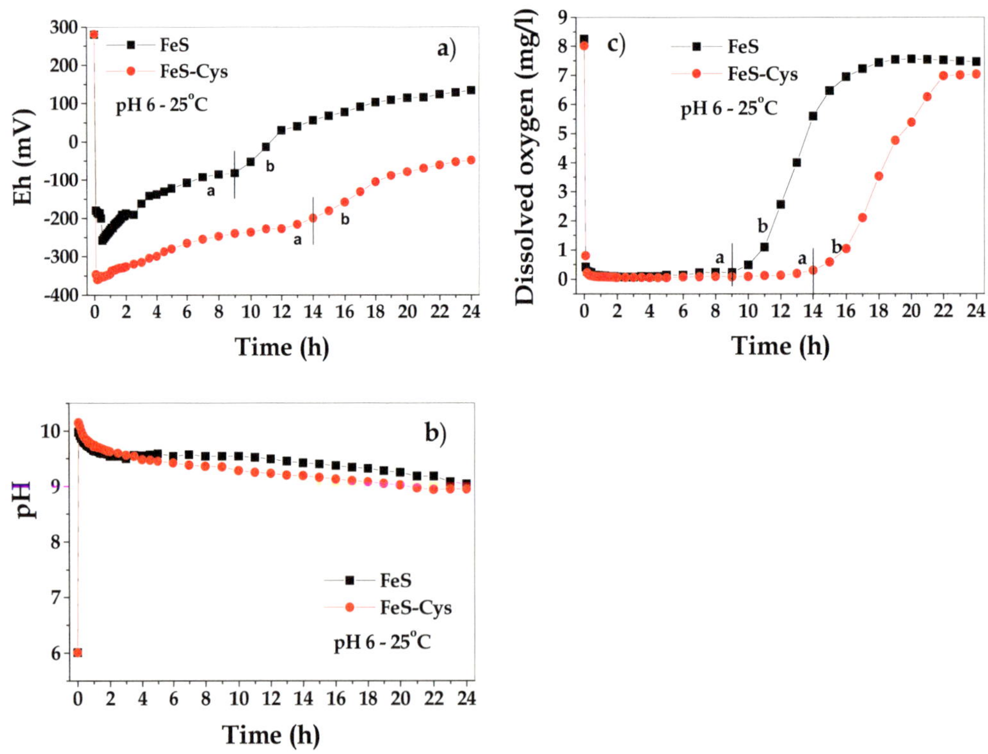

Fig. 16. Eh, pH and DO of the suspension during the oxidation process of mackinawite (long time).

Phase	FeS Consummed O_2 (mol)	FeS-Cys Consummed O_2 (mol)
Initial adsorption + inhibition	5.854 x10-4	7.810 x10-4
Fast oxidation	1.318 x10-4	1.035 x10-4

Table 4. Oxygen consummed during the mackinawite oxidation.

Short chains can easily undergo rearrangement in structures with close packing, reducing the mobility of the oxidant species. As the chains increase in size, it become more permeable, and the bulk solid oxidize faster.

The increase in polysulfide chains occurs through an oxidation process, catalyzed by proton, according to equation 10. Thus, the higher pH indicate the presence of long chains of polysulfide, with higher oxidant diffusion capacity. The lower H^+ concentration during the inhibition phase produced by unmodified mackinawite (Fig. 16b and Fig. 17) in comparison with modified mackinawite, indicates its lower oxidation resistence.

$$\equiv S_m^{2-} + \tfrac{1}{2} O_2 + 2H^+ \rightarrow \equiv S_n^{2-} + H_2O \qquad (10)$$

The end of the inhibition phase is the collapse of the surface layer of polysulfides, which facilitates the diffusion of oxygen, resulting in fast oxidation of the material (phase b). As cysteine does not polymerize, its presence on the modified mackinawite surface should result in small polysulfides chains. This promotes the increased of the oxidation resistance observed.

The mechanism of oxidation of iron sulfides (II) under acidic conditions involves the oxidation of Fe (II) to Fe (III). Researchers have proposed that iron (III) acts as the preferred oxidizing agent for sulfur, ie, the process of electron transfer from sulfur to iron (III) is easier in comparison to transfer to the oxygen (Chiriță and Descostes, 2006; Burton et al, 2009; Kamei and Ohmoto, 2000; Bourdoiseau et al, 2008).

In this work, the FeS consumes less oxygen than the FeS-Cys. To explain the fact that FeS present higher oxidation and lower oxygen consumption, we propose that the polysulfide chains in this material grow through oxidation by Fe (III) (equation 11).

$$\equiv S_m^{2-} + Fe^{3+} \rightarrow \equiv S_n^{2-} + Fe^{2+} \qquad (11)$$

This reaction is minimized in the system FeS-Cys, since part of the iron released is bound to cysteine through the formation of complexes $Fe(Cys)^+$. Thus, a larger amount of oxygen is consumed during the surface oxidation of modified mackinawite.

The concentration of protons as function of time is shown in Fig. 17. The concentration of $[H^+]$ is a result of several reactions that occur in suspension. Therefore, this parameter can only be used as a comparison between the two systems, since the experiments were executed under same conditions, being the cysteine the unique difference. The observations were supported by other analytical techniques results, such as the XRPD pattern.

The surface oxidation of sulfide groups ($\equiv S_m^{2-}$) (equation 8) produces protons, while the oxidation of iron sulfides and oxidation of the polysulfide groups ($\equiv S_n^{2-}$) (equations 9 and 10), consume protons.

Analyzing the concentration of $[H^+]$ in the oxidation process was observed that FeS-Cys releases more $[H^+]$ than FeS, attributed the formation of sulfide groups ($\equiv S_m^{2-}$) and the limited growth of polysulfide chains. This supports the assumption that the oxidation resistance is due of the small polysulfide chains, which can rearrange, keeping the surface layer more impermeable to oxidation for longer time.

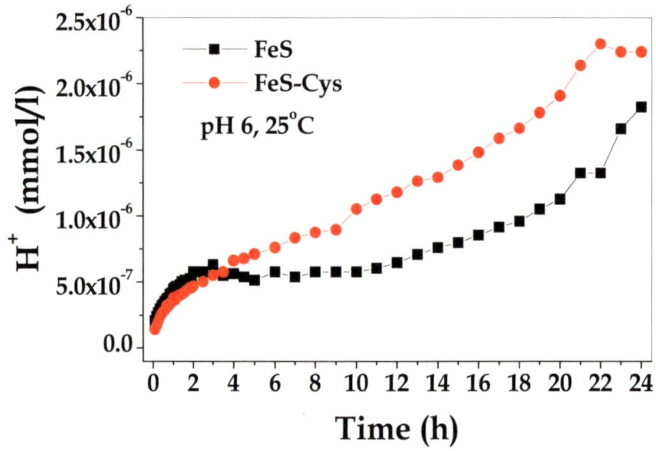

Fig. 17. [H+] concentration during the mackinawite oxidation process (long time), starting after pH maximum.

Mackinawite oxidation resulted in the formation of lepidocrocite and elemental sulfur, as early discussed for experiments of short time. This is in agreement with the features of the crystal structure of mackinawita since the lepidocrocite structure consists of layers of octahedra of iron oxide (III) joined by hydrogen bonding to hydroxyls. This also is in agreement with the oxidation product of the abiotic FeS (Burton et al, 2009).
After 24 hours, both mackinawites (FeS and FeS-Cys) had advanced oxidation states. The XRPD pattern of the solids before (1 and 3) and after (2 and 4) oxidation are shown in the figure 18.

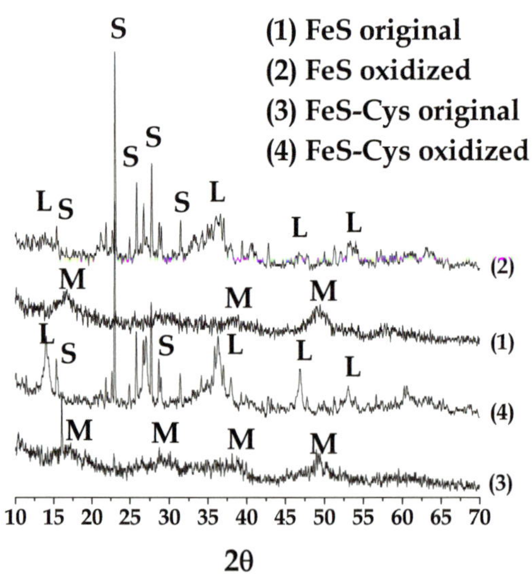

Fig. 18. XRPD pattern of the solids after 24h of oxidation process.

It is clear that the absence of the broad-band 16.5° (2θ) is characteristic of disordered mackinawite. The peaks of lepidocrocite formed by oxidation of FeS-Cys were better defined than for FeS, characteristic of higher crystallinity. This suggests that cysteine can control the microstructure during the formation of lepidocrocite.

3.5 Mercury uptake capacity

The mercury uptake experiments were conducted using 50 ml glass centrifuge tubes. A 5 mmol/l stock solution was prepared by dissolving of $HgCl_2$ (99.9995%, Alfa Aesar) in 32 mmol/l HNO_3 solution (trace metal grade, concentrate) in high pure water (HPLC grade). A Hg(II) 1 mmol/l solution was prepared by dilution of the stock solution, and purged with N_2 ultra high pure during 30 minutes; then, 0.4 g/l of the mackinawite was added into the solution, and magnetic stirred by a PTFE coated stirrer bar during 1, 5, 10, 20, 30, 60, and 90 minutes. After the desired time, the samples were filtered using a Whatmann seringe filter 0.45µm, being completed within 30 seconds after sampling. The 10 ml sample was preserved with addition of 20 µl of HNO_3 solution (trace metal grade, concentrate). The mercury content was determined using cold vapor method CVAAS (mercury Instruments, Lab Analyzer 254) based on EPA method # 7470 A.

Experiments to determine the mercury sequestration capacity of the studied mackinawites was carried through, using solution of $HgCl_2$ dissolved in high pure water. Results showed that unmodified and modified mackinawite is capable to uptake 99.99% and 99.94%, respectively, of mercury in solution within 30 minutes of contact (Fig. 19). It means that the modification with L-cysteine does not showed sensible influence on the mercury uptake; approximately 490 mg Hg/g mackinawite in medium concentration of 1mmol/l Hg (II) was observed, being an excellent material for mercury sequestration.

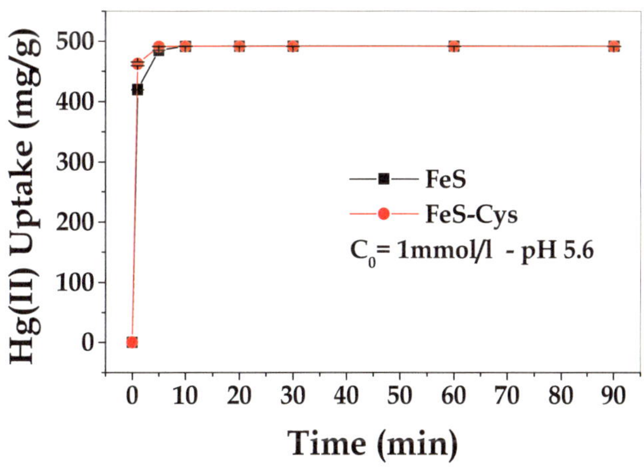

Fig. 19. Mercury uptake by unmodified (FeS) and modified (FeS-Cys) mackinawite.

4. Implications to mercury immobilization in contaminated sediment

The use of mackinawite as a reactive material in an active barrier system (ABS) associated with in-situ capping (ISC) will improve the mercury uptake in a contaminated sediment, since mercury does not have a large chemical affinity to silicate-based materials commonly used in ISC.

In spite of its excellent capability to immobilize mercury, one limitation characteristic to use of the mackinawite in ISC is due it easily be oxidized when applied as a reactive material to natural sediments. The modification of mackinawite, using the amino acid L-cysteine, increases its oxidation resistance, which can make possible the mackinawite aplication as ISC reactive material. The time to build a capping using mackinawite needs to be evaluated, for determine if the increase in the mackinawite oxidation resistance by cysteine is enough to its use.

Results from this study have shown that the modification of mackinawite occured using a simple experimental procedure, the efficiency of L-cysteine in retarding the mackinawite oxidation was 55%, and has a high mercury uptake capability, without sensible difference in comparison to the unmodified mackinawite. These are important features and therefore, modified mackinawite should be tested as a reactive capping material under laboratory conditions using mercury contaminated field sediments.

5. Acknowledgements

This project was funded by a grant (06HQGR0088) from the United States Department of Interior through the Louisiana Water Resources Research Institute. Dr. Chaves was supported by a Chevron postdoctoral fellowship through the College of Engineering at LSU. Dr. Chaves and Dr. Buchler are grateful to Ministry of Education of Brazil, CNPq /CAPES/ PNPD Project 2016/2008.

6. References

Ayyar, R. R., Srinivasan, R. (1965). Crystal structure of L (+) hydrochloride monohydrate. *Current Sc.* Vol. 34, No. 15, pp. 449-450.

Belzile, N., Maki, S., Chen, Y-W., and Goldsack, D. (1997). Inhibition of pyrite oxidation by surface treatment. *Sci. Total Environ.* Vol. 196, No. 2, pp. 177-186.

Behra, P. et al. (2001). XPS and XAS study of the sorption of Hg(II) onto pyrite. *Langmuir,* Vol. 17, pp. 3970-3979.

Bourdoiseau, J. A. et al. (2008). Characterization of mackinawite by Raman spectroscopy: effects of crystallization, drying and oxidation. *Corrosion Sci.* Vol. 50, pp. 3247-3255.

Boursiquot, S. et al (2001) The dry oxidation of tetragonal FeS_{1-x} mackinawite. *Phys. Chem. Minerals* Vol. 28, pp. 600-611.

Burton, E. D. et al. (2009). Iron-monosulfide oxidation in natural sediments: resolving microbially mediated s transformations using xanes, electron microscopy, and selective extractions. *Environ. Sci. Technol.* Vol. 43, pp. 3128-3134.

Chiriță, P., Descostes, M. (2006). Troilite oxidation by hydrogen peroxide. *J. Colloid Interface Sci.* Vol. 299, pp. 260-269.

Chiriță, P. et al. (2008). Oxidation of FeS by oxygen-bearing acidic solutions. *J. Colloid Interface Sci.* Vol. 321, pp. 84-95.

Fontecave, M., and Ollagnier-de-Choudens, S. (2008). Iron–sulfur cluster biosynthesis in bacteria: Mechanisms of cluster assembly and transfer. *Arch. Biochem. Biophys.* Vol. 474, pp. 226-237.

Jeong, H. Y., et al. (2008) Characterization of synthetic nanocrystalline mackinawite: Crystal structure, particle size, and specific surface area. *Geochim. Cosmochim. Acta*, Vol. 72, pp. 493-505.

Kamei, G.; Ohmoto, H. (2000). The kinetics of reactions between pyrite and O2-bearing water revealed from in situ monitoring of DO, Eh and pH in a closed system. *Geochim. Cosmochim. Acta*,Vol. 64, pp. 2585-2601.

Lennie, A. R. et al. (1995). Synthesis and Rietveld crystal structure refinement of mackinawite tetragonal FeS. *Min. Mag.*, Vol. 59, pp. 677-683.

Liu, J. et al. (2007). Observations of mercury fate and transport beneath a sediment cap. *Land Contamination & Reclamation,* Vol. 15, pp. 401-411.

Liu, J. et al. (2008). Immobilization of aqueous Hg(II) by mackinawite (FeS). *J. Hazard. Mater.*, Vol. 157, pp. 432-440.

Liu, J. et al. (2009). Inhibition of Mercury Methylation by Iron Sulfides in an Anoxic Sediment. *Environ. Eng. Sci.*, Vol 26, pp. 833-840.

Moses, C. O. Nordstrom, D. K., Herman, J. S., and Mills, A. L. (1987). Aqueous pyrite oxidation by dissolved oxygen and by ferric iron. *Geochim. Cosmochim. Acta*, Vol. 51, pp. 1561-1571.

Mullet, M. et al. (2002) Surface chemistry and structural properties of mackinawite prepared by reaction of sulfide ions with metallic iron. *Geochim. Cosmochim. Acta* Vol. 66, pp. 829-836.

Nesbitt, H. W, Muir, I. J. (1994). X-ray phoelectron spectroscopic study of a pristine pyrite surface reacted with water vapour and air. *Geochim. Cosmochim. Acta* Vol. 58, (1994), pp. 4667-4679.

Palermo, M. R. (1998). Design considerations for in-situ capping of contaminated sediments. *Wat. Sci. Tech.* Vol. 37, (1998), pp. 315-321.

Peak, D.; Ford, R. D.; Sparks, D. L. (1999). An in situ ATR-FTIR investigation of sulfate bonding mechanisms on goethite. *J. Colloid Interface Sci.*, Vol. 218, pp. 289-299.

Pieulle, L., Guigliarelli, B., Asso, M., Dole, F., Bernadac, A., and Hatchikian C. E. (1995). Isolation and characterization of the pyruvate-ferredoxin oxidoreductase from the sulfate-reducing bacterium Desulfovibrio africanus. *Biochim. Biophys. Acta*, Vol. 1250, pp. 49-59.

Rickard, D. (1997). Kinetics of pyrite formation by the H2S oxidation of iron (II) monosulfide in aqueous solutions between 25 and 125°C: The rate equation. *Geochim. Cosmochim. Acta,* Vol. 61, pp. 115-134.

Rickard, D. (2006).The solubility of FeS. *Geochim. Cosmochim. Acta*, Vol. 70, pp. 5779-5789.

Sing, K. S. W. (1985) IUPAC: Reporting physisorption data for gas/solid systems with special reference to the determination of surface area and porosity (Recommendations 1984). *Pure appl. Chem.*, Vol. 57, pp. 603-619.

Sweeney, R. E., and Kaplan, I. R. (1973) Pyrite framboid formation, laboratory synthesis and marine sediments, Econ. Geol., Vol. 68, pp. 618-634.

Taylor, P., Rummery, T.E., and Owen, D.G. (1979). On the conversion of mackinawite and greigite. J. *Inorg. Nucl. Chem.*, Vol. 41, pp. 595-596.

Widler, A. M. (2002). The adsorption of gold (I) hydrosulphide complexes by iron sulphide surfaces. *Geochim. Cosmochim. Acta*, Vol. 66, pp. 383-402.

Wolthers, M., Charlet, L., van Der Linde, P. R., Rickard, D., and van Der Weijden, C. H. (2005) Surface chemistry of disordered mackinawite. *Geochim. Cosmochim. Acta*, Vol. 69, No. 14, pp. 3469-3481.

Zhang, X., Borda, M., Schoonen, M. A. A., and Strongin, D. R. (2003). Pyrite oxidation inhibition by a cross-linked lipid coating. *Geochem. Trans.* Vol. 4, No. 2, pp. 8-11.